# KEYNOTE guide to topics in your course

# CLIFFS KEYNOTE REVIEWS

# Calculus and Analytic Geometry

*by*

KAJ L. NIELSEN

CLIFF'S NOTES, INC. • LINCOLN, NEBRASKA 68501

ISBN 0-8220-1712-1



L. C. Catalogue Card Number: 70-89836

Printed in the United States of America

# CONTENTS

# TO THE STUDENT

This KEYNOTE is a flexible study aid designed to help you REVIEW YOUR COURSE QUICKLY and USE YOUR TIME TO STUDY ONLY MATERIAL YOU DON'T KNOW.

**FOR GENERAL REVIEW**

Take the SELF-TEST on the first page of any topic and turn the page to check your answers.

Read the SOLUTIONS of any questions you answered incorrectly.

If you are satisfied with your understanding of the material, move on to another topic.

If you feel that you need further review, read the column of BASIC FACTS. Then, to check your comprehension of the material, do the EXERCISES immediately following the group of problems which gave you difficulty. The solutions to the EXERCISES appear on the final page of each topic.

For a more detailed discussion of the material, read the column of ADDITIONAL INFORMATION.

**FOR QUICK REVIEW**

Read the column of BASIC FACTS for a rapid review of the essentials of a topic and then take the SELF-TEST if you want to test your understanding of the material.

**ADDITIONAL HELP FOR EXAMS**

Review terms in the DICTIONARY-INDEX.
Test yourself by taking the sample FINAL EXAM.

# 1 REVIEW OF ALGEBRA

## SELF-TEST

DIRECTIONS: Write your answers in the numbered regions to the right. To check your answers, turn the page. Study the solutions to the problems you missed, and do the exercises following any problem or group of problems in which you had errors.

1 ______
2 ______
3 ______
4 ______
5 ______
6 ______
7 ______
8 ______
9 ______
10 ______
11 ______
12 ______
13 ______
14 ______
15 ______

Simplify.

1. $\dfrac{42a^2(xy^2)^3}{12(ax)^2y^3}$

2. $\sqrt{3} - \sqrt{27} + \sqrt{81}$

3. $\dfrac{x}{1 - \dfrac{1}{1+x}} + \dfrac{x^3}{x - \dfrac{x}{1-x}}$

4. $\dfrac{\sqrt{x} - 5\sqrt{y}}{2\sqrt{x} + 3\sqrt{y}}$

5. $\dfrac{x+y}{x^2 + 2xy + y^2} - \dfrac{x-y}{x^2 - y^2}$

Solve for $x$.

6. $x^2 - 2x + 5 = 0$

7. $\dfrac{2x+3}{3} - \dfrac{x}{4} < 3$

8. $|x - 5| = 2$

9. $|x - 2| < 3$

10. $x^2 - x - 6 < 0$

Find the fourth term.

11. $(1 + x)^7$

12. $(x - 1)^{-1}$

13. $(2 + x)^{\frac{1}{2}}$

Factor each expression.

14. $(2 - x^2)^{\frac{3}{2}} + x^2(2 - x^2)^{\frac{1}{2}}$

15. $a^{\frac{3}{2}} - b^{\frac{3}{2}}$

## SOLUTIONS

1. $\dfrac{42a^2(xy^2)^3}{12(ax)^2y^3}$

$= \dfrac{42a^2x^3y^6}{12a^2x^2y^3} = \dfrac{7}{2}\,xy^3$

2. $\sqrt{3} - \sqrt{27} + \sqrt{81}$

$= \sqrt{3} + 3\sqrt{3} + 9 = 4\sqrt{3} + 9$

3. $\dfrac{x}{1 - \dfrac{1}{1-x}} + \dfrac{x^3}{x - \dfrac{x}{1-x}}$

$= \dfrac{x}{\dfrac{1-x-1}{1-x}} + \dfrac{x^3}{\dfrac{x-x^2-x}{1-x}}$

$= \dfrac{x(1-x)}{-x} + \dfrac{x^3(1-x)}{-x^2}$

$= -1 + x - x + x^2 = x^2 - 1$

4. $\dfrac{\sqrt{x} - 5\sqrt{y}}{2\sqrt{x} + 3\sqrt{y}} \cdot \dfrac{2\sqrt{x} - 3\sqrt{y}}{2\sqrt{x} - 3\sqrt{y}}$

$= \dfrac{2x - 13\sqrt{xy} + 15y}{4x - 9y}$

5. $\dfrac{x+y}{x^2 + 2xy + y^2} - \dfrac{x-y}{x^2 - y^2}$

$= \dfrac{x+y}{(x+y)^2} - \dfrac{(x-y)}{(x-y)(x+y)}$

$= \dfrac{1}{x+y} - \dfrac{1}{x+y} = 0$

6. $x^2 - 2x + 5 = 0$

$x = \dfrac{2 \pm \sqrt{4-20}}{2}$

$= \dfrac{2 \pm 4i}{2} = 1 \pm 2i$

> **EXERCISES I**
>
> i. Complete the square and write in factored form.
>
> $$3x^2 - 12x + 18$$
>
> ii. Reduce to a form $a + bi$ where $i = \sqrt{-1}$.
>
> $$2i - 3i^5 + \sqrt{-25} + 2i^2$$
>
> iii. Find the value of $k$ so that the two roots of
>
> $$2x^2 + 3kx + 9 = 0$$
>
> are equal.

7. $\dfrac{2x+3}{3} - \dfrac{x}{4} < 3$

$8x + 12 - 3x < 36$

$5x < 24;\ x < \frac{24}{5}$

8. $|x - 5| = 2$

$x - 5 = 2$ or $x = 7$

$x - 5 = -2$ or $x = 3$

9. $|x - 2| < 3$

$-3 < x - 2 < 3$

$-1 < x < 5$

10. $x^2 - x - 6 < 0$

Set $x^2 - x - 6 = (x-3)(x+2) = 0$
Then $x = -2$ and $x = 3$ and we test the inequality at a value in $[-2, 3]$, say $x = 0$: $(0)^2 - (0) - 6 = -6 < 0$ and $x^2 - x - 6 < 0$ in $-2 < x < 3$.

> **EXERCISES II**
>
> True or False
>
> i. If $a > b$ and $c < 0$, then $ac < bc$.
>
> ii. If $a > b$, $b > c$, and $x > 0$, then $ax > cx$.
>
> iii. If $a > b$ and $c > d$, then $a + c + x < b + d + x$.

11. $(1+x)^7 = 1^7 + 7x + \binom{7}{2} x^2 + \ldots$

4th term: $\frac{7 \cdot 6 \cdot 5}{1 \cdot 2 \cdot 3} x^3 = 35x^3$

12. $(x-1)^{-1} = x^{-1} - x^{-2}(-1) + \ldots$

4th term: $\binom{-1}{3} x^{-4}(-1)^3$

$= \frac{(-1)(-2)(-3)}{1 \cdot 2 \cdot 3} x^{-4}(-1)$

$= x^{-4}$

13. 4th term of $(2+x)^{\frac{1}{2}}$

$= \frac{\frac{1}{2}\left(-\frac{1}{2}\right)\left(-\frac{3}{2}\right)}{1 \cdot 2 \cdot 3} (2)^{-\frac{5}{2}} x^3$

$= \frac{\frac{3}{8} x^3}{6\sqrt{2^5}} = \frac{x^3}{2^6 \sqrt{2}}$

**EXERCISES III**

i. Find the value of the binomial coefficient

$$\binom{\frac{5}{2}}{4}.$$

ii. Expand $(2-x)^6$.

iii. Find the values of $x$ which satisfy

$$\left| \frac{3-2x}{3+x} \right| < 5.$$

14. $(2-x^2)^{\frac{3}{2}} + x^2(2-x^2)^{\frac{1}{2}}$

$= (2-x^2)^{\frac{1}{2}}(2 - x^2 + x^2)$

$= 2(2-x^2)^{\frac{1}{2}}$

15. $a^{\frac{3}{2}} - b^{\frac{3}{2}}$. Let $u = a^{\frac{1}{2}}$, $v = b^{\frac{1}{2}}$.

$= (u-v)(u^2 + uv + v^2)$

$= (\sqrt{a} - \sqrt{b})(a + \sqrt{ab} + b)$

Solutions to the exercises are on page 6.

## ANSWERS

1. $\frac{7}{2} xy^3$ — 1
2. $4\sqrt{3} + 9$ — 2
3. $x^2 - 1$ — 3
4. $\frac{2x - 13\sqrt{xy} + 15y}{4x - 9y}$ — 4
5. $0$ — 5
6. $1 \pm 2i$ — 6
7. $x < \frac{24}{5}$ — 7
8. 7, 3 — 8
9. $-1 < x < 5$ — 9
10. $-2 < x < 3$ — 10
11. $35x^3$ — 11
12. $x^{-4}$ — 12
13. $\frac{x^3}{2^6\sqrt{2}}$ — 13
14. $2(2-x^2)^{\frac{1}{2}}$ — 14
15. $(\sqrt{a} - \sqrt{b})(a + \sqrt{ab} + b)$ — 15

## BASIC FACTS

**Axioms of Equality**

$E_1$ Reflexive: $a = a$
$E_2$ Symmetric: If $a = b$, then $b = a$
$E_3$ Transitivity: If $a = b$ and $b = c$, then $a = c$
$E_4$ Addition: If $a = b$ and $c = d$, then $a + c = b + d$
$E_5$ Multiplication: If $a = b$ and $c = d$, then $ac = bd$

**Axioms of Comparison**

$O_1$ Trichotomy: $a = b$, $a < b$, or $b < a$
$O_2$ Transitivity: If $a < b$ and $b < c$, then $a < c$
$O_3$ Addition: If $a < b$, then $a + c < b + c$
$O_4$ Multiplication: If $a < b$ and $c > 0$, then $ac < bc$

**Properties of Inequalities**

The sense of an inequality, $a > b$, is *not changed* if both members are:

i. increased by the same number ($O_3$)
ii. decreased by the same number
iii. multiplied by the same positive number ($O_4$)
vi. divided by the same positive number
v. positive and raised to the same power
vi. increased by the corresponding members of an inequality in the same sense
vii. multiplied by the corresponding members of an inequality in the same sense
viii. divided by the corresponding members of an equality in the reversed sense.

The sense of an inequality is *reversed* if:

ix. both members are multiplied by the same negative number
x. the reciprocals of two nonzero elements of like sign are compared.

A number $x$ is between $a$ and $b$ if *both* the inequalities $a < x$ and $x < b$ are true; i.e., $a < x < b$.

## ADDITIONAL INFORMATION

The reader should be familiar with all of the operations encountered in algebra. In this chapter we review only a few of the basic concepts.

**Radicals**

Like radicals have the same index and the same radicand; they may be added or subtracted. The product of two radicals can be obtained if they have the same index: $(\sqrt{3x})(\sqrt{2x}) = \sqrt{6x^2} = x\sqrt{6}$. The division of radicals usually involves removing all radicals from the denominator, a process called rationalizing the denominator: $\frac{\sqrt{5}}{3\sqrt{2}} = \frac{\sqrt{5}}{3\sqrt{2}} \cdot \frac{\sqrt{2}}{\sqrt{2}} = \frac{\sqrt{10}}{3 \cdot 2} = \frac{\sqrt{10}}{6} = \frac{1}{6}\sqrt{10}$. If the radicands are the same but the indices are different, convert to fractional exponents and use the laws of exponents.

$$(\sqrt[6]{a} \div \sqrt[4]{a^2})(\sqrt{a}) = a^{\frac{1}{6}} \cdot a^{-\frac{1}{2}} \cdot a^{\frac{1}{2}} = a^{\frac{1}{6}} = \sqrt[6]{a}$$

If the radicand is negative, an imaginary quantity appears in radical. Change all such radicals to imaginary numbers and work with the quantity $i (i = \sqrt{-1})$.

$$\frac{3 - \sqrt{-2}}{3 + \sqrt{-2}} = \frac{3 - i\sqrt{2}}{3 + i\sqrt{2}} \cdot \frac{3 - i\sqrt{2}}{3 - i\sqrt{2}} = \frac{9 - 6i\sqrt{2} + 2i^2}{9 - 2i^2} = \frac{9 - 6i\sqrt{2}}{9 + 2} = \frac{7 - 6i\sqrt{2}}{11}$$

**Roots**

The fundamental theorems for finding roots of rational integral equations are: (1) every rational integral equation, $f(x) = 0$, has at least one root, and (2) every rational integral equation of the $n^{th}$ degree has $n$ roots and no more.

The formula for the roots of a quadratic equation is derived by completing the square on the general quadratic. We will use this process in our study of the conics (Chapter 10).

Given the general quadratic: $ax^2 + bx + c$

Factor out the coefficient $a$: $a\left(x^2 + \frac{b}{a}x\right) + c$

Add and subtract $\left(\frac{1}{2} \cdot \frac{b}{a}\right)^2$: $a\left[x^2 + \frac{b}{a}x + \left(\frac{b}{2a}\right)^2\right] + c - a\left(\frac{b}{2a}\right)^2$

The bracket is a perfect square: $a\left(x + \frac{b}{2a}\right)^2 + c - \frac{b^2}{4a}$.

**Real Numbers and Intervals**

The totality of real numbers can be associated with a horizontal line on which zero has been selected at any convenient point. The positive numbers lie to the right of the zero, and negative numbers, to the left (Figure 1-1). If $a$ and $b$ are two numbers as shown in Figure 1-2, then the *open interval* from $a$ to $b$ is all the numbers between $a$ and $b$ not including the end points, $]a, b[$.

**Absolute Value, $|a|$**
If $a \geq 0$, then $|a| = a$
If $a < 0$, then $|a| = -a$

$$|ab| = |a| \cdot |b| \text{ and } \left|\frac{a}{b}\right| = \frac{|a|}{|b|},\ b \neq 0$$

**Laws of Exponents**
$a^m \cdot a^n = a^{m+n};\ (a^m)^n = a^{mn}$

$$(ab)^n = a^n b^n;\ \left(\frac{a}{b}\right)^n = \frac{a^n}{b^n},\ b \neq 0$$

$$\frac{a^m}{a^n} = a^{m-n},\ a \neq 0;\ a^0 = 1,\ a \neq 0$$

$$a^{-n} = \frac{1}{a^n},\ a \neq 0;\ \sqrt[n]{a} = a^{\frac{1}{n}}$$

**Factorial Notation**
$n! = 1 \cdot 2 \cdot 3 \cdot 4 \cdot 5 \ldots n$ and $0! = 1$

**The Binomial Theorem**

$$(a + b)^n = a^n + na^{n-1}b + \binom{n}{2} a^{n-2}b^2 + \cdots + \binom{n}{r} a^{n-r}b^r + \cdots + b^n$$

where the binomial coefficient is

$$\binom{n}{r} = \frac{n(n-1)(n-2)\ldots(n-r+1)}{r!}$$

**Typical Products and Their Factors**
$ab + ac = a(b + c)$
$a^2 \pm 2ab + b^2 = (a \pm b)^2$
$a^2 - b^2 = (a + b)(a - b)$
$x^2 + (a + b)x + ab = (x + a)(x + b)$
$x^2 - (a + b)x + ab = (x - a)(x - b)$
$a^3 + b^3 = (a + b)(a^2 - ab + b^2)$
$a^3 - b^3 = (a - b)(a^2 + ab + b^2)$
$a^3 + 3a^2b + 3ab^2 + b^3 = (a + b)^3$

**Quadratic Formula**
If $ax^2 + bx + c = 0$, then the roots are

$$x = \frac{-b \pm \sqrt{b^2 - 4ac}}{2a}.$$

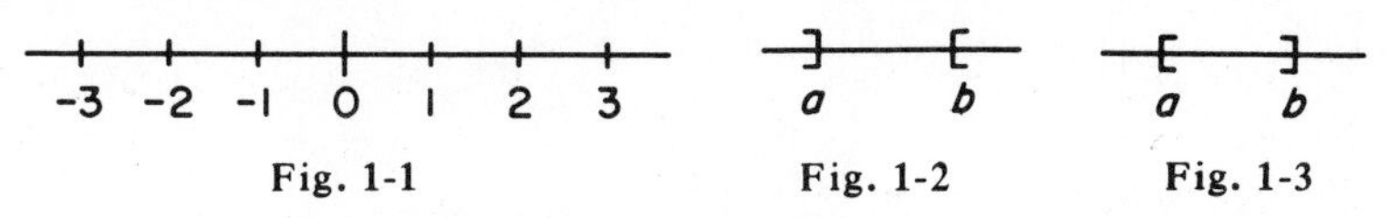

Fig. 1-1 Fig. 1-2 Fig. 1-3

If the end points $a$ and $b$ are included, we have a *closed interval*, $[a, b]$. If $x$ is a number between $a$ and $b$, then we write $a < x < b$ for an open interval and $a \leq x \leq b$ for a closed interval. An interval may also be closed on one end and open on the other: $]a, b]$, $a < x \leq b$; $[a, b[$, $a \leq x < b$. If we consider all the numbers larger than a given number, say 4, $x > 4$, we consider the interval extending to infinity to the right of 4. Although $\infty$ is not a number it is a symbol which is used, and we write $]4, \infty[$ and $4 < x < \infty$. Similarly if $x < 4$, we write $-\infty < x < 4$.

**Absolute Values**

When the absolute value is used in a statement of equality, such as $|x - 3| = 5$, the statement expresses the fact that $x - 3 = 5$ and $x - 3 = -5$. Thus there are two solutions ($x = 8$ and $x = -2$). This can be easily checked since $|8 - 3| = 5$ and $|-2 - 3| = 5$. When an absolute value is used with an inequality, we are considering a range of values. For example, the expression $|x| < 3$ is true for all values of $x$ in the interval from $-3$ to 3 (i.e., $-3 < x < 3$). The inequality $|x - 4| < 7$ is solved by writing it in the form $-7 < x - 4 < 7$ and adding 4 to each term to obtain $-7 + 4 < x < 7 + 4$ or $-3 < x < 11$. An inequality of the form $\left|\frac{x + 2}{x - 3}\right| < 4$, or its equivalent $-4 < \frac{x + 2}{x - 3} < 4$, requires careful consideration. The denominatory may be either positive or negative. If $x - 3 > 0$, we may multiply the terms by $x - 3$ and preserve the sense of the inequalities (Properties of Inequalities iii). Thus $-4(x - 3) < x + 2$ or $2 < x$ and $x + 2 < 4(x - 3)$ or $\frac{14}{3} < x$. The three conditions are $3 < x$ (from $x - 3 > 0$), $2 < x$, and $\frac{14}{3} < x$, all of which are satisfied in the interval $]\frac{14}{3}, \infty[$. On the other hand if $x - 3 < 0$, then a multiplication by this expression reverses the sense of the inequalities (Properties of Inequalities ix), and we have $-4(x - 3) > x + 2$ or $x < 2$, $x < 3$ (from $x - 3 < 0$), and $x + 2 > 4(x - 3)$ or $x < \frac{14}{3}$, all of which are satisfied in the interval $]-\infty, 2[$. The solution to the original inequality is the set of numbers *not* in the closed interval $[2, \frac{14}{3}]$.

NOTE TO STUDENT: Carefully review all the properties of inequalities.

## SOLUTIONS TO THE EXERCISES

I

i. $3x^2 - 12x + 18$
$= 3(x^2 - 4x + 6)$
$= 3(x^2 - 4x + 4 + 6 - 4)$
$= 3(x^2 - 4x + 4) + 3(2)$
$= 3(x - 2)^2 + 6$

ii. $2i - 3i^5 + \sqrt{-25} + 2i^2$
$= 2i - 3i + 5i - 2$
$= -2 + 4i$

iii. The roots of $2x^2 + 3kx + 9 = 0$ are equal if the discriminant is equal to zero.

$$(3k)^2 - 4(2)9 = 0$$
$$9k^2 - (8)9 = 0;\ k^2 = 8$$
$$k = \pm\sqrt{8} = \pm 2\sqrt{2}$$

II

i. A multiplication by a negative quantity reversed the sense of the inequality. *True*

ii. Under the given conditions $ax > bx$ and $bx > cx$, therefore $ax > cx$. *True*

iii. Since the sense of an inequality is preserved when equalities of the same sense are added and when the same quantity is added to both members, the inequality should read

$$a + c + x > b + d + x.$$

The given statement is *False.*

III

i. $\binom{\frac{5}{2}}{4} = \dfrac{\left(\frac{5}{2}\right)\left(\frac{3}{2}\right)\left(\frac{1}{2}\right)\left(-\frac{1}{2}\right)}{1\cdot 2\cdot 3\cdot 4}$

$\binom{\frac{5}{2}}{4} = -\dfrac{5}{2^7} = -\dfrac{5}{128}$

ii. $(2 - x)^6 = 2^6 + 6(2)^5(-x)$
$+ \binom{6}{2}(2)^4(-x)^2$
$+ \binom{6}{3}(2)^3(-x)^3$
$+ \binom{6}{4}(2)^2(-x)^4$
$+ \binom{6}{5}(2)(-x)^5$
$+ (-x)^6$
$= 2^6 - 6(2)^5 x$
$+ 15(2)^4 x^2$
$- 20(2)^3 x^3$
$+ 15(2)^2 x^4$
$- 6(2) x^5 + x^6$
$= 64 - 192x$
$+ 240x^2$
$- 160x^3 + 60x^4$
$- 12x^5 + x^6$

iii. $\left|\dfrac{3 - 2x}{3 + x}\right| < 5$

$$-5 < \frac{3 - 2x}{3 + x} < 5$$

See Additional Information for a detailed explanation of this type of problem.

If $3 + x > 0$, $x > -3$, $x > -6$,

and $x > -\frac{12}{7}$. These three conditions are all satisfied if $x > -\frac{12}{7}$.

If $3 + x < 0$, $x < -3$, $x < -6$, and $x < -\frac{12}{7}$. These conditions are all satisfied if $x < -6$. The given inequality is satisfied by all values outside the interval $\left[-6, -\frac{12}{7}\right]$.

# 2 FUNCTIONS, LIMITS, AND CONTINUITY

## SELF-TEST

DIRECTIONS: Write your answers in the numbered regions to the right. To check your answers, turn the page. Study the solutions to the problems you missed, and do the exercises following any problem or group of problems in which you had errors.

1. If $f(x) = 3x^2 - 8x + 5$, find $f(0)$, $f(1)$, $f(-1)$; $f(x + \Delta x)$.

2. If $f(x) = x^3 - x^2 + x - 1$, find $f(1)$, $f(-1)$, $f(a)$, $f(x + 1)$.

3. If $f(x) = \dfrac{x + 3}{x^2 - x - 6}$, find $f(0)$, $f(-2)$, $f(x - 1)$.

4. If $f(x) = 10^x$, find $f(2)$, $f(-x + 1) - f(-x)$.

5. If $f(x) = 2 \sin x - 3 \cos x$, find $f(0)$, $f\left(\frac{\pi}{4}\right)$.

In 6–14, evaluate the limit.

6. $\lim\limits_{x \to 0} (3x^2 - 5x + 2)$

7. $\lim\limits_{x \to 1} \dfrac{x^2 + x}{2x - 1}$

8. $\lim\limits_{x \to 2} \dfrac{x^2 - 4}{x - 2}$

9. $\lim\limits_{x \to -1} \dfrac{3 - x}{x^2 - x}$

10. $\lim\limits_{x \to 3^+} \dfrac{1}{x - 3}$

11. $\lim\limits_{x \to \infty} \dfrac{x^2 + 2x + 1}{x^3}$

12. $\lim\limits_{x \to \infty} \dfrac{3x^2 + x}{2x^2 - 3x}$

13. $\lim\limits_{h \to 0} \dfrac{(x + h)^2 - x^2}{h}$

14. $\lim\limits_{h \to 0} \dfrac{3(x + h)^3 - 3x^3}{h}$

15. Is the following function continuous in its domain?

$f(x) = \dfrac{1}{x + 1}$ for $0 < x < 3$

1 ___
2 ___
3 ___
4 ___
5 ___
6 ___
7 ___
8 ___
9 ___
10 ___
11 ___
12 ___
13 ___
14 ___
15 ___

## SOLUTIONS

1. $f(x) = 3x^2 - 8x + 5$
$f(0) = 0 - 0 + 5 = 5$
$f(1) = 3 - 8 + 5 = 0$
$f(-1) = 3 + 8 + 5 = 16$
$f(x + \Delta x)$
$= 3(x + \Delta x)^2 - 8(x + \Delta x) + 5$
$= 3x^2 - 8x + 5 + 6x\Delta x + 3\overline{\Delta x}^2 - 8\Delta x$

2. $f(x) = x^3 - x^2 + x - 1$
$f(1) = 1 - 1 + 1 - 1 = 0$
$f(-1) = -1 - 1 - 1 - 1 = -4$
$f(a) = a^3 - a^2 + a - 1$
$f(x+1) = (x+1)^3 - (x+1)^2 + (x+1) - 1$
$= x^3 + 2x^2 + 2x$

3. $f(x) = \dfrac{x+3}{x^2 - x - 6}$

$f(0) = \dfrac{0+3}{0-0-6} = -\dfrac{1}{2}$

$f(-2) = \dfrac{-2+3}{4+2-6} = \dfrac{1}{0}$, undefined

$f(x-1) = \dfrac{x-1+3}{(x-1)^2 - (x-1) - 6}$

$= \dfrac{x+2}{x^2 - 3x - 4}$

4. $f(x) = 10^x$; $f(2) = 10^2 = 100$
$f(-x+1) - f(-x)$
$= 10^{-x+1} - 10^{-x}$
$= (10^{-x})\,10 - 10^{-x}$
$= 10^{-x}(10 - 1) = 9(10^{-x})$

5. $f(x) = 2\sin x - 3\cos x$
$f(0) = 2\sin 0 - 3\cos 0$
$= 0 - 3 = -3$

$f\left(\frac{\pi}{4}\right) = 2\dfrac{\sqrt{2}}{2} - 3\dfrac{\sqrt{2}}{2} = -\dfrac{\sqrt{2}}{2}$

6. $\lim\limits_{x\to 0}(3x^2 - 5x + 2)$

$= 3\lim\limits_{x\to 0} x^2 - 5\lim\limits_{x\to 0} x + \lim\limits_{x\to 0} 2$

$= 3(0) - 5(0) + 2 = 2$

| EXERCISES I |
|---|
| i. If $xy = y^2 - 6$, find $y$ when $x = 1$. |
| ii. If $f(x) = \dfrac{1}{x}$, find $f(a+h) - f(a)$. |
| iii. If $xy = 2y + 3$, find $y$ when $x = 2$. |

7. $\lim\limits_{x\to 1}\dfrac{x^2 + x}{2x - 1} = \dfrac{\lim\limits_{x\to 1}(x^2 + x)}{\lim\limits_{x\to 1}(2x - 1)}$

$= \dfrac{1+1}{2-1} = 2$

8. $\lim\limits_{x\to 2}\dfrac{x^2 - 4}{x - 2} = \lim\limits_{x\to 2}\dfrac{(x-2)(x+2)}{(x-2)}$

$= \lim\limits_{x\to 2}(x+2) = 4$

9. $\lim\limits_{x\to -1}\dfrac{3 - x}{x^2 - x} = \dfrac{3+1}{1+1} = 2$

10. $\lim\limits_{x\to 3^+}\dfrac{1}{x-3} = +\infty$, since

$f(5) = \frac{1}{2}$, $f(4) = 1$, $f\left(\frac{7}{2}\right) = 2$,

etc.

11. $\lim\limits_{x\to\infty}\dfrac{x^2 + 2x + 1}{x^3}$

$= \lim\limits_{x\to\infty}\dfrac{1}{x} + \lim\limits_{x\to\infty}\dfrac{2}{x^2} + \lim\limits_{x\to\infty}\dfrac{1}{x^3}$

$= 0 + 0 + 0 = 0$

12. $\lim\limits_{x\to\infty}\dfrac{3x^2 + x}{2x^2 - 3x} = \lim\limits_{x\to\infty}\dfrac{3 + \frac{1}{x}}{2 - \frac{3}{x}}$

$= \dfrac{3}{2}$

13. $\lim\limits_{h\to 0}\dfrac{(x+h)^2 - x^2}{h}$

$= \lim\limits_{h\to 0}\dfrac{x^2 + 2hx + h^2 - x^2}{h}$

$= \lim\limits_{h\to 0}(2x + h) = 2x$

**14.** $\lim_{h\to 0} \frac{3(x+h)^3 - 3x^3}{h}$

$= \lim_{h\to 0} \frac{3(x^3 + 3x^2h + 3xh^2 + h^3 - x^3)}{h}$

$= \lim_{h\to 0} 3(3x^2 + 3xh + h^2) = 9x^2$

**EXERCISES II**

i. Find $\lim_{x\to 2} \frac{x^2-4}{x^3-8}$.

ii. Find $\lim_{x\to 1} \sqrt{\frac{2x+3}{5x-2}}$.

iii. Find $\lim_{h\to 0} \frac{1}{h}\left(\frac{1}{x+h} - \frac{1}{x}\right)$.

iv. Find $\lim_{x\to 0} x\sqrt{1+\frac{1}{x^2}}$.

**15.** $f(x) = \frac{1}{x+1}$, $0 < x < 3$

We calculate

| $x$ | 0 | 1 | 2 | 3 |
|---|---|---|---|---|
| $f(x)$ | 1 | $\frac{1}{2}$ | $\frac{1}{3}$ | $\frac{1}{4}$ |

and it is evident that $f(x)$ exists for all values of $x$ in $0 < x < 3$. The function is continuous.

**EXERCISES III**

Is the following function continuous in its domain?

i. $f(x) = \begin{cases} \frac{x^2-4}{x^2+2x-8} \end{cases}$

$0 < x < 3,\ x \neq 2$

$f(2) = \frac{2}{3}$

Solutions to the exercises are on page 12.

## ANSWERS

1. 5, 0, 16,
   $3x^2 - 8x + 5 + 6x\Delta x + 3\overline{\Delta x}^2 - 8\Delta x$ — 1
2. 0, −4, $a^3 - a^2 + a - 1$,
   $x^3 + 2x^2 + 2x$ — 2
3. $-\frac{1}{2}$, undefined,
   $(x+2) \div (x^2 - 3x - 4)$ — 3
4. 100, $9(10^{-x})$ — 4
5. $-3, -\frac{\sqrt{2}}{2}$ — 5
6. 2 — 6
7. 2 — 7
8. 4 — 8
9. 2 — 9
10. $+\infty$ — 10
11. 0 — 11
12. $\frac{3}{2}$ — 12
13. $2x$ — 13
14. $9x^2$ — 14
15. Yes — 15

## BASIC FACTS

A **function** is a set of ordered pairs such that no two ordered pairs have the same first element. The **domain** of the function is the set of all the first elements of the ordered pairs. The **range** of the function is the set of all the second elements.

If $x$ and $y$ are two variables and for each value of $x$ a *single* value of $y$ may be determined, then we say that $y$ is a function of $x$. In this situation $x$ is called the **independent variable** and $y$ is called the **dependent variable.** The function is the association that determines the specific value of $y$ when $x$ is assigned a specific value. The correspondence between $x$ and $y$ may be shown by a formula, a graph, a table of values, or a statement of a principle. The notation used to express a function is usually a letter, and $f$, $g$, $F$, $G$, $\phi$, etc. are frequently used. If $f$ is a function that associates values of $x$ with values of the dependent variable $y$, then the symbol $f(x)$, read $f$ of $x$, is used to denote the value of $y$ that corresponds to the assigned value of $x$.

To obtain the value of a function at a specified value of the independent variable, replace the independent variable with this value and perform the indicated operation. The **limit of f(x)** as $x$ tends to a constant $a$ is a constant $L$ if for each positive number $\varepsilon$ there is a positive number $\delta$ such that $|f(x) - L| < \varepsilon$ whenever $0 < |x - a| < \delta$. The notation is

$$\lim_{x \to a} f(x) = L.$$

### Properties of Limits

1. If $\lim_{x \to a} f(x) = L_1$ and $\lim_{x \to a} f(x) = L_2$, then $L_1 = L_2$.

2. If $f(x) = c$, then $\lim_{x \to a} f(x) = c$

## ADDITIONAL INFORMATION

The notion of a **function** is very fundamental to a study of the calculus. We have already defined it as the correspondence which permits us to obtain the second element, $y$, of an ordered pair $(x, y)$, once the first element, $x$, has been assigned. A crucial thing to remember is that for each given value of $x$, there is *one* and *only one* value of $y$. In our study most of the functions will be defined by equations, and it is possible to write equations which do not determine a unique value of $y$ for each $x$. Consider, for example, the equation $x^2 + y^2 = 25$. If we solve for $y$ to obtain $y = \pm\sqrt{25 - x^2}$, we see that for each $x$ there are two values of $y$; e.g. $(0,5)$ and $(0,-5)$. This correspondence between $x$ and $y$ is called a **relation**, and we shall also make extensive studies of relations.

### Graphing

The **graph of a function** is a very informative tool. If the function is defined by an equation connecting $x$ and $y$, we can obtain the graph by considering a rectangular coordinate system, which permits us to associate each point in the plane with a unique ordered pair of numbers $(x, y)$. The coordinate system is shown in Figure 2–1, where two perpendicular lines, $X$ and $Y$, intersect at the origin, $O$.

Any point $P(x, y)$ has coordinates $x$ and $y$. The coordinate axes divide the plane into four quadrants, and the signs of the coordinates in each are as follows.

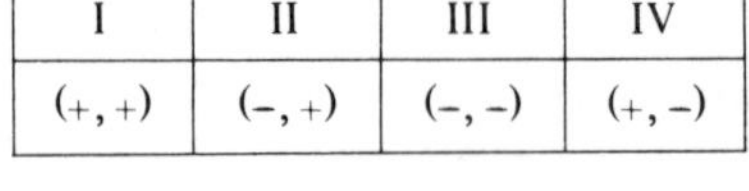

| I | II | III | IV |
|---|---|---|---|
| (+, +) | (–, +) | (–, –) | (+, –) |

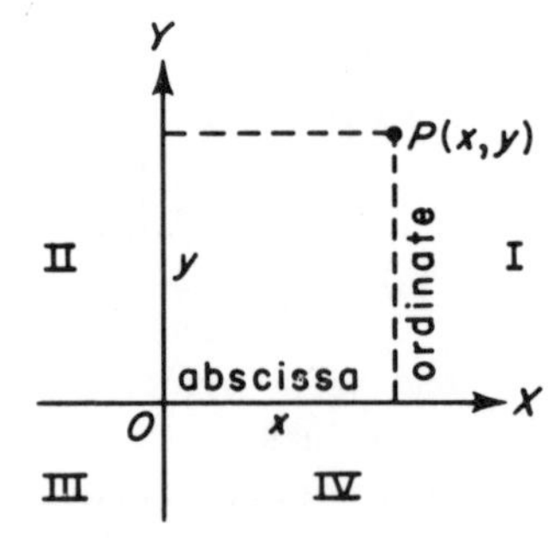


Fig. 2–1

To *plot* a point $P(x, y)$ is to locate its position from the values of $x$ along the horizontal and $y$ along the vertical. There is a *one-to-one correspondence* between the points in the plane and the set of ordered pairs $(x, y)$.

An equation in $x$ and $y$, such as $y = x^2 - x - 2$, has an infinite number of solutions, the ordered pair consisting of a value for $x$ and the corresponding value for $y$, which satisfy the equation. A table of these pairs follows.

| $x$ | –3 | –2 | –1 | 0 | 1 | 2 | 3 |
|---|---|---|---|---|---|---|---|
| $y$ | 10 | 4 | 0 | –2 | –2 | 0 | 4 |

The totality of such pairs is called the **solution set** of the equation, and if we associate these pairs with a set of points in the plane we have the **locus** of the equation. The geometric representation of the locus is called the **graph** of the equation. It should be clear that every point whose coordinates satisfy the equation lies on the locus, and, conversely, the coordinates of every point on the locus satisfy the equation.

3. If $u$, $v$, and $w$ are functions of $x$ and if $\lim_{x \to a} u = A$, $\lim_{x \to a} v = B$, and $\lim_{x \to a} w = C$, then

   a) $\lim_{x \to a} (u + v - w) = A + B - C$

   b) $\lim_{x \to a} (uvw) = ABC$

   c) $\lim_{x \to a} \frac{u}{v} = \frac{A}{B}$ if $B \neq 0$.

4. If $c$ is a constant and $c \neq 0$,

| | |
|---|---|
| $\lim_{x \to 0} \frac{c}{x} = \infty$ | $\lim_{x \to \infty} cx = \infty$ |
| $\lim_{x \to \infty} \frac{x}{c} = \infty$ | $\lim_{x \to \infty} \frac{c}{x} = 0$ |

5. If $\lim_{x \to a} f(x) = A$ and if

   a) $n$ is an odd integer or

   b) $n$ is an even integer and $A \geq 0$

   then $\lim_{x \to a} \sqrt[n]{f(x)} = \sqrt[n]{A}$.

**Continuity**

A function $f$ is said to be continuous at the number $a$ if (i) $f(a)$ is defined and (ii) if $\lim_{x \to a} f(x) = f(a)$.

If $f$ is continuous at every point in an interval $[c,d]$, it is said to be *continuous on the interval.*

A function which is not continuous at $a$ is said to be *discontinuous* at $a$.

The values in the above table can be used to locate a few of the points on the locus. If we then join these points with a smooth curve, we have a portion of the graph of the given equation (Figure 2-2).

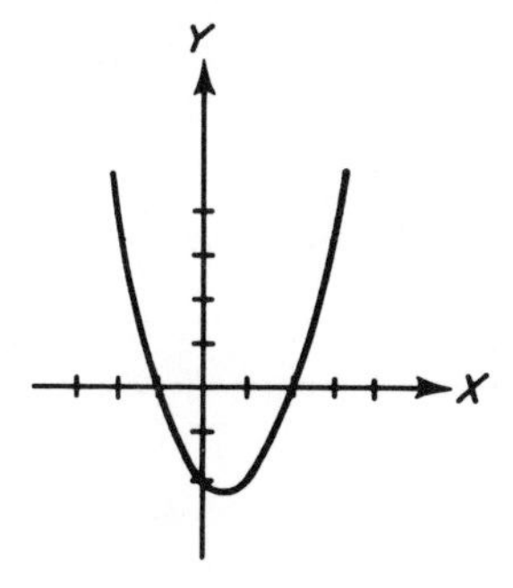


Fig. 2-2

In sketching the graph of an equation, some valuable assistance can be obtained from the characteristics of the resulting curve. The value of $x$ for which the curve crosses the $X$-axis is called the $x$-intercept; similarly, the value of $y$ for which the curve crosses the $Y$-axis is called the **$y$-intercept**. The intercepts may be obtained by: (1) setting $y$ equal to zero in the equation and solving for $x$, and then (2) setting $x$ equal to zero and solving for $y$. **Symmetry** is another useful characteristic. A curve can be symmetric with respect to a line or a point. To test for symmetry perform the following steps.

1. Replace $y$ by $-y$ in the equation. If the equation is unchanged, the curve is symmetric with respect to the $X$-axis.

2. Replace $x$ by $-x$ in the equation. If the equation is unchanged, the curve is symmetric with respect to the $Y$-axis.

3. Replace $x$ by $-x$ and $y$ by $-y$ in the equation. If the equation is unchanged, the curve is symmetric with respect to the origin, and the points $(a, b)$ and $(-a, -b)$ both lie on the locus.

**Continuity**

The graph of an equation which defines a function illustrates the concept of continuity very clearly. Consider, for example, the function defined by $y = \frac{x}{x-1}$. Calculate a table of values.

| $x$ | $-3$ | $-2$ | $-1$ | $0$ | $1$ | $2$ | $3$ |
|---|---|---|---|---|---|---|---|
| $y$ | $\frac{3}{4}$ | $\frac{2}{3}$ | $\frac{1}{2}$ | $0$ | - | $2$ | $\frac{3}{2}$ |

We note that at $x = 1$ the function is not defined and is therefore discontinuous at this point. We calculate some additional values.

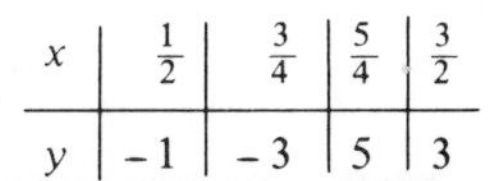

| $x$ | $\frac{1}{2}$ | $\frac{3}{4}$ | $\frac{5}{4}$ | $\frac{3}{2}$ |
|---|---|---|---|---|
| $y$ | $-1$ | $-3$ | $5$ | $3$ |

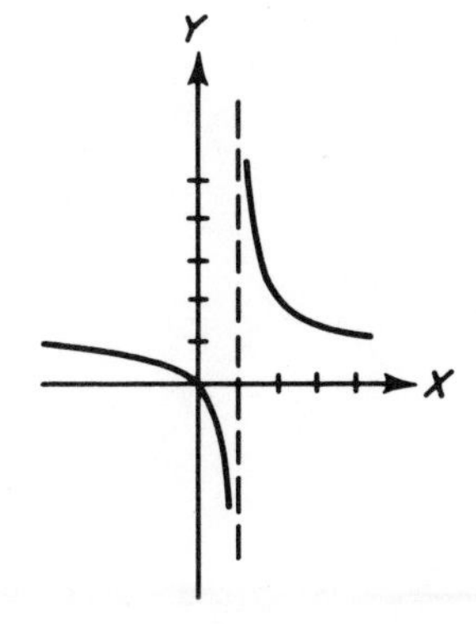


Fig. 2-3

The graph is shown in Figure 2-3.

The line $x = 1$ (dotted in Figure 2-3) is called a vertical asymptote. As $x$ approaches 1, the value of $y$ increases (decreases) beyond bound and the distance between the curve and the asymptote becomes infinitesimal.

## SOLUTIONS TO EXERCISES

I

i. $xy = y^2 - 6$
If $x = 1$, $y = y^2 - 6$ or
$y^2 - y - 6$
$= (y - 3)(y + 2) = 0$
$y = 3$ and $y = -2$

ii. If $f(x) = \frac{1}{x}$, then

$f(a + h) - f(a)$

$= \frac{1}{a + h} - \frac{1}{a}$

$= \frac{a - a - h}{a(a + h)}$

$= -\frac{h}{a^2 + ah}$

iii. Given $xy = 2y + 3$,
if $x = 2$, then $2y = 2y + 3$
which simplifies to $0 = 3$.
Since this statement is false there are no values of $y$ which satisfy the equation when $x = 2$.

II

i. $\lim_{x\to2} \frac{x^2 - 4}{x^3 - 8}$

Direct substitution yields $\frac{0}{0}$.

$\frac{x^2 - 4}{x^3 - 8}$

$= \frac{(x - 2)(x + 2)}{(x - 2)(x^2 + 2x + 4)}$

$= \frac{x + 2}{x^2 + 2x + 4}$

$\lim_{x\to2} \frac{x^2 - 4}{x^3 - 8}$

$= \lim_{x\to2} \frac{x + 2}{x^2 + 2x + 4}$

$= \frac{4}{4 + 4 + 4} = \frac{4}{12} = \frac{1}{3}$

ii. $\lim_{x\to1} \sqrt{\frac{2x + 3}{5x - 2}} = \sqrt{\frac{5}{5 - 2}}$

$= \sqrt{\frac{5}{3}}$

iii. $\lim_{h\to0} \frac{1}{h}\left(\frac{1}{x + h} - \frac{1}{x}\right)$

$\frac{1}{h}\left(\frac{1}{x + h} - \frac{1}{x}\right)$

$= \frac{1}{h}\left(\frac{x - x - h}{x(x + h)}\right)$

$= -\frac{1}{x^2 + xh}$

$\lim_{h\to0} \frac{1}{h}\left(\frac{1}{x + h} - \frac{1}{x}\right)$

$= -\frac{1}{x^2}$

iv. $\lim_{x\to0} x\sqrt{1 + \frac{1}{x^2}}$

$= \lim_{x\to0} \sqrt{x^2 + 1}$

$= \sqrt{1} = 1$

III

i. $f(x) = \begin{cases} \frac{x^2 - 4}{x^2 + 2x - 8} \\ f(2) = \frac{2}{3} \end{cases}$

$0 < x < 3,\ x \neq 2$

The first expression has values of $f(x)$ for $x$ in the interval from 0 to 3 except at $x = 2$. We consider
$\lim_{x\to2} f(x)$

$= \lim_{x\to2} \frac{(x - 2)(x + 2)}{(x - 2)(x + 4)}$

$= \lim_{x\to2} \frac{x + 2}{x + 4}$

$= \frac{2 + 2}{2 + 4} = \frac{4}{6} = \frac{2}{3}.$

Since this is the value of $f(2)$ assigned to $f(x)$ by the second expression the function is continuous.

# 3 DIFFERENTIATING ALGEBRAIC FORMS

## SELF-TEST

DIRECTIONS: Write your answers in the numbered regions to the right. To check your answers, turn the page. Study the solutions to the problems you missed, and do the exercises following any problem or group of problems in which you had errors.

In **1-10** find $\frac{dy}{dx}$.

**1.** $y = 2x^5 - 3x^2$

**2.** $y = ax^4 - bx^{-2}$

**3.** $y = (2x^2 - x)^2$

**4.** $y = 2x^3(a + bx)^2$

**5.** $y = \sqrt{1 - 2x}$

**6.** $y = x\sqrt{a - x}$

**7.** $y = \frac{a - x}{a + x}$

**8.** $y = \frac{b}{a}\sqrt{a^2 - x^2}$

**9.** $y = \frac{x^2}{2 + x}$

In **10-14**, find the value of the derivative at the given value of the variable.

**10.** $y = (x + 2)(x - 3)$, $x = 3$

**11.** $y = x^2(1 - x)(2x - 3)$, $x = 2$

**12.** $y = \frac{3x - 2}{x^2 - 1}$, $x = 2$

**13.** $y = (2x^2 - 3x)^3$, $x = \frac{1}{2}$

**14.** $y = \frac{\sqrt{1 - x^2}}{x}$, $x = \frac{1}{2}$

1 ______

2 ______

3 ______

4 ______

5 ______

6 ______

7 ______

8 ______

9 ______

10 ______

11 ______

12 ______

13 ______

14 ______

## SOLUTIONS

The derivatives are found by using the formulas given in the Basic Facts. You should especially be familiar with the chain rule.

**1.** $y = 2x^5 - 3x^2$
$y' = 2D_x(x^5) - 3D_x(x^2)$
$= 10x^4 - 6x$

**2.** $y = ax^4 - bx^{-2}$
$y' = aD_x(x^4) - bD_x(x^{-2})$
$= 4ax^3 + 2bx^{-3}$

**3.** $y = (2x^2 - x)^2 = v^2$
where $v = 2x^2 - x$
$y' = 2(2x^2 - x)D_x(2x^2 - x)$
$= 2(2x^2 - x)(4x - 1)$
$= 16x^3 - 12x + 2x$

**4.** $y = 2x^3(a + bx)^2 = (u)(v)^2$
where $u = 2x^3$ and $v = a + bx$
$y' = 2x^3D_x(a + bx)^2 + 2(a + bx)^2D_xx^3$
$= 2x^3[2(a + bx)b] + 2(a + bx)^2 3x^2$
$= 4bx^3(a + bx) + 6x^2(a + bx)^2$
$= 2x^2(a + bx)(2bx + 3a + 3bx)$
$= 2x^2(a + bx)(5bx + 3a)$

**5.** $y = \sqrt{1 - 2x} = (1 - 2x)^{\frac{1}{2}} = v^{\frac{1}{2}}$

$y' = \frac{1}{2}(1 - 2x)^{-\frac{1}{2}}D_x(1 - 2x)$

$= \frac{1}{2}(1 - 2x)^{-\frac{1}{2}}(-2)$

$= -\dfrac{1}{\sqrt{1 - 2x}}$

**6.** $y = x\sqrt{a - x} = x(a - x)^{\frac{1}{2}}$

$y' = (a - x)^{\frac{1}{2}} + x\left(\frac{1}{2}\right)(a - x)^{-\frac{1}{2}}(-1)$

$= (a - x)^{\frac{1}{2}} - \dfrac{x}{2\sqrt{a - x}}$

$= \dfrac{2a - 2x - x}{2\sqrt{a - x}}$

$= \dfrac{2a - 3x}{2\sqrt{a - x}}$

**7.** $y = \dfrac{a - x}{a + x} = \dfrac{u}{v}$

$y' = \dfrac{(a + x)D_x(a - x) - (a - x)D_x(a + x)}{(a + x)^2}$

$= \dfrac{(a + x)(-1) - (a - x)(1)}{(a + x)^2}$

$= -\dfrac{2a}{(a + x)^2}$

**8.** $y = \dfrac{b}{a}\sqrt{a^2 - x^2} = c\,v^{\frac{1}{2}}$

$y' = \dfrac{b}{a}\left(\dfrac{1}{2}\right)(a^2 - x^2)^{-\frac{1}{2}}(-2x)$

$= -\dfrac{b}{a}\,\dfrac{x}{\sqrt{a^2 - x^2}}$

**9.** $y = \dfrac{x^2}{2 + x}$

$y' = \dfrac{(2 + x)(2x) - x^2(1)}{(2 + x)^2}$

$= \dfrac{4x + 2x^2 - x^2}{(2 + x)^2}$

$= \dfrac{4x + x^2}{(2 + x)^2}$

EXERCISES I

Find $y'$ in each problem.

i. $y = x(2 - x^2)^{\frac{3}{2}}$

ii. $y = \dfrac{3}{x^2}$

iii. $y = \dfrac{x^2 - 2x}{2x}$

iv. $y = \dfrac{3x^2}{\sqrt[3]{x}} - 9\sqrt[3]{x^4}$

v. $y = (ax - b)\sqrt{cx - d}$

**10.** $y = (x + 2)(x - 3)$
$y' = x - 3 + x + 2 = 2x - 1$
$y'(3) = (2)(3) - 1 = 5$

**11.** $y = x^2(1 - x)(2x - 3)$
$y' = 2x(1 - x)(2x - 3)$
$+ x^2(2x - 3)(-1) + x^2(1 - x)(2)$
$= 15x^2 - 8x^3 - 6x$
$y'(2) = 60 - 64 - 12 = -16$

12. $y = \dfrac{3x - 2}{x^2 - 1}$

$$y' = \frac{(x^2 - 1)(3) - (3x - 2)2x}{(x^2 - 1)^2}$$

$$= \frac{4x - 3x^2 - 3}{(x^2 - 1)^2}$$

$$y'(2) = \frac{8 - 12 - 3}{9} = -\frac{7}{9}$$

**EXERCISES II**

Find the derivative of $y = f(x)$ by applying the 5-step rule.

i. $y = ax^3 - x$

ii. $y = 2x^{-2}$

13. $y = (2x^2 - 3x)^3$

$$y' = 3(2x^2 - 3x)^2(4x - 3)$$

$$y'\left(\tfrac{1}{2}\right) = 3\left(\tfrac{1}{2} - \tfrac{3}{2}\right)^2(2 - 3) = -3$$

14. $y = \dfrac{\sqrt{1 - x^2}}{x} = \dfrac{u^{\frac{1}{2}}}{x}$

$$y' = \frac{x\left(\frac{1}{2}\right)(1 - x^2)^{-\frac{1}{2}}(-2x) - (1 - x^2)^{\frac{1}{2}}}{x^2}$$

$$y' = -\frac{1}{x^2\sqrt{1 - x^2}}$$

$$y'\left(\frac{1}{2}\right) = -\frac{1}{\frac{1}{4}\sqrt{\frac{3}{4}}} = -\frac{8}{\sqrt{3}} = -\frac{8\sqrt{3}}{3}$$

**EXERCISES III**

Find the value of $y' = f(x)$ at the indicated value.

i. $y = x^{-3} + x^{-1}$, $x = \frac{1}{2}$

ii. $y = \sqrt{4 - x^2}$, $x = 1$

iii. $y = x^2 - x - 6$, $x = \frac{1}{2}$

iv. $y = x\sqrt{x} - x\sqrt[3]{x}$, $x = 4$

v. $y = (x^2 + 3)^3 - (x^2 + 3)^2$ at $x = 1$.

Solutions to the exercises are on page 18.

**ANSWERS**

1. $10x^4 - 6x$
2. $4ax^3 + 2bx^{-3}$
3. $16x^3 - 12x + 2x$
4. $2x^2(a + bx)(5bx + 3a)$
5. $-\dfrac{1}{\sqrt{1 - 2x}}$
6. $\dfrac{2a - 3x}{2\sqrt{a - x}}$
7. $-\dfrac{2a}{(a + x)^2}$
8. $-\dfrac{bx}{a\sqrt{a^2 - x^2}}$
9. $\dfrac{4x + x^2}{(2 + x)^2}$
10. 5
11. $-16$
12. $-\frac{7}{9}$
13. $-3$
14. $-\dfrac{8\sqrt{3}}{3}$

## BASIC FACTS

**Increment.** The difference between two consecutive values of a variable is called the increment of the variable and is denoted by $h$ or $\Delta x$. If we have $y = f(x)$ and change $x$, then there is a corresponding change in $y$, and we call $\Delta y$ the increment of $y$.

**The Derivative.** If $f$ is a function of the independent variable $x$ defined by the equation $y = f(x)$, the derivative of the function $f$, denoted by $y'$, is defined by the formula

$$y' = \lim_{h \to 0} \frac{f(x+h) - f(x)}{h}$$

where $x$ remains fixed as $h$ tends to zero and providing that the limit exists.

**Notation.** The following symbols are in common use to denote the derivative of a function.

$$D_x y,\ D_x f(x),\ y',\ f'(x),\ \frac{dy}{dx},\ \frac{df(x)}{dx}$$

The process of finding a derivative is called **differentiation.** Derivatives may be obtained by a method called the **five-step rule.**

**Formulas for the Algebraic Forms**

The derivative of a constant is zero.

1. $$\frac{dc}{dx} = 0$$

The derivative of the product of a constant and a function is equal to the product of the constant and the derivative of the function.

2. $$\frac{d}{dx}(cv) = c\,\frac{dv}{dx}$$

The derivative with respect to $x$ of a power function of $x$ is equal to the product of the index of the power and the function with the index of the power reduced by one.

## ADDITIONAL INFORMATION

The formulas for the derivative of an algebraic form are derived from the definition by applying a **five-step rule.**

**Theorem 3-1.** If $f(x) = c$, a constant for all $x$, then $f'(x) = 0$ for all $x$.

*Step 1.* Write the formula for the functions at $x$: $f(x) = c$

*Step 2.* Write the formula for the function at $x + h$: $f(x+h) = c$

Note: Since $f(x) = c$ for all $x$, $f(x) = c$ and $f(x+h) = c$.

*Step 3.* Subtract $f(x)$ from $f(x+h)$: $f(x+h) - f(x) = c - c = 0$

*Step 4.* Divide by $h$ and simplify: $\dfrac{f(x+h) - f(x)}{h} = \dfrac{0}{h} = 0$

*Step 5.* Take the limit as $h \longrightarrow 0$: $\lim\limits_{h \to 0} \dfrac{\Delta f}{h} = \lim\limits_{h \to 0} 0 = 0.$

This is the definition of $f'(x)$. In step five of the above proof we made use of the abbreviated notation for the incremental change in the function $y$, defined by $y = f(x)$, due to an incremental change in the independent variable $x$.

**Notation to Remember**

Increment of $x$: $h = \Delta x$.

Increment of the function:

$$\Delta f = f(x+h) - f(x)$$

Derivative: $\lim\limits_{h \to 0} \dfrac{\Delta f}{h} = f'(x)$

Keep this notation firmly in mind as we apply the five-step rule to the remaining theorems. Also to test your comprehension of this rule work the problems in Exercise II.

**Theorem 3-2.** If $c$ is a constant and $f(x)$ has a derivative $f'(x)$, then $g(x) = cf(x)$ has the derivative $g'(x) = cf'(x)$.

*Step 1.* $g(x) = cf(x)$ *Step 2.* $g(x+h) = cf(x+h)$

*Step 3.* $\Delta g = g(x+h) - g(x) = cf(x+h) - cf(x) = c\Delta f$

*Step 4.* $\dfrac{\Delta g}{h} = c\,\dfrac{\Delta f}{h}$ *Step 5.* $g'(x) = \lim\limits_{h \to 0} c\,\dfrac{\Delta f}{h} = c \lim\limits_{h \to 0} \dfrac{\Delta f}{h} = cf'(x)$

In simplifying the expressions in Step 5, we make use of the properties of limits discussed in the last chapter. In particular we used "the limit of a product is the product of the limits" and "the limit of a constant is a constant."

**Theorem 3-3.** If $n$ is a positive integer and $f(x) = x^n$, then $f'(x) = nx^{n-1}$

*Step 1.* $f(x) = x^n$ *Step 2.* $f(x+h) = (x+h)^n$

*Step 3.* $\Delta f = f(x+h) - f(x) = (x+h)^n - x^n$

Applying the binomial theorem (see Chapter 1):

$$\Delta f = x^n + nx^{n-1}h + \binom{n}{2} x^{n-2}h^2 + \cdots + h^n - x^n$$

The first and last terms cancel to yield:

$$\Delta f = nx^{n-1}h + \binom{n}{2} x^{n-2}h^2 + \binom{n}{3} x^{n-3}h^3 + \cdots + h^n$$

where every term on the right has $h$ as a factor.

3. $\dfrac{d}{dx}(x^n) = nx^{n-1}$

3a. $\dfrac{dx}{dx} = 1$

3b. $\dfrac{d}{dx}(v^n) = nv^{n-1}\dfrac{dv}{dx}$

The derivatives of the sum of functions is equal to the sum of the derivatives of the functions.

4. $\dfrac{d}{dx}(u + v - w) = \dfrac{du}{dx} + \dfrac{dv}{dx} - \dfrac{dw}{dx}$

The derivative of the product of two functions is equal to the product of the first function and the derivative of the second plus the product of the second function and the derivative of the first.

5. $\dfrac{d}{dx}(uv) = u\dfrac{dv}{dx} + v\dfrac{du}{dx}$

The derivative of a quotient of two functions is equal to the product of the denominator and the derivative of the numerator minus the product of the numerator and the derivative of the denominator, all divided by the square of the denominator.

6. $\dfrac{d}{dx}\left(\dfrac{u}{v}\right) = \dfrac{v\dfrac{du}{dx} - u\dfrac{dv}{dx}}{v^2}$

6a. $\dfrac{d}{dx}\left(\dfrac{u}{c}\right) = \dfrac{1}{c}\dfrac{du}{dx}$

If $y = f(v)$ and $v = g(x)$, then

7. $\dfrac{dy}{dx} = \dfrac{dy}{dv} \cdot \dfrac{dv}{dx}$.

This is known as the **Chain Rule**.

If $y = f(x)$, then

8. $\dfrac{dx}{dy} = \dfrac{1}{\dfrac{dy}{dx}}$.

The **value of a derivative** is found by substitution. Insert the given value of the variable into the expression for the derivative.

*Step 4.* $\dfrac{\Delta f}{h} = nx^{n-1} + \binom{n}{2}x^{n-2}h + \binom{n}{3}x^{n-3}h^2 + \cdots + h^{n-1}$

Note the terms that have $h$ as a factor.

*Step 5.* $f'(x) = \lim\limits_{h\to 0}\dfrac{\Delta f}{h} = nx^{n-1} + 0 + 0 + \cdots + 0 = nx^{n-1}$

**Theorem 3-4.** If $u(x)$ has derivative $u'(x)$ and $v(x)$ has derivative $v'(x)$, then $f(x) = u(x)v(x)$ has derivative $f'(x) = u(x)v'(x) + v(x)u'(x)$.

*Step 1.* $f(x) = u(x)v(x)$

*Step 2.* $f(x+h) = u(x+h)v(x+h) = (u + \Delta u)(v + \Delta u)$

In the last product $f(x+h)$ can be written $f(x) + \Delta f$ (see "notation to remember") and, using the more precise functional notation (see Chapter 2), $u$ and $v$ replace $u(x)$ and $v(x)$.

*Step 3.* $\Delta f = f(x+h) - f(x) = (u + \Delta u)(v + \Delta v) - uv$
$= u\Delta v + v\Delta u + \Delta u\Delta v$

*Step 4.* $\dfrac{\Delta f}{h} = u\dfrac{\Delta v}{h} + v\dfrac{\Delta u}{h} + \Delta u\dfrac{\Delta v}{h}$

*Step 5.* $f'(x) = \lim\limits_{h\to 0}\dfrac{\Delta f}{h} = u\lim\limits_{h\to 0}\dfrac{\Delta v}{h} + v\lim\limits_{h\to 0}\dfrac{\Delta u}{h} + \lim\limits_{h\to 0}\Delta u\dfrac{\Delta v}{h}$
$= u(x)v'(x) + v(x)u'(x) + 0$

The last term $\Delta u\dfrac{\Delta v}{h}$ tends to zero as $h$ tends to zero because $u(x)$ is a continuous function and $\Delta u = u(x+h) - u(x)$ tends to zero as $h \longrightarrow 0$, $\dfrac{\Delta v}{h}$ becomes $v'(x)$ and then $(0)[v'(x)] = 0$.

**Theorem 3-5.** (Chain Rule). If $f$, $g$, and $v$ are functions such that $f(x) = g[v(x)]$ and $g$ and $v$ are differentiable, then $f$ is differentiable and $f'(x) = g'(v)v'(x) = \dfrac{df}{dv}\dfrac{dv}{dx}$.

*Step 1.* $f(x) = g[v(x)]$

*Step 2.* $f(x+h) = g[v(x+h)] = g[v(x) + v(x+h) - v(x)]$

In the delta notation this becomes

$f + \Delta f = g(v + \Delta v)$.

*Step 3.* $\Delta f = g(v + \Delta v) - g(v)$

Multiply by $\dfrac{\Delta v}{\Delta v}$ and get $\left[\dfrac{g(v - \Delta v) - g(v)}{\Delta v}\right]\Delta v$

$\Delta f = [g'(v) - G(\Delta v)]\Delta v$

The last form is obtained since the function $g$ is differentiable [$g'(v)$ exists] and by definition of the derivative $[g(v + \Delta v) - g(v)]/\Delta v = g'(v) +$ a term $[G(\Delta v)]$ which tends to zero as $\Delta v$ tends to zero; thus we define $G(0) = 0$.

*Step 4.* $\dfrac{\Delta f}{h} = [g'(v) + G(\Delta v)]\dfrac{\Delta v}{h}$

*Step 5.* Since $\Delta v \longrightarrow 0$, $G(\Delta v) \longrightarrow 0$, and $\dfrac{\Delta v}{h} \longrightarrow v'(x)$ as $h \longrightarrow 0$, we have $\lim\limits_{h\to 0}\dfrac{\Delta f}{h} = f'(x) = g'(v)v'(x)$.

## SOLUTIONS TO EXERCISES

I

**i.** $y = x(2 - x^2)^{\frac{3}{2}}$

$y' = (2 - x^2)^{\frac{3}{2}}$

$+ x\left(\frac{3}{2}\right)(2 - x^2)^{\frac{1}{2}}(-2x)$

$= (2 - x^2)^{\frac{3}{2}}$

$-3x^2(2 - x^2)^{\frac{1}{2}}$

$= (2 - x^2)^{\frac{1}{2}}(2 - 4x^2)$

**ii.** $y = \frac{3}{x^2} = 3x^{-2}$

$y' = 3(-2)x^{-3} = -\frac{6}{x^3}$

**iii.** $y = \frac{x^2 - 2x}{2x} = \frac{1}{2}x - 1$

$y' = \frac{1}{2}$

**iv.** $y = \frac{3x^2}{\sqrt[3]{x}} - 9\sqrt[3]{x^4}$

$= 3x^{\frac{5}{3}} - 9x^{\frac{4}{3}}$

$y' = 3\left(\frac{5}{3}\right)x^{\frac{2}{3}} - 9\left(\frac{4}{3}\right)x^{\frac{1}{3}}$

$= 5x^{\frac{2}{3}} - 12x^{\frac{1}{3}}$

$= \sqrt[3]{x}\,(5x^2 - 12)$

**v.** $y = (ax - b)\sqrt{cx - d}$

$y' = a\sqrt{cx - d}$

$+ (ax - b)\left(\frac{1}{2}\right)(cx - d)^{-\frac{1}{2}}(c)$

$= a\sqrt{cx - d} + \frac{c(ax - b)}{2\sqrt{cx - d}}$

II

**i.** $y = ax^3 - x$

$y + \Delta y = a(x + h)^3 - (x + h)$

$\Delta y = a(x^3 + 3x^2h + 3xh^2 + h^3)$

$- x - h - ax^3 + x$

$= 3ax^2h + 3axh^2 + ah^3 - h$

$\frac{\Delta y}{h} = 3ax^2 + 3axh + ah^2 - 1$

$\lim_{h \to 0} \frac{\Delta y}{h} = 3ax^2 - 1$

**ii.** $y = 2x^{-2} = \frac{2}{x^2}$

$y + \Delta y = \frac{2}{(x + h)^2}$

$\Delta y = \frac{2}{(x + h)^2} - \frac{2}{x^2}$

$= \frac{2(x^2 - x^2 - 2xh - h^2)}{x^2(x + h)^2}$

$\frac{\Delta y}{h} = \frac{-4x - 2h}{x^2(x + h)^2}$

$\lim_{h \to 0} \frac{\Delta y}{h} = -\frac{4x}{x^4} = -\frac{4}{x^3}$

III

**i.** $y = x^{-3} + x^{-1}$

$y' = -3x^{-4} - 1x^{-2}$

$= -\frac{3}{x^4} - \frac{1}{x^2}$

$y'\left(\frac{1}{2}\right) = -3(2)^4 - (2)^2$

$= -(48 + 4) = -52$

**ii.** $y = \sqrt{4 - x^2}$

$y' = \frac{1}{2}(4 - x^2)^{-\frac{1}{2}}(-2x)$

$= -x(4 - x^2)^{-\frac{1}{2}}$

$y'(1) = -\frac{1}{\sqrt{3}} = -\frac{\sqrt{3}}{3}$

**iii.** $y = x^2 - x - 6$

$y' = 2x - 1$

$y'\left(\frac{1}{2}\right) = 2\left(\frac{1}{2}\right) - 1 = 0$

**iv.** $y = x\sqrt{x} - x\sqrt[3]{x}$

$= x^{\frac{3}{2}} - x^{\frac{4}{3}}$

$y' = \frac{3}{2}x^{\frac{1}{2}} - \frac{4}{3}x^{\frac{1}{3}}$

$y'(4) = \frac{3}{2}\sqrt{4} - \frac{4}{3}\sqrt[3]{4}$

$= 3 - \frac{4}{3}\sqrt[3]{4}$

**v.** $y = (x^2 + 3)^3 - (x^2 + 3)^2$

$y' = 3(x^2 + 3)^2(2x)$

$-2(x^2 + 3)(2x)$

$= 6x(x^2 + 3)^2 - 4x(x^2 + 3)$

$y'(1) = 6(4)^2 - 4(4)$

$= 4^2(6 - 1) = 16 \cdot 5$

$= 80.$

# 4 HIGHER DERIVATIVES AND GEOMETRIC INTERPRETATION

## SELF-TEST

DIRECTIONS: Write your answers in the numbered regions to the right. To check your answers, turn the page. Study the solutions to the problems you missed, and do the exercises following any problem or group of problems in which you had errors.

In **1-4** find $\dfrac{d^2y}{dx^2} = y''$

1. $y = 3x^4 - 6x^2 + 2x$
2. $y = x^3 - 9x^2 + x$
3. $y = (2x + 1)^{\frac{3}{2}}$
4. $yx + y^2 = 2x^2$
5. Obtain $y'$ by implicit differentiation. Solve for $y_i = f_i(x)$ and obtain $f_i'(x)$. Show that $f_i'(x)$ satisfies $y'$.

$$2x^2 + y^2 = 4$$

In **6-9** find the value of $\dfrac{d^3y}{dx^3}$ at the specified value of the variable.

6. $y = x^5 - 2x^4$ at $x = 2$
7. $y = \sqrt{2x - 1}$ at $x = 1$
8. $y = \dfrac{a}{x}$ at $x = a$
9. $y = 3x^2 + 2x - x^{-2}$ at $x = 2$

Find the equations of the tangent line to each curve in **10-12** at the points indicated.

10. $y = x^3 - 9x$, $(2, -10)$
11. $y = \sqrt[3]{x}$, $(8, 2)$
12. $y = \dfrac{2}{x}(1 - x)$, $(1, 0)$

Find the equations of the normal line to each curve in **13-14** at the points indicated.

13. $y = \sqrt{x}$, $(9, 3)$
14. $x^2 + y^2 = 4$, $(1, \sqrt{3})$

| |
|---|
| 1 |
| 2 |
| 3 |
| 4 |
| 5 |
| 6 |
| 7 |
| 8 |
| 9 |
| 10 |
| 11 |
| 12 |
| 13 |
| 14 |

## SOLUTIONS

1. $y = 3x^4 - 6x^2 + 2x$
$y' = 12x^3 - 12x + 2$
$y'' = 36x^2 - 12$

2. $y = x^3 - 9x^2 + x$
$y' = 3x^2 - 18x + 1$
$y'' = 6x - 18$

3. $y = (2x + 1)^{\frac{3}{2}}$
$y' = \frac{3}{2}(2x + 1)^{\frac{1}{2}}(2)$
$y'' = 3\left(\frac{1}{2}\right)(2x + 1)^{-\frac{1}{2}}$
$= \frac{3}{2}(2x + 1)^{-\frac{1}{2}}$

4. $yx + y^2 = 2x^2$
$xy' + y + 2yy' = 4x$
$(x + 2y)y' + y = 4x$
$(1 + 2y')y' + (x + 2y)y'' + y' = 4$
$y'' = \dfrac{4 - 2y'^2 - 2y'}{x + 2y}$

5. $2x^2 + y^2 = 4$
$4x + 2yy' = 0,\ y' = -\dfrac{2x}{y}$
$y_i = \pm\sqrt{4 - 2x^2}$
$y_i' = \pm\frac{1}{2}(4 - 2x^2)^{-\frac{1}{2}}(-4x)$
$= \mp 2x(4 - 2x^2)^{-\frac{1}{2}}$
Substitute $y_i$ into $y'$
$y' = -\dfrac{2x}{\pm\sqrt{4 - 2x^2}}$
$= \mp 2x(4 - 2x^2)^{-\frac{1}{2}}$

### EXERCISES I

Obtain $y'$ by implicit differentiation. Solve for $y_i = f_i(x)$ and obtain $f_i'(x)$. Show that each $f_i'(x)$ satisfies $y'$.

i. $x^{\frac{2}{3}} + y^{\frac{2}{3}} = a^{\frac{2}{3}}$
($a$ = constant)

ii. $xy + y^2 = x^2$

iii. Find $y''$

$y = x^2(x - 3)^3$

6. $y = x^5 - 2x^4$ at $x = 2$
$y' = 5x^4 - 8x^3$
$y'' = 20x^3 - 24x^2$
$y''' = 60x^2 - 48x$
$y'''(2) = 240 - 96 = 144$

7. $y = \sqrt{2x - 1}$ at $x = 1$
$y' = \frac{1}{2}(2x - 1)^{-\frac{1}{2}}(2)$
$y'' = -\frac{1}{2}(2x - 1)^{-\frac{3}{2}}(2)$
$y''' = \frac{3}{2}(2x - 1)^{-\frac{5}{2}}(2)$
$y'''(1) = 3$

8. $y = \dfrac{a}{x}$ at $x = a$
$y' = -ax^{-2},\ y'' = 2ax^{-3}$
$y''' = -6ax^{-4};\ y'''(a) = -6a^{-3}$

9. $y = 3x^2 + 2x - x^{-2}$ at $x = 2$
$y' = 6x + 2 + 2x^{-3}$
$y'' = 6 - 6x^{-4}$
$y''' = 24x^{-5};\ y'''(2) = \frac{24}{32} = \frac{3}{4}$

10. $y = x^3 - 9x;\ (2, -10)$
$y' = 3x^2 - 9$
$y'(2) = m = 12 - 9 = 3$
$y + 10 = 3(x - 2),\ y = 3x - 16$

### EXERCISES II

True or False

i. The $n$th derivative of $y = x^n$ is $n!$

ii. If $s$ is a function of $t$, $s = f(t)$, the second derivative of $s$ is $\dfrac{d^2s}{dx^2}$.

iii. The equation of the tangent to the line $y = x$ at $P_0(x_0, y_0)$ is $y = x$.

11. $y = \sqrt[3]{x},\ (8, 2)$
$y' = \frac{1}{3}x^{-\frac{2}{3}}$
$y'(8) = m = \frac{1}{3}\left(\frac{1}{4}\right) = \frac{1}{12}$
$y - 2 = \frac{1}{12}(x - 8),\ 12y = x + 16$

12. $y = \frac{2}{x}(1 - x),\ (1, 0)$
$y' = -2x^{-2},\ y'(1) = -2$
$y - 0 = -2(x - 1),\ y + 2x = 2$

**13.** $y = \sqrt{x}$, $(9, 3)$

$y' = \frac{1}{2\sqrt{x}}$, $y'(9) = \frac{1}{6}$

$m_N = -6$; $y - 3 = -6(x - 9)$

$y + 6x = 57$

**14.** $x^2 + y^2 = 4$, $(1, \sqrt{3})$

$2x + 2yy' = 0$, $y' = -\frac{x}{y}$

$m = -\frac{1}{\sqrt{3}}$, $m_N = \sqrt{3}$

$y - \sqrt{3} = \sqrt{3}(x - 1)$, $y = \sqrt{3}x$

EXERCISES III

Find the lengths of the subtangent, subnormal, tangent, and normal to the curve $C$ at $P_0$.

**i.** $y = x^3 - 2x^2 - x + 3$, $P_0(2, 1)$

**ii.** $y = -\sqrt{3x + 1}$, $P_0(5, -4)$

**iii.** $y = \frac{x^2}{2x + 1}$, $P_0\left(2, \frac{4}{5}\right)$

Solutions to the exercises are on page 24.

**ANSWERS**

| | |
|---|---|
| **1.** $36x^2 - 12$ | 1 |
| **2.** $6x - 18$ | 2 |
| **3.** $\frac{3}{2}(2x + 1)^{-\frac{1}{2}}$ | 3 |
| **4.** $\frac{4 - 2y'^2 - 2y'}{x + 2y}$ | 4 |
| **5.** $y' = -\frac{2x}{y}$ <br> $y'_i = \mp 2x\sqrt{4 - 2x^2}$ | 5 |
| **6.** 144 | 6 |
| **7.** 3 | 7 |
| **8.** $-6a^{-3}$ | 8 |
| **9.** $\frac{3}{4}$ | 9 |
| **10.** $3x - 16$ | 10 |
| **11.** $12y = x + 16$ | 11 |
| **12.** $y + 2x = 2$ | 12 |
| **13.** $y + 6x = 57$ | 13 |
| **14.** $y = \sqrt{3}x$ | 14 |

## BASIC FACTS

The derivative of a function $f$ is also a function $g$, and may also have a derivative. Thus if $f'(x) = g(x)$ exists and

$$\lim_{h \to 0} \frac{g(x+h) - g(x)}{h} = g'(x)$$

exists, then $g'(x) = f''(x)$ and is called a **second derivative.** The process can be continued, and we have **third derivatives, fourth derivatives,** etc. The notations for the second derivative are

$$f''(x),\ \frac{d^2y}{dx^2},\ D_x^2\,y,\ y''.$$

For the $n$th derivative we write

$$f^{(n)}(x),\ \frac{d^ny}{dx^n},\ D_x^{(n)}y,\ y^{(n)}.$$

### Implicit Differentiation

If a relation between the variables $x$ and $y$ is written in the form $f(x, y) = 0$, we shall say that the associated functions $y_i = f_i(x)$ are defined *implicitly*. It is not necessary to solve the equation $f(x, y) = 0$ for $y$ explicitly in terms of $x$ before differentiating. We can apply the chain rule indicating the derivative of $y$ with respect to $x$ by $y'$ and solve the resulting equation for $y'$. Thus, for example, if we are given $x^2 + y^2 - 9 = 0$ and apply the rules of differentiation directly to obtain $2x + 2yy' = 0$, we can then solve for $y'$ and write $y' = -\frac{x}{y}$ providing $y \neq 0$. The value of $y'$ at a given point $(x_0, y_0)$ can then be found by direct substitution of $x_0$ and $y_0$. If only $x_0$ is given, then $y_0$ must be calculated from $f(x, y) = 0$ if $y$ appears in the derivative $y'$.

### Tangents and Normals

**Definition.** The tangent to the curve defined by the equation $y = f(x)$ at a point $P_0(x_0, y_0)$ is the line through $P_0$ with slope $f'(x_0)$. The equation of the

## ADDITIONAL INFORMATION

Consider the curve $C$ defined by the function $y = f(x)$ and two points $P_0(x_0, y_0)$ and $P(x_0 + h, y + \Delta y)$ where $y + \Delta y = f(x_0 + h)$. The line through the points $P_0$ and $P$ is called the secant line and makes an angle $\theta$ with the $X$-axis.

By definition $\tan\theta = \frac{\Delta y}{h}$ (see Figure 4-1). As $h$ tends to zero, the point $P$ moves along the curve and tends to the point $P_0$. The secant line rotates about the point $P_0$, and the limiting line becomes tangent to the curve at $P_0$ when $P$ coincides with $P_0$. The angle $\theta$ then approaches the angle $\alpha$ which is the angle that the tangent line at $P_0$ makes with the $X$-axis. Thus we have

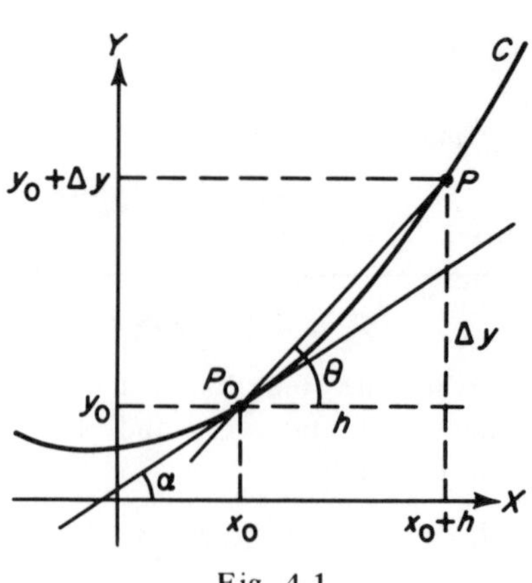


Fig. 4-1

$\lim_{h \to 0} \frac{\Delta y}{h} = f'(x_0) = \lim_{h \to 0} \tan\theta = \tan\alpha$. By definition this is the slope of the tangent line at $P_0$, and we see that the value of the derivative of $y = f(x)$ at $P_0$ is equal to the slope of the tangent line to the curve $C$ at $P_0$. The equation of a line through a point $P_0(x_0, y_0)$ with a given slope $m$ is $y - y_0 = m(x - x_0)$. Consequently, the equation of the tangent line to $C$ at $P_0$ is given by $y - y_0 = m(x - x_0)$ with $m = f'(x_0)$.

Since the **normal** to the curve $C$ at $P_0$ is a line through $P_0$ perpendicular to the tangent line, the slope of the normal is $-\frac{1}{m}$ and the equation of the normal is $y - y_0 = -\frac{1}{m}(x - x_0)$ where $m = f'(x_0)$.

In Figure 4-2, $P_0(x_0, y_0)$ is a point on the curve $C$ defined by $y = f(x)$ such that there is an oblique **tangent** to $C$ at $P_0$. $T$ is the intersection point of the tangent line and the $X$-axis, point $M(x_0, 0)$ is the intersection point of the perpendicular from $P_0$ to the $X$-axis, and the point $N$ is the intersection point of the normal and the $X$-axis. The length of the tangent is $\overline{TP_0}$. The length of the normal is $\overline{P_0N}$. $\overline{TM}$ is called the **subtangent.** $\overline{MN}$ is called the **subnormal.**

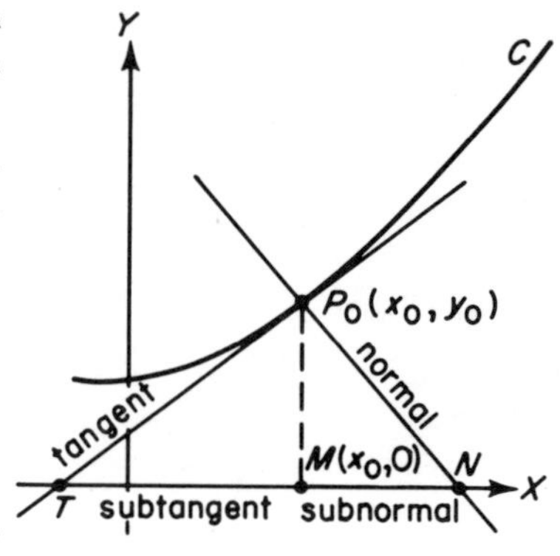


Fig. 4-2

$$\text{length of subtangent: } |TM| = \left|\frac{y_0}{m}\right|$$

$$\text{length of subnormal: } |MN| = |my_0|$$

tangent line: $y - y_0 = m(x - x_0)$ where $m = f'(x_0)$.

**Definition.** The **normal** to the curve defined by the equation $y = f(x)$ at $P_0(x_0, y_0)$ is the line through $P_0$ perpendicular to the tangent line at $P_0$. The equation of the normal line is

$$y - y_0 = -\frac{1}{m}(x - x_0)$$

where $m = f'(x_0)$.

**Definition.** An **oblique tangent** to a curve is one that is not parallel to either coordinate axis.

length of tangent: $|TP| = \left|\dfrac{y_0\sqrt{1 + m^2}}{m}\right|$

length of normal: $|PN| = \left|y_0\sqrt{1 + m^2}\right|$

**Implicit Differentiation**

When a function $f$ is defined by an equation of the form $y = f(x)$, we have an explicit expression for the dependent variable $y$ in terms of the independent variable $x$. On the other hand an equation of the form $f(x, y) = 0$ which also relates the variables $x$ and $y$ may define more than one function. This type of expression is then said to define the functions implicitly. In some cases the equation $f(x, y) = 0$ may be changed and solved explicitly for one variable in terms of the other. Consider, for example, the equation $y^2 - 3xy - 4x^2 = 0$. The expression on the left side may be factored, and we can write $(y - 4x)(y + x) = 0$. If we set each factor equal to zero and solve for $y$, we have two functions defined explicitly by $y_1 = 4x$ and $y_2 = -x$. Each of these functions has a derivative, $y_1' = 4$ and $y_2' = -1$. Let us now apply the rules of differentiation directly to the relation $y^2 - 3xy - 4x^2 = 0$, using the notation $y'$ for the derivative of $y$ with respect to $x$, to obtain the equation $2yy' - 3y - 3xy' - 8x = 0$. (Check the derivative of each term carefully and note the use of the chain rule.) Solve this equation for $y'$ through the steps:

$$(2y - 3x)y' = 3y + 8x \quad \text{and} \quad y' = \frac{3y + 8x}{2y - 3x}.$$

This is the general derivative of the relation $f(x, y) = 0$ with respect to $x$ and contains the derivative of each function expressed by the equation $f(x, y) = 0$. Consider the function $y_1 = 4x$, and replace $y$ by this value in the general derivative $y_1' = \dfrac{3(4x) + 8x}{2(4x) - 3x} = \dfrac{12x + 8x}{8x - 3x} = \dfrac{20x}{5x} = 4$ which is the same as the derivative $y_1'$ obtained from the equation $y_1 = 4x$. Similarly, $y_2' = \dfrac{3(-x) + 8x}{2(-x) - 3x} = \dfrac{5x}{-5x} = -1$. Implicit differentiation of this kind is valid for all values of the variables for which the resulting derivative exists. The relation $f(x, y) = 4x^2 + y^2 - 4 = 0$ has the general derivative $8x + 2yy' = 0$ or $y' = -\dfrac{8x}{2y} = -\dfrac{4x}{y}$ which is not defined if $y = 0$. The relation defines two functions $y_1 = 2\sqrt{1 - x^2}$ and $y_2 = -2\sqrt{1 - x^2}$, each of which is zero when $x = \pm 1$. Thus the derivative $y'$ does not exist when $y = 0$, or the equivalent $y' = f'(x)$ does not exist at $x = 1$ or $x = -1$.

## SOLUTIONS TO EXERCISES

I

i. $x^{\frac{2}{3}} + y^{\frac{2}{3}} = a^{\frac{2}{3}}$

$\frac{2}{3}x^{-\frac{1}{3}} + \frac{2}{3}y^{-\frac{1}{3}}y' = 0,\ y' = -\sqrt[3]{\frac{y}{x}}$

$y_i = \pm\left(a^{\frac{2}{3}} - x^{\frac{2}{3}}\right)^{\frac{3}{2}}$

$y_i' = \pm\frac{3}{2}\left(a^{\frac{2}{3}} - x^{\frac{2}{3}}\right)^{\frac{1}{2}}\left(-\frac{2}{3}\right)x^{-\frac{1}{3}}$

$= \mp x^{-\frac{1}{3}}\left(a^{\frac{2}{3}} - x^{\frac{2}{3}}\right)^{\frac{1}{2}}$

Substitute $y_i$ into $y'$:

$y' = -x^{-\frac{1}{3}}y^{\frac{1}{3}}$

$= \mp x^{-\frac{1}{3}}\left[\left(a^{\frac{2}{3}} - x^{\frac{2}{3}}\right)^{\frac{3}{2}}\right]^{\frac{1}{3}}$

$= \mp x^{\frac{1}{3}}\left(a^{\frac{2}{3}} - x^{\frac{2}{3}}\right)^{\frac{1}{2}}$

ii. $xy + y^2 = x^2$

$y + xy' + 2yy' = 2x$

$(x + 2y)y' = 2x - y$

$y' = \frac{2x - y}{x + 2y}$

$y_i = \frac{-x \pm\sqrt{x^2 + 4x^2}}{2}$

$y_i = \frac{x}{2}(-1 \pm\sqrt{5})$

$y_i' = \frac{1}{2}(-1 \pm\sqrt{5})$

Substitute $y_i$ into $y'$.

$$y' = \frac{2x - \frac{x}{2}(-1 \pm\sqrt{5})}{x + x(-1 \pm\sqrt{5})}$$

$$= \frac{4 - (-1 \pm\sqrt{5})}{2(1 - 1 \pm\sqrt{5})}$$

$$= \frac{5 \mp\sqrt{5}}{\pm 2\sqrt{5}} = \frac{5}{\pm 2\sqrt{5}} \mp \frac{\sqrt{5}}{\pm 2\sqrt{5}}$$

$$= \frac{1}{2}\left(-1 \pm\frac{5}{\sqrt{5}}\right) = \frac{1}{2}(-1 \pm\sqrt{5})$$

iii. $y = x^2(x - 3)^3$

$y' = 2x(x - 3)^3 + x^2(3)(x - 3)^2(1)$

$= (x - 3)^2(2x^2 - 6x + 3x^2)$

$y'' = 2(x - 3)(1)(5x^2 - 6x) + (x - 3)^2(10x - 6)$

$= (x - 3)(10x^2 - 12x + 10x^2 - 36x + 18)$

$= (x - 3)(20x^2 - 48x + 18)$

II

i. True ii. False iii. True

III

i. $y = x^3 - 2x^2 - x + 3,\ P_0(2, 1)$

$y' = 3x^2 - 4x - 1$

$m = 3(4) - 4(2) - 1 = 3$

$|TM| = \left|\frac{y_0}{m}\right| = \frac{1}{3}$

$|MN| = |my_0| = 3$

$|TP| = \left|\frac{y_0\sqrt{1 + m^2}}{m}\right| = \frac{\sqrt{10}}{3}$

$|PN| = |y_0\sqrt{1 + m^2}| = \sqrt{10}$

ii. $y = -\sqrt{3x + 1},\ P_0(5, -4)$

$y' = -\frac{1}{2}(3x + 1)^{-\frac{1}{2}}(5)$

$m = -\frac{3}{2}(16)^{-\frac{1}{2}} = \frac{3}{2}\left(\frac{1}{4}\right) = -\frac{3}{8}$

$\frac{1}{m} = -\frac{8}{3};\ \sqrt{1 + m^2} = \frac{1}{8}\sqrt{73}$

$|TM| = \left|(-4)\left(-\frac{8}{3}\right)\right| = \frac{32}{3}$

$|MN| = \left|(-4)\left(-\frac{3}{8}\right)\right| = \frac{3}{2}$

$|TP| = \left|(-4)\left(-\frac{8}{3}\right)\left(\frac{\sqrt{73}}{8}\right)\right| = \frac{4}{3}\sqrt{73}$

$|PN| = \left|(-4)\frac{73}{8}\right| = \frac{1}{2}\sqrt{73}$

iii. $y = \frac{x^2}{2x + 1},\ P_0\left(2, \frac{4}{5}\right)$

$y' = \frac{(2x + 1)2x - x^2(2)}{(2x + 1)^2}$

$= \frac{2x(x + 1)}{(2x + 1)^2}$

$m = \frac{(4)(3)}{25} = \frac{12}{25};\ \frac{1}{m} = \frac{25}{12}$

$\sqrt{1 + m^2} = \frac{\sqrt{769}}{25}$

$|TM| = \left|\left(\frac{4}{5}\right)\left(\frac{25}{12}\right)\right| = \frac{5}{3}$

$|MN| = \left|\left(\frac{4}{5}\right)\left(\frac{12}{25}\right)\right| = \frac{48}{125}$

$|TP| = \left|\left(\frac{5}{3}\right)\frac{\sqrt{769}}{25}\right| = \frac{\sqrt{769}}{15}$

$|PN| = \left|\left(\frac{4}{5}\right)\frac{\sqrt{769}}{25}\right| = \frac{4\sqrt{769}}{125}$

# 5 APPLICATIONS OF THE DERIVATIVE

## SELF-TEST

DIRECTIONS: Write your answers in the numbered regions to the right. To check your answers, turn the page. Study the solutions to the problems you missed, and do the exercises following any problem or group of problems in which you had errors.

1. Show that the rate of change of the area of a circle with respect to its radius is equal to the circumference.

In problems 2–5 for the given equation find $v$ and $a$. Also find the value of $t \geq 0$ at which the velocity is zero and the corresponding value of the acceleration.

2. $s = 48t - 16t^2$

3. $s = \frac{1}{2}gt^2 + v_0t + s_0$, where $g$, $v_0$, $s_0$ are constants.

4. $s = t^4 - 16t^2$

5. A ball rolls down an inclined plane. At the end of $t$ seconds, its distance (in feet) from the starting point is given by $s = 4t^2 - 2t$. Find its acceleration at the end of 3 seconds.

In problems 6 and 7 find the relative maxima and minima and the points of inflection of the curves defined by the given equations.

6. $y = 3 + 8x + 4x^2$

7. $y = x^{\frac{1}{2}} + x^{-\frac{1}{2}}$

8. A box with a square base and lid is to contain 640 cu. feet. If the bottom costs 32¢, the lid 48¢, and the sides 32¢ per square foot, what are the dimensions for minimum cost?

9. If $n_1$ and $n_2$ are two positive numbers whose sum is 5, find the maximum value of $n_1^2 n_2^3$.

10. A poster board contains 24 square feet. If the margins at the top and bottom are to be 6 in., and those on the sides 4 in., what are the dimensions of the board for maximum printing area?

| |
|---|
| 1 |
| 2 |
| 3 |
| 4 |
| 5 |
| 6 |
| 7 |
| 8 |
| 9 |
| 10 |

## SOLUTIONS

1. Area of a circle: $A = \pi r^2 \; \frac{dA}{dr} = 2\pi r$ which is the circumference of a circle.

2. $s = 48t - 16t^2$
$v = s' = 48 - 32t$
$a = s'' = -32$
$v = 48 - 32t = 0$ at $t = \frac{48}{32} = \frac{3}{2}$
At $t = \frac{3}{2}$ we have $a = -32$.

3. $s = \frac{1}{2}gt^2 + v_0 t + s_0$
$v = s' = gt + v_0$; $a = s'' = g$
$v = gt + v_0 = 0$ at $t = -\frac{v_0}{g}$
At $t = -\frac{v_0}{g}$, we have $a = g$.

4. $s = t^4 - 16t^2$
$v = s' = 4t^3 - 32t$
$a = s'' = 12t^2 - 32$
$v = 4t(t^2 - 8) = 0$ at $t = 0, 2\sqrt{2}$
At $t = 0$, we have $a = -32$.
At $t = 2\sqrt{2}$, we have $a = 64$.

**EXERCISES I**

For the given equation find $t$ for $v = 0$ $(t \geq 0)$ and determine $a$ at this value of $t$.

i. $s = 12t - t^3$

ii. $s = (4 - t^2)^{\frac{3}{2}}, \; 0 \leq t \leq 2$

5. $s = 4t^2 - 2t$
$v = \frac{ds}{dt} = 8t - 2$
$a = \frac{dv}{dt} = 8$
The acceleration at the end of three seconds is 8 ft/sec$^2$.

6. $y = 3 + 8x + 4x^2$
$y' = 8 + 8x = 0$ at $x = -1$
$y'' = 8$ and $y''(-1) = 8$
$y(-1) = 3 - 8 + 4 = -1$
Therefore there is a minimum at $(-1,-1)$. Since $y''$ is a constant, there is no point of inflection.

**EXERCISES II**

i. Two boats start at the same point. One sails due east starting at 10 A.M. at a constant rate of 20 miles per hour. The other sails due south starting at 11 A.M. at the constant rate of 9 mph. How fast are they separating at noon?

ii. The rectilinear motion of a particle is given by $s = \sqrt{t+1}$. Show that the acceleration is negative and proportional to the cube of the velocity.

7. $y = x^{\frac{1}{2}} + x^{-\frac{1}{2}}$
$y' = \frac{1}{2}x^{-\frac{1}{2}} - \frac{1}{2}x^{-\frac{3}{2}}$
$= \frac{1}{2\sqrt{x}}\left(\frac{x-1}{x}\right) = 0$ at $x = 1$
$y'' = -\frac{1}{4}x^{-\frac{3}{2}} + \frac{3}{4}x^{-\frac{5}{2}}$
$= \frac{1}{4}x^{-\frac{3}{2}}\left(\frac{3-x}{x}\right) = 0$
at $x = 3$ and $y''' \neq 0$
$y(1) = 1 + 1 = 2$
$y''(1) = \frac{1}{4}(1)(2) = \frac{1}{2}$
$y(3) = \sqrt{3} + \frac{\sqrt{3}}{3} = \frac{4\sqrt{3}}{3}$
We have a minimum at $(1,2)$ and a point of inflection at $\left(3, \frac{4\sqrt{3}}{3}\right)$.

8. The variable dimensions of the base and top are $x$ and $x$ and the sides are $x$ and $z$ where $z$ is the height. The volume is $x^2 z = 640$ or $z = 640\,x^{-2}$. The cost of the box is
$c = 32x^2 + 48x^2 + 4(32\,xz)$
$= 80x^2 + 128(640\,x^{-1})$
$\frac{dc}{dx} = 160x - 128(640)x^{-2}.$
Set the derivative equal to zero
$x^{-2}(160)(x^3 - 512) = 0$ and
$x^3 - 512 = 0$ or $x = 8$.
At $x = 8$: $\frac{d^2c}{dx^2} = 160 + 2(128)(640)x^{-3} > 0$
and $z = 640x^{-2} = \frac{640}{64} = 10$. The dimensions are $8' \times 8' \times 10$ .

### EXERCISES III

In i and ii find the relative maxima and minima and the points of inflection.

i. $y = x^4 - 2x^2 + 2$

ii. $y = x^4 + 1$

**9.** $n_1 + n_2 = 5$

$y = n_1^2 n_2^3 = n_1^2(5 - n_1)^3$

$$\frac{dy}{dn_1} = 2n_1(5 - n_1)^3 + 3n_1^2(5 - n_1)^2(-1)$$
$$= (5 - n_1)^2(10n_1 - 2n_1^2 - 3n_1^2)$$
$$= 5n_1(5 - n_1)^2(2 - n_1) = 0$$

at $n_1 = 0, 5, 2$ and the corresponding $n_2 = 5, 0, 3$. The maximum $y$ is at $n_1 = 2$ and $n_2 = 3$ when $y = 2^2 3^3 = 108$.

**10.** Draw the figure shown. Note that the dimensions given in inches have been changed into feet.

Area of board = $xy = 24$

Printing area = $A = \left(x - \frac{2}{3}\right)(y - 1)$

Since $y = 24x^{-1}$,

we have

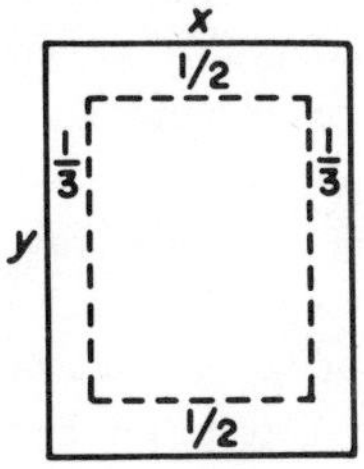

**Fig. 5-1**

$$A = \left(x - \tfrac{2}{3}\right)(24x^{-1} - 1)$$
$$= 24 - 16x^{-1} - x + \tfrac{2}{3}.$$

$\frac{dA}{dx} = 16x^{-2} - 1$ which equals zero at $x = 4$. At $x = 4$, $\frac{d^2A}{dx^2} = -32x^{-3} < 0$ and

$y = 24x^{-1} = \frac{24}{4} = 6$.

Dimensions are 4' × 6'.

Solutions to the exercises are on page 30.

## ANSWERS

1. $\frac{dA}{dt} = 2\pi r$
2. $t = \frac{3}{2}$; $a = -32$
3. $t = -\frac{v_0}{g}$; $a = g$
4. $t = 0$, $a = -32$; $t = 2\sqrt{2}$, $a = 64$
5. 8 ft/sec$^2$
6. $(-1, -1)$, minimum
7. $(1, 2)$, min.; $\left(3, \frac{4\sqrt{3}}{3}\right)$, infl.
8. 8' × 8' × 10'
9. 108
10. 4' × 6'

1

2

3

4

5

6

7

8

9

10

## BASIC FACTS

If a particle moves along a straight line and the distance from a fixed point is $s$, then the position of the particle at any time $t$ can be expressed as a function $s = f(t)$. The **average velocity** of the particle is defined as the ratio of the distance traveled $\Delta s$ to the elapsed time, $\Delta t$. The **instantaneous velocity** is defined as the limit approached by the average velocity as $\Delta t$ approaches zero. $v = \lim_{\Delta t \to 0} \frac{\Delta s}{\Delta t} = f'(t)$. The velocity can be either positive or negative depending upon the direction in which the particle is moving. The term *speed* is used to indicate the magnitude of the velocity, $|v|$. The acceleration of the particle is the rate of change of the velocity, $a = \lim_{\Delta t \to 0} \frac{\Delta v}{\Delta t} = \frac{dv}{dt} = f''(t)$.

### Guide in Solving Rate Problems

1. Draw a figure illustrating the problem. Denote by $x$, $y$, $z$, etc. the quantities which vary with time.
2. Make a list of the given and the desired quantities.
3. Obtain a relation between the variables which will be true at any instant of time.
4. Differentiate with respect to time.
5. Substitute the known quantities and solve for the unknowns.

### Definitions of Extrema

A **relative maximum** value of a function is one that is *greater than* any value immediately preceding and following.

A **relative minimum** value of a function is one that is *less than* any value immediately preceding and following.

The **relative extrema** of a function $f$, defined by the equation $y = f(x)$, exist only at the values of $x$ for which $f'(x) = 0$.

If $f'(x) = 0$ and $f''(x)$ is negative

## ADDITIONAL INFORMATION

### Rate Problems

In linear motion the distance is equal to the average velocity times the time, $s = \bar{v}t$. If the velocity is also changing with time, we have $v = g(t)$, and the rate of change of $v$ is the acceleration. If the distance can be expressed as a function of $t$, $s = f(t)$, and, if we consider the motion along a horizontal line to the right as positive, we can then describe the motion in terms of moving right and left. Consider the motion defined by $s = 16t^2 - 48t + 100$. The velocity is $v = f'(t) = 32t - 48$, and the acceleration is $a = f''(t) = 32$. Since the acceleration is a positive constant, the velocity is continuously increasing. The velocity is zero at $t = \frac{48}{32} = \frac{3}{2}$; obtained from the equation $32t - 48 = 0$. Consider the following values.

| $t$ | 0 | 1 | 1.5 | 2 |
|---|---|---|---|---|
| $s$ | 100 | 68 | 64 | 66 |
| $v$ | −48 | −16 | 0 | 16 |

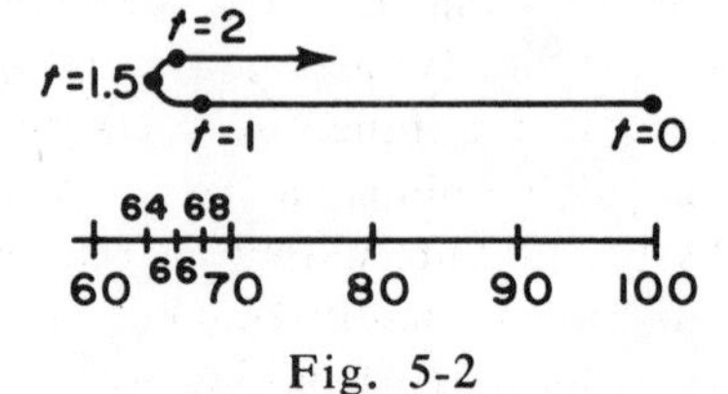


Fig. 5-2

The particle starts at $s = 100$ with a negative velocity causing it to move to the left. The positive acceleration diminishes the magnitude of the negative velocity until it is zero at $t = 1.5$ and $s = 64$. The velocity then becomes positive, and the particle starts to move to the right. It will continue to move to the right at ever increasing speed.

The form of the function $f(t)$ is determined by considering the physics of the moving body and taking measurements to determine the constant. Thus, a body falling freely under the action of gravity only (in a vacuum, near the surface of the earth) moves according to the function $s = \frac{1}{2}gt^2$ where $g$, the acceleration of gravity, is approximately 32.2 feet per second.

### Maxima and Minima

**Extreme Value Theorem.** If $f$ is a continuous function defined on the closed interval $[a, b]$, there is at least one value, $x_1$, in $[a, b]$ where $f$ has a largest value and at least one value, $x_2$, where $f$ has a smallest value.

**Rolle's Theorem.** If $f$ is continuous in the interval $[a, b]$, has a derivative at every value of $x$ between $a$ and $b$, and $f(a) = f(b) = 0$, then there is at least one value, $x_1$, between $a$ and $b$ such that $f'(x_1) = 0$.

**Theorem of the Mean.** If $f$ is a continuous function of $x$ in $[a, b]$ and $f'(x)$ exists for each $x$ in $]a, b[$, then there is at least one value $x_1$ in $]a, b[$ such that $f'(x_1) = \frac{f(b) - f(a)}{b - a}$.

The right member is the slope of a line through two points on the curve and the left member is the slope of the tangent line to the curve at $x = x_1$. (See Chapter 4.) Therefore, geometrically the theorem states that there is at least one point on the curve defined by $f$ between the point $P(a, f(a))$ and $Q(b, f(b))$ at which the tangent to the curve is parallel to the secant through $P$ and $Q$.

for $x = x_1$, then $f$ has a **relative maximum** at $x = x_1$.

If $f'(x) = 0$ and $f''(x)$ is positive for $x = x_1$, then $f$ has a **relative minimum** at $x = x_1$.

The curve $C$, defined by the equation $y = f(x)$, has a **point of inflection** at $x = x_1$ if $f''(x_1) = 0$ and $f'''(x_1) \neq 0$. The curve $C$ is **concaved upward** at any point where $f''(x)$ is positive and **concaved downward** at any point where $f''(x)$ is negative.

The roots of the equations $f'(x) = 0$ are called the **critical values** of $x$ for the function $f$. The corresponding values of $f(x)$ are called critical values of the function, and the corresponding points on the curve defined by $y = f(x)$ are called critical points.

The most interesting part of a curve is that interval of the domain of the function which includes all the critical values of $x$. The graph of the curve should always exhibit these critical points.

**Procedure for Locating Extrema**

1. Find the expression for the first and second derivatives of the function.
2. Set the first derivative equal to zero and solve the resulting equation for the real roots which are the critical values of the variable.
3. Test the second derivative at each of the critical values—$f''(x_0) < 0$, maximum; $f''(x_0) > 0$, minimum. If the second derivative is also zero at a critical value, use the theorem given in the additional information.

**Definitions.** If $f(x_2) > f(x_1)$ for $x_2 > x_1$, then the function $f$ is said to be *increasing*. If $f(x_2) < f(x_1)$ for $x_2 > x_1$, the function is said to be *decreasing*.

**Theorem.** If $f'(x) > 0$ for each $x$ in $]a, b[$, then $f$ is increasing on $]a, b[$; if $f'(x) < 0$ for each $x$ in $]a, b[$, then $f$ is decreasing on $]a, b[$.

The fact that $f'(x_0) = 0$ is not sufficient to determine that $f$ has a relative extreme at $x_0$. We use the value of $f''(x_0)$ to test the situation. However, if $f''(x_0) = 0$, the test fails and we use the above theorem.

$$\text{If } f'(x) \begin{cases} > 0 \text{ for } x \text{ in } [a, x_0[ \\ < 0 \text{ for } x \text{ in } ]x_0, b] \end{cases} \text{ then } f(x_0) \text{ is a maximum.}$$

$$\text{If } f'(x) \begin{cases} < 0 \text{ for } x \text{ in } [a, x_0[ \\ > 0 \text{ for } x \text{ in } ]x_0, b] \end{cases} \text{ then } f(x_0) \text{ is a minimum.}$$

This should be intuitively clear to the reader. In the first case the function changes from an increasing function to a decreasing function at $x_0$, and in the second case we have the opposite condition.

**Definition.** If the curve of $f$ always remains *above* the tangent line to the curve at each point of an interval, the curve is said to be **concaved upward** on the interval. If the curve remains *below* the tangent line, it is said to be **concaved downward.**

The tangent line crosses the curve at the point of **inflection.** At this point, the curve changes its direction of concavity from upward to downward or vice versa.

**General Directions for Extrema Problems.**

(a) Draw a figure illustrating the problem.
(b) Establish the function whose extreme value is desired and express it as a function of one variable or a relation between two variables.
(c) Apply the procedure for finding the extrema. (Be certain to find the absolute maximum or minimum for the interval involved.)

The following properties are useful.

1. The maximum and minimum values of a continuous function must occur alternately.
2. If $c$ is a positive constant, $cf(x)$ is a maximum or a minimum for those and only those values of $x$ that make $f(x)$ a maximum or a minimum. If $c$ is negative, $cf(x)$ is a maximum when $f(x)$ is a minimum and conversely.
3. If $c$ is any constant, $f(x)$ and $c + f(x)$ have maximum and minimum values for the same values of $x$.

## SOLUTIONS TO EXERCISES

### I

i. $s = 12t - t^3$
$v = s' = 12 - 3t^2$
$a = s'' = -6t$
$v = 12 - 3t^2 = 0$ at $t = 2$ ($t$ is limited to values $\geq 0$.)
At $t = +2$, we have $a = -12$.

ii. $s = (4 - t^2)^{\frac{3}{2}}$

$$v = s' = \frac{3}{2}(4 - t^2)^{\frac{1}{2}}(-2t)$$
$$= -3t(4 - t^2)^{\frac{1}{2}}$$
$$a = s'' = -3(4 - t^2)^{\frac{1}{2}} - 3t\left(\frac{1}{2}\right)(4 - t^2)^{-\frac{1}{2}}(-2t)$$
$$= \frac{6t^2 - 12}{\sqrt{4 - t^2}}$$

$v = 0$ at $t = 0$ and $t = 2$, $t \geq 0$
At $t = 0$, $a = -\frac{12}{2} = -6$.
At $t = 2$, $a = \frac{12}{0}$ is undefined.

### II

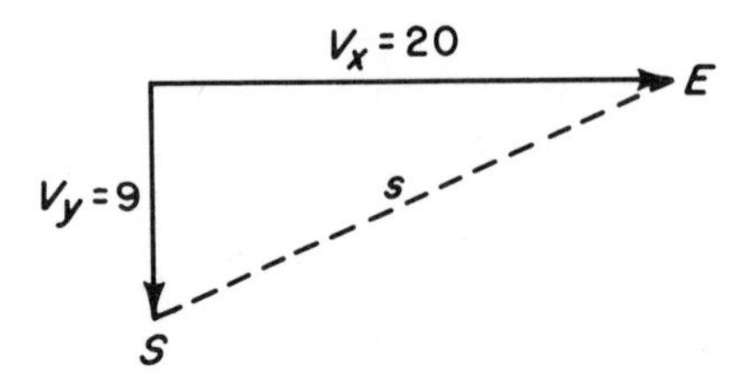


Fig. 5-3

i. Draw the figure shown above. Let $x$ be the distance travelled by ship going east; $y$, the distance of ship going south; and $s$ the distance between the ships. Then $s^2 = x^2 + y^2$. Each of these distances is a function of time, and $2s\frac{ds}{dt} = 2x\frac{dx}{dt} + 2y\frac{dy}{dt}$. At noon $x = 40$, $y = 9$, $v_x = 20$, and $v_y = 9$. $s = \sqrt{x^2 + y^2} = \sqrt{1681} = 41$, and

$$v_s = \frac{ds}{dt} = \frac{1}{s}[xv_x + yv_y]$$
$$= \frac{1}{41}[40(20) + 9(9)] = \frac{881}{41}.$$

ii. $s = \sqrt{t + 1} = (t + 1)^{\frac{1}{2}}$

$$v = \frac{ds}{dt} = \frac{1}{2}(t + 1)^{-\frac{1}{2}}$$
$$a = \frac{dv}{dt} = -\frac{1}{2}\left(\frac{1}{2}\right)(t + 1)^{-\frac{3}{2}}$$
$$= -\frac{1}{4}(t + 1)^{-\frac{3}{2}}$$
$$v^3 = \left[\frac{1}{2}(t + 1)^{-\frac{1}{2}}\right]^3$$
$$= \frac{1}{8}(t + 1)^{-\frac{3}{2}}$$

$\therefore a = kv^3$ with $k = -2$.

### III

i. $y = x^4 - 2x^2 + 2$
$y' = 4x^3 - 4x = 0$
at $x = 0, \pm 1$
$y'' = 12x^2 - 4 = 0$
at $x = \pm\frac{\sqrt{3}}{3}$

Form the table.

| $x$ | 0 | $\pm 1$ | $\pm\frac{\sqrt{3}}{3}$ |
|---|---|---|---|
| $y$ | 2 | 1 | $\frac{13}{9}$ |
| $y'$ | 0 | 0 | – |
| $y''$ | $-4$ | 8 | 0 |
| | max. | min. | infl. |

ii. $y = x^4 + 1$
$y' = 4x^3$
$y'' = 12x^2$
$y''' = 24x$
All are zero at $x = 0$ and $y = 1$.
At $x = -1$:
$y' = -4 < 0$.
At $x = 1$:
$y' = +4 > 0$.
Thus $(0, 1)$ is a minimum, and there is no point of inflection.

Fig. 5-4

# 6 THE DIFFERENTIAL; APPROXIMATIONS

## SELF-TEST

DIRECTIONS: Write your answers in the numbered regions to the right. To check your answers, turn the page. Study the solutions to the problems you missed, and do the exercises following any problem or group of problems in which you had errors.

In **1–3** find the differential $dy$ for each function.

1. $y = x^2 + 16$
2. $y = \sqrt{x}$
3. $y = \dfrac{4x}{x^2 - 4}$

In **4** and **5** find $dy$, $\Delta y$ and $\Delta y - dy$ for the indicated values of $x$ and $\Delta x$.

4. $y = \frac{1}{2}x^2 + x$, at $x = 2$ and $\Delta x = \frac{1}{2}$
5. $y = \dfrac{3}{x^2}$, at $x = 2$ and $\Delta x = \frac{1}{3}$
6. Use the differential to approximate the value of $\sqrt[3]{0.009}$.

In **7–9** find $\dfrac{dy}{dt}$.

7. $y = (x^2 + 2x)^3$ and $x = \sqrt{t^2 - 2t}$
8. $y = x^2 + 3x$ and $x = \dfrac{t}{t^2 - 2t}$
9. $y = \sqrt{x}$, and $x = \sqrt{1 - r}$, and $r = t^2$
10. Compare the differential of the area of a circle of radius $r$ with the area of a circular ring whose inner radius is $r$ and whose width is $\Delta r$.

1

2

3

4

5

6

7

8

9

10

## SOLUTIONS

1. $y = x^2 + 16$
$y' = 2x$ and $dy = 2x\,dx$

2. $y = \sqrt{x}$
$$y' = \frac{1}{2\sqrt{x}} \text{ and } dy = \frac{1}{2\sqrt{x}}\,dx$$

3. $$y = \frac{4x}{x^2 - 4},\quad y' = -\frac{4x^2 + 16}{(x^2 - 4)^2}$$
$$dy = -\frac{(4x^2 + 16)\,dx}{(x^2 - 4)^2}$$

> **EXERCISES I**
>
> Use differentials to find the derivative $\dfrac{dy}{dx}$.
>
> i. $x^2 + y^2 = 25$
>
> ii. $(x - 3)^2 - (y - 2)^2 = 2x^2y^2$
>
> iii. $x^2 - y^2 = z^2$, $z = f(z)$
>
> iv. $y^2 = \sqrt{x + 1}\,\sqrt{x - 1}$

Remember $dy = y'\,dx$ and $\Delta y = f(x + h) - y$.

4. $y = \frac{1}{2}x^2 + x$, $x = 2$, $\Delta x = \frac{1}{2}$
$$dy = (x + 1)\,dx = 3\left(\tfrac{1}{2}\right) = \tfrac{3}{2}$$
$$\Delta y = \tfrac{1}{2}(x + \Delta x)^2 + (x + \Delta x) - y$$
$$= \tfrac{1}{2}x^2 + x\Delta x + \tfrac{1}{2}\overline{\Delta x}^2 + x + \Delta x - \tfrac{1}{2}x^2 - x$$
$$= (x + 1)\,\Delta x + \tfrac{1}{2}\overline{\Delta x}^2$$
$$= \tfrac{3}{2} + \tfrac{1}{8} = \tfrac{13}{8}$$
$$\Delta y - dy = \tfrac{13}{8} - \tfrac{3}{2} = \tfrac{1}{8}$$

5. $y = \dfrac{3}{x^2}$, $x = 2$, $\Delta x = \frac{1}{3}$
$$dy = \frac{(-2)\,3}{x^3}\,dx = -\tfrac{6}{8}\left(\tfrac{1}{3}\right) = -\tfrac{1}{4}$$
$$\Delta y = \frac{3}{(x + \Delta x)^2} - \frac{3}{x^2}$$
$$= \frac{-6x\,\Delta x - 3\,\overline{\Delta x}^2}{x^2\,(x + \Delta x)^2}$$
$$= \frac{(-12)\left(\frac{1}{3}\right) - 3\left(\frac{1}{9}\right)}{4\left(2 + \frac{1}{3}\right)^2} = -\frac{39}{196}$$
$$\Delta y - dy = -\left(\frac{39}{196} - \frac{1}{4}\right) = \frac{5}{98}$$

> **EXERCISES II**
>
> In i and ii, use the differential to approximate the values.
>
> i. $\sqrt{101}$
>
> ii. $\sqrt{15} + \sqrt[4]{15}$
>
> iii. Find $\Delta y - dy$ at $x = 1$ and $\Delta x = 0.1$ for $y = \sqrt{5 - x}$

6. $\sqrt[3]{0.009}$, $y = \sqrt[3]{x}$
$dy\,(.008) = f'(.008)\,\Delta x$
with $\Delta x = .001$
$$= \tfrac{1}{3}(.008)^{-\frac{2}{3}}\,(.001)$$
$$= \frac{1}{3\,(.04)}\,(.001) = \frac{1}{120}$$
$$y\,(.009) = y\,(.008) + dy\,(.008)$$
$$= .2 + \frac{1}{120}$$
$$= .2 + .00833$$
$$= .20833$$

7. $y = (x^2 + 2x)^3$ and $x = \sqrt{t^2 - 2t}$
$$\frac{dy}{dx} = 3\,(x^2 + 2x)^2\,(2x + 2)$$
$$= 6\,(x^2 + 2x)^2\,(x + 1)$$
$$\frac{dx}{dt} = \tfrac{1}{2}\,(t^2 - 2t)^{-\frac{1}{2}}\,(2t - 2)$$
$$= (t^2 - 2t)^{-\frac{1}{2}}\,(t - 1)$$
$$\frac{dy}{dt} = 6\,(x^2 + 2x)^2\,(x + 1)\,\frac{t - 1}{\sqrt{t^2 - 2t}}$$

8. $y = x^2 + 3x$ and $x = \dfrac{t}{t^2 - 2t}$
$$\frac{dy}{dx} = 2x + 3$$

$$\frac{dx}{dt} = \frac{d}{dt}\left[\frac{1}{t-2}\right] = -\frac{1}{(t-2)^2}$$

$$\frac{dy}{dt} = -\frac{2x+3}{(t-2)^2}$$

9. $y = \sqrt{x},\ x = \sqrt{1-r},\ r = t^2$

$$\frac{dy}{dx} = \frac{1}{2\sqrt{x}},\quad \frac{dx}{dr} = -\frac{1}{2}(1-r)^{-\frac{1}{2}}$$

$$\frac{dr}{dt} = 2t,\quad \frac{dy}{dt} = \frac{dy}{dx}\frac{dx}{dr}\frac{dr}{dt}$$

$$\frac{dy}{dt} = -\frac{t}{2\sqrt{x}\sqrt{1-r}}$$

EXERCISES III

Given that $z = 2y^2$, $y = \dfrac{x}{x+1}$ and $x = \dfrac{1}{t-1}$

i. Find $\dfrac{dz}{dt}$ using the chain rule.

ii. Express $\dfrac{dz}{dt}$ as a function of $t$.

iii. Express $z$ as a function of $t$.

iv. Find $\dfrac{dz}{dt}$ by using the results of iii.

10. Area of a circle: $A_C = \pi r^2$

$dA_C = 2\pi r dr$

Area of ring: $A_r = A_1 - A_2$

$A_r = \pi(r + \Delta r)^2 - \pi r^2$

$= \pi(r^2 + 2r\Delta r + \overline{\Delta r}^2 - r^2)$

$= \pi(2r\Delta r + \overline{\Delta r}^2)$

Since $\Delta r = dr$ we have

$A_r = dA_C + \pi\overline{\Delta r}^2$.

Solutions to the exercises are on page 36.

## ANSWERS

|  | |
|---|---|
| 1. $2x\,dx$ | 1 |
| 2. $\dfrac{1}{2\sqrt{x}}\,dx$ | 2 |
| 3. $-\dfrac{(4x^2+16)\,dx}{(x^2-4)^2}$ | 3 |
| 4. $\frac{3}{2}, \frac{13}{8}, \frac{1}{8}$ | 4 |
| 5. $-\frac{1}{4}, -\frac{39}{196}, \frac{5}{98}$ | 5 |
| 6. 0.20833 | 6 |
| 7. $6(x^2+2x)^2(x+1)\dfrac{t-1}{\sqrt{t^2-2t}}$ | 7 |
| 8. $-\dfrac{2x+3}{(t-2)^2}$ | 8 |
| 9. $-\dfrac{t}{2\sqrt{x}\sqrt{1-r}}$ | 9 |
| 10. $A_r = dA_C + \pi\overline{\Delta r}^2$ | 10 |

## BASIC FACTS

**Definition.** The differential of $y$ (denoted by $dy$) is that function of two variables, $x$ and $h$, defined by

$$dy = f'(x)\,h$$

where $y = f(x)$ and $h$ is an arbitrary increment in the independent variable $x$.

In our previous notation we have used $h = \Delta x$, and if we consider the function $y = f(x) = x$ then we have

$$dx = f'(x)\,h = h$$

since $f'(x) = 1$ when $f(x) = x$. Thus there are three equivalent forms for the differential

$$\begin{aligned} dy &= f'(x)\,h \\ &= f'(x)\,\Delta x \\ &= f'(x)\,dx \end{aligned}$$

The last of these is the one that is used most frequently. A simple division by the differential $dx$ results in the familiar notation for the derivative

$$\frac{dy}{dx} = f'(x) \quad \text{if} \quad dx \neq 0.$$

**Theorem.** If $y = f(x)$, then it is always true that

$$dy = f'(x)\,dx$$

whether or not $x$ is an independent variable.

Recall the definition for the increment of the function $y = f(x)$ as

$$\Delta y = f(x + h) - f(x)$$

and we see that $dy$ and $\Delta y$ are not the same. However, the limit of the difference between $\Delta y$ and $dy$ divided by $h$ is zero; i.e.,

$$\lim_{h \to 0} \frac{\Delta y - dy}{h} = 0.$$

Consequently, if $h$ is sufficiently small, $dy$ is approximately equal to $\Delta y$. This approximation is used in

## ADDITIONAL INFORMATION

The difference between the increment $\Delta y$ of a function defined by $y = f(x)$ and the differential $dy$ can be better understood by considering the graph of $y = f(x)$ as shown in Figure 6-1. Let $P$ and $Q$ be two points with coordinates $P(x, y)$ and $Q(x + h, y + \Delta y)$. Let $S$ be the point with coordinates $S(x + h, y)$ and $T$ be the point of intersection of the tangent line to the curve of $y = f(x)$ at the point $P$ with the line $QS$. From the figure we see that $\Delta y = \overline{SQ}$ and $\overline{PS} = h$. The slope of the tangent line $\overline{PT}$ is equal to $f'(x)$ by definition and is also the ratio $\frac{\overline{TS}}{\overline{PS}} = \frac{\overline{TS}}{h}$ so that we have $\overline{TS} = f'(x)\,h$ which by definition is the differential $dy$. Thus geometrically $dy = \overline{TS}$. Using these line segments, we have

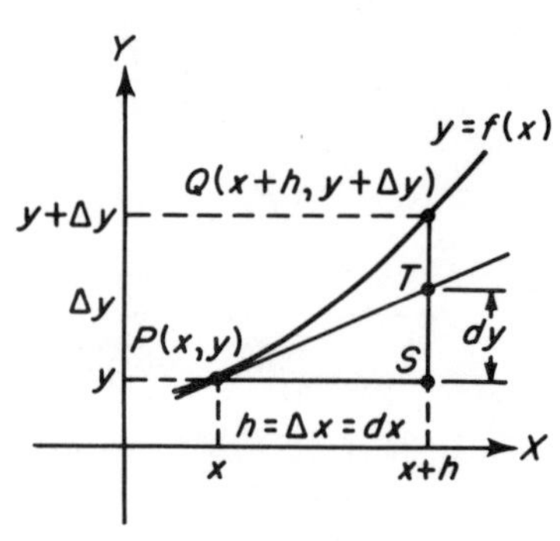

Fig. 6-1

$$\Delta y - dy = \overline{QS} - \overline{TS} = \overline{QT}.$$

As the point $Q$ moves along the curve towards $P$, we see that $\overline{QT}$ gets smaller and smaller and if $h$ is small enough $\Delta y$ and $dy$ are approximately equal, $\Delta y \doteq dy$ (the symbol $\doteq$ means "approximately equal to").

Let us give a rigorous proof of the statement

$$\lim_{h \to 0} \frac{\Delta y - dy}{h} = 0.$$

*Proof.* By the definition of the derivative $\lim_{h \to 0} \frac{\Delta y}{h} = f'(x)$ and by the definition of the differential $\frac{dy}{h} = f'(x)$. Thus

$$\lim_{h \to 0} \frac{\Delta y - dy}{h} = \lim_{h \to 0} \frac{\Delta y}{h} - \lim_{h \to 0} \frac{dy}{h} = f'(x) - f'(x) = 0.$$

**Errors of Measurement**

> Absolute Error
>
> $E_a$ = true value − approx. value $= y - y_1 = \Delta y$
>
> Relative Error
>
> $E_r = \dfrac{\text{absolute error}}{\text{true value}} = \dfrac{E_a}{y} = \dfrac{\Delta y}{y}$
>
> Percentage Error = (relative error) 100 = $E_r(100)$

Because it is easier to calculate $dy$ than $\Delta y$, $dy$ is used to approximate the errors made in measurements. Consider, for example,

calculating **errors of measurement** and in approximating the values of functions.

The elementary formulas for the derivatives can now be written in terms of differentials.

$$d(cu) = c\, du$$
$$d(u + v) = du + dv$$
$$d(uv) = udv + vdu$$
$$d\left(\frac{u}{v}\right) = \frac{vdu - udv}{v^2}$$
$$d(u^n) = n u^{n-1}\, du$$

The use of differentials in the notation for the derivatives gives us a further realization of the truth of the chain rule for

$$\frac{dy}{dx} = \frac{dy}{du} \cdot \frac{du}{dx}$$

if $du \neq 0$ and $dx \neq 0$ is certainly true by simple algebra.

the calculation of the approximate and maximum percentage error in determining the volume of a sphere by both $\Delta y$ and $dy$. The diameter is measured to be $10 \pm 0.05$ inches. From the given date we have

$$V = \tfrac{4}{3}\pi r^3 = \tfrac{\pi}{6} d^3 \text{ with } d = 10 \text{ and } \Delta d = h = 0.05. \text{ Then}$$

$$dV = V'h = \tfrac{\pi}{2} d^2 h = \tfrac{\pi}{2}(100)(.05) = 2.5\,\pi$$

Let the true volume be $V = \frac{\pi}{6}(10)^3 = \frac{500}{3}\pi$, then the approximate percentage error (using $dy$) is

$$100\, E_r = \frac{dV}{V}(100) = \frac{2.5\,\pi}{500\,\pi}(3)(100) = \mathbf{1.5\%}$$

The true error $\Delta V = V(10.05) - V(10) = \frac{\pi}{6}(10.05)^3 - \frac{\pi}{6}(10)^3$

$$= \tfrac{\pi}{6}[1015.075125 - 1000] = \tfrac{\pi}{6}(15.075)$$

The percentage error (using $\Delta y$) is

$$100\, E_r = \frac{\Delta V}{V}(1\,00) = \frac{15.075\,\pi\,(6)}{6\,(1000)\,\pi}(100) = \mathbf{1.5075\ \%}$$

Differentials may also be used to find the approximate value of a function. Examples of this are given in Exercises II.

Differentials are frequently used to find the derivatives of functions which are given in implicit form. Consider, for example, the equation $u^3 + u^2 v - v^2 + 2v = 0$. Let us first find the differentials of each term.

$$3u^2 du + 2uv\, du + u^2\, dv - 2v\, dv + 2\, dv = 0$$

Next factor the differentials.

$$(3u^2 + 2uv)\, du + (u^2 - 2v + 2)\, dv = 0$$

If we divide by $dv$ and solve, we obtain

$$\frac{du}{dv} = -\frac{u^2 - 2v + 2}{3u^2 + 2uv}.$$

We can also divide by $du$. These derivatives are defined everywhere except where the denominators are zero.

## SOLUTIONS TO EXERCISES

I

i. $x^2 + y^2 = 25$
Take the differential:
$2x\,dx + 2y\,dy = 0$
Divide by $dx$ and solve for the derivative:

$$\frac{dy}{dx} = -\frac{x}{y}$$

ii. $(x-3)^2 - (y-2)^2 = 2x^2y^2$
$2(x-3)\,dx - 2(y-2)\,dy$
$= 4xy^2dx + 4x^2ydy$
$(x - 3 - 2xy^2)\,dx = (y - 2 + 2x^2y)dy$

$$\frac{dy}{dx} = \frac{x - 2xy^2 - 3}{y + 2x^2y - 2}$$

iii. $x^2 - y^2 = z^2$
$2xdx - 2ydy = 2zdz$
$xdx - zdz = ydy$

$$\frac{dy}{dx} = \frac{x}{y} - \frac{z}{y}\frac{dz}{dx}$$

iv. $y^2 = \sqrt{x+1}\,\sqrt{x-1}$
$= \sqrt{x^2-1}$
$2y\,dy = \frac{1}{2}(x^2-1)^{-\frac{1}{2}}\,2x\,dx$

$$\frac{dy}{dx} = \frac{x}{2y\sqrt{x^2-1}}$$

II

i. $y = \sqrt{x};\ x = 100;\ \Delta x = 1$
$\sqrt{x + \Delta x} \doteq y + dy$
$dy = \frac{1}{2}(x)^{-\frac{1}{2}}\,dx$
$= \frac{1}{2}(100)^{-\frac{1}{2}}\,(1) = \frac{1}{20}$

$\sqrt{100+1} = \sqrt{100} + \frac{1}{20} = 10.05$

ii. $\sqrt{15} + \sqrt[4]{15}$
$y = \sqrt{x} + \sqrt[4]{x}$
$dy = \left(\frac{1}{2}x^{-\frac{1}{2}} + \frac{1}{4}x^{-\frac{3}{4}}\right)h$
$y(15) = y(16) + dy(16)$
with $h = -1$

$y(15) = (4+2) + \left(\frac{1}{8} + \frac{1}{32}\right)(-1)$
$= 6 - \frac{5}{32} = 6 - .15625$
$= 5.84375$

iii. $y = \sqrt{5-x},\ x = 1,\ \Delta x = .1$

$$dy = \frac{(-1)}{2\sqrt{5-x}}\,dx$$

$$= -\frac{1}{2\,(2)}\left(\frac{1}{10}\right) = -\frac{1}{40}$$

$\Delta y = \sqrt{5 - x - x} - \sqrt{5-x}$
$= \sqrt{5 - 1 - .1} - \sqrt{5-1}$
$= \sqrt{3.9} - 2$
$= 1.9748 - 2 = -0.0252$
$\Delta y - dy = -0.0252 + 0.025$
$= -0.0002$

III

i. $z = 2y^2,\ y = \dfrac{x}{x+1},\ x = \dfrac{1}{t-1}$

$$\frac{dz}{dt} = \frac{dz}{dy}\cdot\frac{dy}{dx}\cdot\frac{dx}{dt}$$
$= 4y\,(x+1)^{-2}\,[-(t-1)^{-2}]$
$= -4y\,(x+1)^{-2}\,(t-1)^{-2}$

ii. Since $x = (t-1)^{-1}$, we have

$$(x+1)^{-2} = \left(\frac{1}{t-1} + 1\right)^{-2}$$
$$= \left(\frac{t-1}{t}\right)^2.$$

$$y = \frac{x}{x+1} = \frac{1}{t-1} \cdot \frac{t}{t-1} = \frac{1}{t}$$

$$\frac{dz}{dt} = -4\,\frac{1}{t}\left(\frac{t-1}{t}\right)^2\frac{1}{(t-1)^2} = -4\,t^{-3}$$

iii. $z = 2y^2 = 2\left(\dfrac{x}{x+1}\right)^2$

$\dfrac{x}{x+1} = \dfrac{1}{t}$ (See ii.); $z = 2\,\dfrac{1}{t^2} = 2t^{-2}$

iv. $\dfrac{dz}{dt} = -4t^{-3}$

# 7 INTEGRATION, THE ANTIDERIVATIVE

## SELF-TEST

DIRECTIONS: Write your answers in the numbered regions to the right. To check your answers, turn the page. Study the solutions to the problems you missed, and do the exercises following any problem or group of problems in which you had errors.

| |
|---|
| 1 |
| 2 |
| 3 |
| 4 |
| 5 |
| 6 |
| 7 |
| 8 |
| 9 |
| 10 |

In Problems **1–5** perform the indicated integrations.

**1.** $\int (4x^3 - 3x^2 + 5)\,dx$

**2.** $\int x\sqrt{x}\,dx$

**3.** $\displaystyle\int \frac{4x^2 - \sqrt{x}}{x}\,dx$

**4.** $\int 3x\sqrt{3 - x^2}\,dx$

**5.** $\displaystyle\int \frac{(x + 1)\,dx}{\sqrt{x^2 + 2x}}$

In Problems **6–9** find the value of the antiderivative $y = f(x)$ for the given conditions.

**6.** Find $f(3)$ if $y' = x^2 - 5$ and $f(2) = \frac{2}{3}$.

**7.** Find $f\left(\frac{4}{5}\right)$ if $y' = x\sqrt{1 - x^2}$ and $f(1) = \frac{84}{125}$.

**8.** Find $f(3)$ if $y' = (x^3 + 3x)(x^2 + 1)$ and $f(1) = -\frac{1}{3}$.

**9.** Find $f(4)$ if $y' = \sqrt{3x^2}$ and $f(2) = 0$.

**10.** Find a general form for $f(t)$ if

$f''(t) = -\frac{1}{2}gt$.

## SOLUTIONS

1. $\int (4x^3 - 3x^2 + 5)\, dx$
$= 4 \int x^3\, dx - 3 \int x^2 dx + 5 \int dx$
$= 4 \left(\frac{x^4}{x}\right) - 3 \left(\frac{x^3}{3}\right) + 5x + c$
$= x^4 - x^3 + 5x + c$

2. $\int x\sqrt{x}\, dx = \int x^{\frac{3}{2}}\, dx$
$= x^{\frac{5}{2}} \div \frac{5}{2} + c$
$= \frac{2}{5}\, x^{\frac{5}{2}} + c$

3. $\int \frac{4x^2 - \sqrt{x}}{x}\, dx = \int \left(4x - x^{-\frac{1}{2}}\right) dx$
$= 4 \left(\frac{x^2}{2}\right) - \frac{x^{\frac{1}{2}}}{\frac{1}{2}} + c$
$= 2x^2 - 2x^{\frac{1}{2}} + c$

4. $\int 3x\sqrt{3 - x^2}\, dx$
Let $u = 3 - x^2$, then
$du = -2x\, dx$
and we write the integral as
$-\frac{3}{2} \int (3 - x^2)^{\frac{1}{2}} (-2x\, dx)$
$= -\frac{3}{2} (3 - x^2)^{\frac{3}{2}} \div \frac{3}{2} + c$
$= -(3 - x^2)^{\frac{3}{2}} + c.$

5. $\int \frac{(x + 1)\, dx}{\sqrt{x^2 + 2x}}$
See Additional Information section for a detailed discussion of this problem.

6. $y' = x^2 - 5 = f'(x)$
$y = \int (x^2 - 5)\, dx$
$= \frac{x^3}{3} - 5x + c$
$f(2) = \frac{2}{3} = \frac{2^3}{3} - 5(2) + c$
and $c = 8$
$f(3) = \frac{27}{3} - 15 + 8 = 2$

7. $y' = x\sqrt{1 - x^2} = f'(x)$
$y = -\frac{1}{2} \int (1 - x^2)^{\frac{1}{2}} (-2x\, dx)$
$= -\frac{1}{2} (1 - x^2)^{\frac{3}{2}} \left(\frac{2}{3}\right) + c$
$= -\frac{1}{3} (1 - x^2)^{\frac{3}{2}} + c$
$f(1) = \frac{84}{125} = 0 + c = c$
$f\left(\frac{4}{5}\right) = -\frac{1}{3} \left(1 - \frac{16}{25}\right)^{\frac{3}{2}} + \frac{84}{125}$
$= -\frac{1}{3} \left(\frac{9}{25}\right)^{\frac{3}{2}} + \frac{84}{125}$
$= -\frac{1}{3} \left(\frac{27}{125}\right) + \frac{84}{125}$
$= -\frac{9}{125} + \frac{84}{125}$
$= \frac{75}{125} = \frac{3}{5}$

8. $y' = (x^3 + 3x)(x^2 + 1) = f'(x)$
$y = \frac{1}{3} \int (x^3 + 3x)(3x^2 + 3)\, dx$
$= \frac{1}{3} (x^3 + 3x)^2 \left(\frac{1}{2}\right) + c$
$f(1) = \frac{1}{6} (4)^2 + c = -\frac{1}{3}$
and $c = -3$
$f(3) = \frac{1}{6} (27 + 9)^2 - 3$
$= \frac{1}{6} (36)^2 - 3 = 213$

9. $y' = \sqrt{3x^2} = \sqrt{3}\, x = f'(x)$
$y = \frac{\sqrt{3}}{2} x^2 + c$
$f(2) = \frac{\sqrt{3}}{2} (4) + c = 0$
Then $c = -2\sqrt{3}$.
$f(4) = \frac{\sqrt{3}}{2} (16) - 2\sqrt{3}$
$= 8\sqrt{3} - 2\sqrt{3} = 6\sqrt{3}$

EXERCISES I

Perform the integration.

i. $\int (x^2 + x)(2x)\, dx$

ii. $\int \frac{x^2 - 3}{\sqrt{x}}\, dx$

iii. $\int \left(5\sqrt{x} - \frac{3}{\sqrt{x}}\right) x\, dx$

iv. $\int (x^3 - 4x^2)(3x^2 - 8x)\, dx$

v. $\int (x^2 + \sqrt{x})(4x + x^{-\frac{1}{2}})\, dx$

EXERCISES II

Find $f(2)$.

i. $f'(x) = x^2 - \sqrt{x};\ f(1) = 0$

ii. $f'(x) = \sqrt{5 + x};\ f(-1) = \frac{22}{3}$

iii. $f'(x) = \frac{1}{x^2};\ f(1) = 3$

iv. $f'(x) = (x^3 - 5)^5\, x^2$

$f(1) = \frac{5}{6}\,(4^4)$

10. $f''(t) = -\frac{1}{2}\,gt$

$$f'(t) = -\tfrac{1}{2}\,g \int t\,dt = -\tfrac{1}{2}\,g\,\frac{t^2}{2} + c_1 = -\tfrac{1}{4}\,gt^2 + c_1$$

$$f(t) = -\tfrac{1}{4}\,g \int t^2\,dt + c_1 \int dt = -\tfrac{1}{4}\,g\,\frac{t^3}{3} + c_1 t + c_2 = -\tfrac{1}{12}\,gt^3 + c_1 t + c_2$$

EXERCISES III

i. Find $f(x)$ if $f''(x) = 24x$.

ii. Find $f(2)$ if $f''(x) = 18x - 8$, $f'(1) = 3$, and $f(1) = 7$.

Solutions to the exercises are on page 42.

**ANSWERS**

|  | |
|---|---|
| 1. $x^4 - x^3 + 5x + c$ | 1 |
| 2. $\frac{2}{5}\,x^{\frac{5}{2}} + c$ | 2 |
| 3. $2x^2 - 2x^{\frac{1}{2}} + c$ | 3 |
| 4. $-(3 - x^2)^{\frac{3}{2}} + c$ | 4 |
| 5. $(x^2 - 2x)^{\frac{1}{2}} + c$ | 5 |
| 6. 2 | 6 |
| 7. $\frac{3}{5}$ | 7 |
| 8. 213 | 8 |
| 9. $6\sqrt{3}$ | 9 |
| 10. $-\frac{1}{12}\,gt^3 + c_1 t + c_2$ | 10 |

## BASIC FACTS

If we are given the derivative, $f'(x)$, of a function, $f(x)$, of the independent variable $x$ and have a process for finding $f(x)$, we are performing an operation which is the inverse of differentiation.

**Definition.** The function $f(x)$ is an **antiderivative** of $f'(x)$.

Since the derivative of a constant is zero, the antiderivative is not unique; that is, $f(x) + 2$ and $f(x) + 3$ are both antiderivatives of $f'(x)$. However, these functions differ only by a constant. Consequently, we can express the antiderivative of $f'(x)$ by including an arbitrary constant, $f(x) + c$.

**Axiom.** If $f'(x) = g(x)$, then the antiderivative of $g(x)$ is $f(x) + c$ where $c$ is an arbitrary constant.

We could adopt an inverse notation for the antiderivative, $D_x^{-1}[f'(x)] = f(x) + c$. However, classical mathematics has used the integral sign and the process is more commonly called integration or finding the indefinite integral.

**Definition.** If $f'(x) = g(x)$, then $\int g(x)\,dx = f(x) + c$ is called the **indefinite integral** of $g(x)$. The function $g(x)$ is called the **integrand** and $c$ is called the **constant of integration.**

The differential $dx$ is used to indicate the variable of integration. We recall from Chapter 6 that if $y = f(x)$ then $dy = f'(x)\,dx$, and the process of finding the indefinite integral is also an operation on the differential.

The following elementary formulas are direct consequences of the properties of derivatives.

1. $\int(du + dv - dw) = \int du + \int dv - \int dw$
2. $\int a\,du = a\int du$
   where $a$ is a constant.
3. $\int dx = x + c$
4. $\int u^n du = \dfrac{u^{n+1}}{n+1} + c,\ n \neq -1$

**Warning.** If $u = f(x)$, then Formula **4** can be applied only if $du = f'(x)\,dx$.

## ADDITIONAL INFORMATION

**The Process of Integration**

Since, for the present, we have defined integration as the inverse of differentiation, it is clear that the derivative of an integral is the integrand; that is,

$$\frac{d}{dx}\left(\int f(x)\,dx\right) = f(x) \quad \text{and} \quad d\int f(x)\,dx = f(x)\,dx.$$

Let $u$ and $v$ be functions of the independent variable $x$, then we have $\dfrac{d}{dx}\displaystyle\int (u + v)\,dx = u + v$ and

$$\frac{d}{dx}\left(\int u\,dx + \int v\,dx\right) = \frac{d}{dx}\int u\,dx + \frac{d}{dx}\int v\,dx = u + v.$$

Since the derivatives of the functions are the same, the functions themselves can differ at the most by a constant. Consequently, we have Formula **1**, $\int (u + v)\,dx = \int u\,dx + \int v\,dx$. The truth of Formula **2**, $\int a\,du + a\int du$, can be shown in the same manner. Formulas **3** and **4** are verified by differentiating the right member. Consider, for example, Formula **4**.

$$\int u^n du = \frac{u^{n+1}}{n+1} + c, \quad n \neq -1$$

$$\text{Differentiating: } \frac{d}{du}\left(\frac{u^{n+1}}{n+1} + c\right) = \frac{n+1}{n+1}u^{n+1-1} + 0 = u^n.$$

To find the indefinite integral we arrange the integrand to take the form of one of the standard formulas and apply the properties. Consider, for example, the integral

$$\int \frac{3 + 4x - 5x^2}{\sqrt{x}}\,dx = f(x).$$

The function representing this integral is found in the following manner.

$$\text{Rewrite the integrand} = 3x^{-\frac{1}{2}} + 4x^{\frac{1}{2}} - 5x^{\frac{3}{2}}$$

$$\text{Apply Formula 1: } f(x) = \int 3x^{-\frac{1}{2}}\,dx + \int 4x^{\frac{1}{2}}\,dx - \int 5x^{\frac{3}{2}}\,dx$$

$$\text{Apply Formula 2:} \quad = 3\int x^{-\frac{1}{2}}\,dx + 4\int x^{\frac{1}{2}}\,dx - 5\int x^{\frac{3}{2}}\,dx$$

$$\text{Apply Formula 4:} \quad = 3\,\frac{x^{\frac{1}{2}}}{\frac{1}{2}} + c_1 + 4\,\frac{x^{\frac{3}{2}}}{\frac{3}{2}} + c_2 - 5\,\frac{x^{\frac{5}{2}}}{\frac{5}{2}} + c_3$$

$$= 6x^{\frac{1}{2}} + c_1 + \tfrac{8}{3}x^{\frac{3}{2}} + c_2 - 2x^{\frac{5}{2}} + c_3$$

Combine the constants of integration $c_1 + c_2 + c_3 = C$:

$$\int \frac{3 + 4x - 5x^2}{\sqrt{x}}\,dx = 6x^{\frac{1}{2}} + \tfrac{8}{3}x^{\frac{3}{2}} - 2x^{\frac{5}{2}} + C.$$

If $u$ is a function of $x$, then Property 4 can be applied only if we also have $du = f'(x)dx$ present. The problem is one of writing the expression under the integral sign in the proper grouping. Consider, for example, $\int(x^2 + 3x)^{\frac{1}{2}}(2x + 3)dx$. If we let $u = x^2 + 3x$, then $du =$

**Definition.** The equation $y' = f(x)$ is called an **ordinary differential equation of the first order and first degree.**

It may also be written in terms of differentials, $dy = f(x)\,dx$ and $dy - f(x)\,dx = 0$.

**Definition.** A solution of an ordinary differential equation in two variables is a relation between the two variables which satisfies the differential equation.

For an ordinary differential equation of the first order and first degree, the solution is the antiderivative or indefinite integral. Thus if $y' = f(x)$, the solution is $y = \int f(x)\,dx = g(x) + c$.

**Definition.** The equation $y'' = f(x)$ is called an **ordinary differential equation of the second order and first degree.**

The solution of this equation may be obtained by finding the indefinite integral

$y' = \int f(x)\,dx = g(x) + c$, and then

finding the indefinite integral

$$y = \int [g(x) + c_1]\,dx = h(x) + c_1 x + c_2.$$

Note that the solution has two arbitrary constants $c_1$ and $c_2$.

$\frac{d}{dx}(x^2 + 3x)dx = (2x + 3)dx$, and we can write the integral in the form

$\int u^{\frac{1}{2}}\,du = u^{\frac{3}{2}} \div \frac{3}{2} + C$, or in terms of $x$ we have $\frac{2}{3}(x^2 + 3x)^{\frac{3}{2}} + C$.

The differential $du$ may be altered by a constant multiplier using the reciprocal of the multiplier outside the integral sign since $\int u\,du = \frac{1}{k}\int u\,k\,du$ where $k$ is a constant. This is equivalent to multiplying by the identity element $\left(\frac{1}{k}\right) k = 1$. This procedure is necessary in many problems and is a source of difficulty to some students. We shall illustrate by an example.

Given: $\int \frac{(x + 1)\,dx}{\sqrt{x^2 + 2x}}$. Let $u = x^2 + 2x$, then

$du = (2x + 2)\,dx = 2(x + 1)\,dx$. We see that the numerator of the integrand differs from $du$ by a multiple of 2. We therefore multiply the numerator of the integrand by 2 and the integral by $\frac{1}{2}$. The sequential steps of the solution are

$$\int \frac{(x + 1)\,dx}{\sqrt{x^2 + 2x}} = \frac{1}{2}\int (x^2 + 2x)^{-\frac{1}{2}}(2x + 2)\,dx = \frac{1}{2}(x^2 + 2x)^{\frac{1}{2}} \div \frac{1}{2} + c$$
$$= (x^2 + 2x)^{\frac{1}{2}} + c.$$

By the **solution of a differential equation** we mean exactly the same as the normal definition of "solution" for any mathematical equation. The only thing that may be new to the student is that the solution is now expressed as a functional relation instead of a number or an algebraic expression. The solution of the simple differential equation $y' = f(x)$ is the antiderivative $y = g(x) + c$ where $g'(x) = f(x)$. Thus if we substitute $y' = g'(x)$ into the equation, we obtain an identity $f(x) = f(x)$. To illustrate this point let us show that $y = 2x^3 + Ax + B$ is a solution of the second order differential equation $y'' = 12x$. Let us differentiate $y$ to find $y' = 6x^2 + A$ and then differentiate again to obtain $y'' = 12x$. If we substitute this into the differential equation, we obtain $12x = 12x$, an identity, and the equation is satisfied.

The solutions of the simple equations $y' = f(x)$, $y'' = f(x)$, and $y^{(n)} = f(x)$ are obtained by repeated integrations to find $y = g(c_i, x)$. Consider, for example, the equation $y'' = 6x + 4$. Integrating both sides we obtain

$$y' = 3x^2 + 4x + c_1.$$

Integrating again we obtain

$$y = x^3 + 2x^2 + c_1 x + c_2.$$

The two arbitrary constants are $c_1$ and $c_2$.

If a particular value of the function $y = f(x)$ is known, then the arbitrary constant $c$ may be determined. There is a particular solution to $y' = \sqrt{x}$ if $y(2) = \frac{\sqrt{2}}{3}$. We first integrate to obtain $y = \frac{2}{3}x^{\frac{3}{2}} + c$. Now find $y(2) = \frac{2}{3}(2)^{\frac{3}{2}} + c = \frac{2}{3}(2)\sqrt{2} + c = \frac{4}{3}\sqrt{2} + c$ and set this equal to $\frac{\sqrt{2}}{3}$. The resulting equation is $\frac{4}{3}\sqrt{2} + c = \frac{\sqrt{2}}{3}$ or $c = \frac{\sqrt{2}}{3} - \frac{4\sqrt{2}}{3} = -\frac{3\sqrt{2}}{3} = -\sqrt{2}$. The particular solution is $y = \frac{2}{3}x\sqrt{x} - \sqrt{2}$.

## SOLUTIONS TO EXERCISES

### I

i. $\int (x^2 + x)(2x)\,dx = I$

This is not of the form $u^2 du$.

$$I = \int (2x^3 + 2x^2)\,dx$$
$$= 2\int x^3 dx + 2\int x^2 dx$$
$$= \frac{x^4}{2} + \frac{2x^3}{3} + c$$

ii. $\displaystyle\int \frac{x^2 - 3}{\sqrt{x}} = \int (x^{\frac{3}{2}} - 3x^{-\frac{1}{2}})\,dx$

$$= \left(x^{\frac{5}{2}} + \tfrac{5}{2}\right) - \left(3x^{\frac{1}{2}} + \tfrac{1}{2}\right) + c$$
$$= \tfrac{2}{5}x^{\frac{5}{2}} - 6x^{\frac{1}{2}} + c$$

iii. $\displaystyle\int \left(5\sqrt{x} - \frac{3}{\sqrt{x}}\right) x dx = \int (5x^{\frac{3}{2}} - 3x^{\frac{1}{2}})\,dx$

$$= 5x^{\frac{5}{2}} \div \tfrac{5}{2} - 3x^{\frac{3}{2}} \div \tfrac{3}{2} + c$$
$$= 2x^{\frac{5}{2}} - 2x^{\frac{3}{2}} + c$$

iv. $\int (x^3 - 4x^2)(3x^2 - 8x)\,dx = I$

Let $u = x^3 - 4x^2$, $du = (3x^2 - 8x)\,dx$

$$I = \int u du = \tfrac{1}{2}u^2 + c$$
$$= \tfrac{1}{2}(x^3 - 4x^2)^2 + c$$

v. $\int (x^2 + \sqrt{x})(4x + x^{-\frac{1}{2}})\,dx = I$

Factor 2 from second factor.

$$I = 2\int (x^2 + \sqrt{x})\left(2x + \tfrac{1}{2}x^{-\frac{1}{2}}\right)dx$$
$$= \frac{2(x^2 + \sqrt{x})^2}{2} + c$$
$$= x^4 + 2x^2\sqrt{x} + x + c$$

### II

i. $f'(x) = x^2 - \sqrt{x}$, $f(1) = 0$

$$f(x) = \int (x^2 - \sqrt{x})\,dx$$
$$= \tfrac{1}{3}x^3 - \tfrac{2}{3}x^{\frac{3}{2}} + c$$
$$f(1) = \tfrac{1}{3} - \tfrac{2}{3} + c = 0,\ c = \tfrac{1}{3}$$
$$f(2) = \tfrac{1}{3}(8) - \tfrac{2}{3}(2\sqrt{2}) + \tfrac{1}{3}$$
$$= 3 - \tfrac{4}{3}\sqrt{2}$$

ii. $f'(x) = \sqrt{5 + x}$, $f(-1) = \frac{22}{3}$

$$f(x) = \int \sqrt{5 + x}\,dx$$
$$= \tfrac{2}{3}(5 + x)^{\frac{3}{2}} + c$$
$$f(-1) = \tfrac{2}{3}(5 - 1)^{\frac{3}{2}} + c = \tfrac{22}{3}$$
$$c = \tfrac{22}{3} - \tfrac{16}{3} = 2$$
$$f(2) = \tfrac{2}{3}(5 + 2)^{\frac{3}{2}} + 2 = \tfrac{14}{3}\sqrt{7} + 2$$

iii. $f'(x) = x^{-2}$, $f(1) = 3$

$$f(x) = \int x^{-2}\,dx = -x^{-1} + c$$
$$f(1) = -1^{-1} + c = 3,\ c = 4$$
$$f(2) = -2^{-1} + 4 = -\tfrac{1}{2} + 4 = \tfrac{7}{2}$$

iv. $f'(x) = (x^3 - 5)^5 x^2$, $f(1) = \frac{5}{6}(4^4)$

$$f(x) = \int (x^3 - 5)^5 x^2\,dx$$
$$= \tfrac{1}{3}\int (x^3 - 5)^5 3x^2\,dx$$
$$= \frac{(x^3 - 5)^6}{18} + c$$
$$f(1) = \frac{(-4)^6}{18} + c = \tfrac{5}{6}(4^4)$$
$$c = \frac{4^4}{18}(15 - 16) = -\frac{4^4}{18}$$
$$f(2) = \frac{(8 - 5)^6}{18} - \frac{4^4}{18}$$
$$= \tfrac{1}{18}(3^6 - 4^4) = \tfrac{473}{18}$$

### III

i. $f''(x) = 24x$

$$f'(x) = \int 24x\,dx = 12x^2 + A$$
$$f(x) = \int (12x^2 + A)\,dx$$
$$= 4x^3 + Ax + B$$

ii. $f''(x) = 18x - 8$, $f'(1) = 3$, $f(1) = 7$

$$f'(x) = \int (18x - 8)\,dx$$
$$= 9x^2 - 8x + A$$
$$f'(1) = 9 - 8 + A = 3,\ A = 2$$
$$f(x) = \int (9x^2 - 8x + 2)\,dx$$
$$= 3x^3 - 4x^2 + 2x + B$$
$$f(1) = 3 - 4 + 2 + B = 7$$
$$B = 7 - 1 = 6$$
$$f(x) = 3x^3 - 4x^2 + 2x + 6$$
$$f(2) = 24 - 16 + 4 + 6 = 18$$

# 8 APPLICATIONS OF THE DEFINITE INTEGRAL

## SELF-TEST

DIRECTIONS: Write your answers in the numbered regions to the right. To check your answers, turn the page. Study the solutions to the problems you missed, and do the exercises following any problem or group of problems in which you had errors.

| |
|---|
| 1 |
| 2 |
| 3 |
| 4 |
| 5 |
| 6 |
| 7 |
| 8 |
| 9 |

In **1–4** evaluate the definite integrals.

1. $\int_1^3 (3x^2 + 2x)\,dx$

2. $\int_2^\infty \frac{1}{\sqrt{x^3}}\,dx$

3. $\int_a^b \sqrt{a - x}\,dx$

4. $\int_{-2}^2 x^3\,dx$

In **5–7** find the area bounded by the $x$-axis, the curve defined by $y = f(x)$, and the indicated ordinates.

5. $y = 6x^2 - x$, $x = 0$ and $x = 4$

6. $y = 2x - 4$, $x = -1$ and $x = 3$

7. $y = x^3 - 4x^2 + 3x$, $x = 0$ and $x = 3$

8. Find the area bounded by $y = \frac{x^2}{4}$ and $y = 2\sqrt{x}$.

9. The cross section of a tank is an isosceles trapezoid whose upper base is 8 ft, lower base is 6 ft, and the altitude is 3 ft. Find the total force on one end of the tank if it is filled with water ($w = 62.4$ lb/ft$^3$).

## SOLUTIONS

1. $\int_1^3 (3x^2 + 2x)\,dx = [x^3 + x^2]_1^3$

$= (3)^3 + (3)^2 - [(1)^3 + (1)^2]$

$= 27 + 9 - 1 - 1 = 34$

2. $\int_2^\infty \frac{1}{\sqrt{x^3}}\,dx = \lim_{b\to\infty} \int_2^b x^{-\frac{3}{2}}\,dx$

$= \lim_{b\to\infty} \left[x^{-\frac{1}{2}}(-2)\right]_2^b$

$\lim_{b\to\infty}\left(\frac{-2}{\sqrt{b}} - \frac{-2}{\sqrt{2}}\right) = 0 + \sqrt{2} = \sqrt{2}$

3. $\int_a^b \sqrt{a-x}\,dx$

$= \frac{1}{-1}\int_a^b (a-x)^{\frac{1}{2}}(-1)\,dx$

$= -\left[\frac{2}{3}(a-x)^{\frac{3}{2}}\right]_a^b$

$= -\ \frac{2}{3}(a-b)^{\frac{3}{2}} - \frac{2}{3}(a-a)^{\frac{3}{2}}$

$= -\frac{2}{3}(a-b)^{\frac{3}{2}}$

4. $\int_{-2}^2 x^3 dx = \left[\frac{x^4}{4}\right]_{-2}^2$

$= \frac{1}{4}[(2)^4 - (-2)^4]$

$= \frac{1}{4}(16 - 16) = 0$

**EXERCISES I**

Evaluate the integrals.

i. $\int_{-3}^3 (x^2 - x - 6)\,dx$

ii. $\int_0^r \sqrt{r^2 - x^2}\,x dx$

iii. $\int_1^\infty x^{-2}\,dx$

5. $y = 6x^2 - x,\ x = 0,\ x = 4$
The curve $y = f(x)$ lies above the $X$-axis in the interval $[0,4]$.

$A = \int_a^b y dx = \int_0^4 (6x^2 - x)\,dx$

$= \left[2x^3 - \frac{1}{2}x^2\right]_0^4$

$= 2(64) - \frac{1}{2}(16) - 0 = 120$

6. $y = 2x - 4,\ x = -1,\ x = 3$
The curve $y = f(x)$ crosses the $X$-axis at $2x - 4 = 0$ or $x = 2$. In the interval $[-1,2]$ the curve is below the $X$-axis and in $[2,3]$ the curve is above the $X$-axis.

$A = -\int_{-1}^2 (2x - 4)\,dx + \int_2^3 (2x - 4)\,dx$

$= -[x^2 - 4x]_1^2 + [x^2 - 4x]_2^3$

$= -(4 - 8 - 1 - 4) + (9 - 12 - 4 + 8)$

$= -(-9) + 1 = 10$

7. $y = x^3 - 4x^2 + 3x,\ x = 0,\ x = 3$
The curve crosses the $X$-axis at $x^3 - 4x^2 + 3x = 0$ or $x = 0,1$, and 3. For $0 \le x \le 1,\ y \ge 0$ and for $1 \le x \le 3,\ y \le 0$.
$A = A_1[0 \le x \le 1] + A_2[1 \le x \le 3]$

$\int_a^b (x^3 - 4x^2 + 3x)\,dx$

$= \left[\frac{x^4}{4} - \frac{4x^3}{3} + \frac{3x^2}{2}\right]_a^b$

$A_1 = \left[\frac{x^4}{4} - \frac{4x^3}{3} + \frac{3x^2}{2}\right]_0^1 = \frac{5}{12}$

$A_2 = -\left[\frac{x^4}{4} - \frac{4x^3}{3} + \frac{3x^2}{2}\right]_1^3 = \frac{32}{12}$

$A = A_1 + A_2 = \frac{5}{12} + \frac{32}{12} = \frac{37}{12}$

8. $y = \frac{x^2}{4}$ and $y = 2\sqrt{x}$

Draw the two curves which intersect at $x^2 = 8\sqrt{x}$ or $x = 0,\ x = 4$ for which $y = 0$ and $y = 4$.

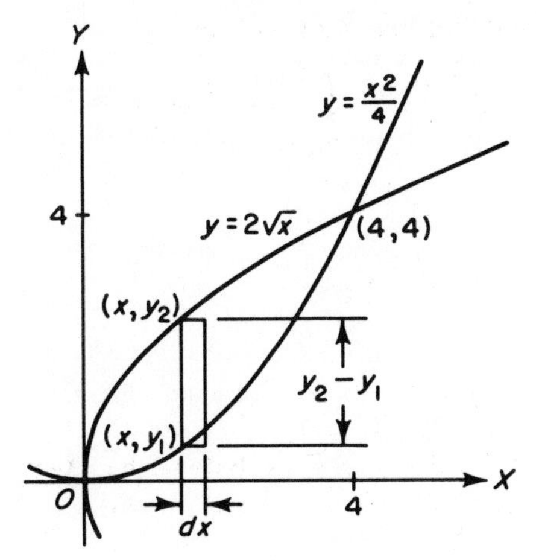


Fig. 8-1

$$A = \int_0^4 \left(2\sqrt{x} - \frac{x^2}{4}\right)dx$$

$$= \left[\frac{4}{3}x^{\frac{3}{2}} - \frac{x^3}{12}\right]_0^4 = \frac{16}{3}$$

EXERCISES II

Find the specified areas.

i. Area bounded by $y = x^2 - x - 6$, the $X$-axis, $x = -3$, and $x = 3$. (Compare with Exercise I-i.)

ii. Area bounded by the curves $y = x$ and $y = x^3$.

iii. Area bounded by the curves $y = x - 3$ and $x = y^2 + 1$.

9. Draw the trapezoid and choose the axes as shown in Figure 8–2.

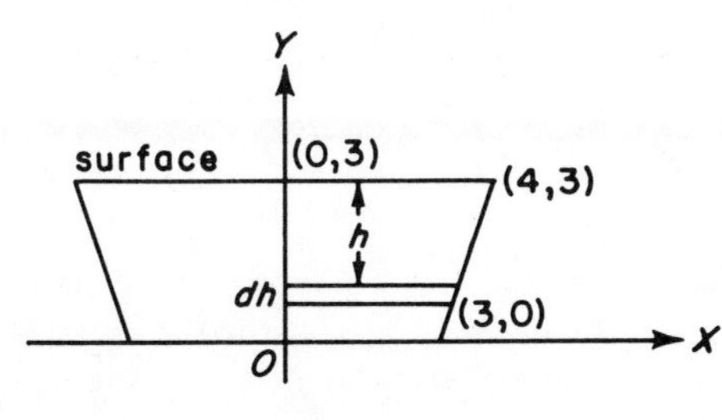


Fig. 8–2

**ANSWERS**

|  | |
|---|---|
| 1. 34 | 1 |
| 2. $\sqrt{2}$ | 2 |
| 3. $-\frac{2}{3}(a-b)^{\frac{3}{2}}$ | 3 |
| 4. 0 | 4 |
| 5. 120 | 5 |
| 6. 10 | 6 |
| 7. $\frac{37}{12}$ | 7 |
| 8. $\frac{16}{3}$ | 8 |
| 9. 1872 lb | 9 |

Since this figure is symmetrical about the $Y$-axis, we can consider the force on the right half and multiply by 2. The element of the force is $dF = hxdh$ and we shall sum with respect to $h$. First express $x$ in terms of $y$ by obtaining the equation of the line through $P_1(3,0)$ and $P_2$ $(4,3)$.

$$y = 3x - 9 \text{ or } x = \frac{y+9}{3}$$

Then to change this equation into $h$ use the relationship

$$y = 3 - h \text{ and } x = \tfrac{1}{3}(12 - h).$$

$$\tfrac{1}{2}F = \frac{w}{3}\int_0^3 (12-h)\,h\,dh$$

$$= 20.8\left[6h^2 - \frac{h^3}{3}\right]_0^3 = 936$$

Therefore $F = 1872$ lb.

## BASIC FACTS

### Definition

$$\int_a^b f(x)\,dx = F(x)\Big|_a^b = F(b) - F(a)$$

is called a **definite integral.** The interval $[a,b]$ is called the **interval of integration.** The value $x = a$ is called the **lower limit**; $x = b$, the **upper limit.**

To find the value of a definite integral first integrate to find $F(x)$, then substitute $x = b$ to find $F(b)$, substitute $x = a$ to find $F(a)$, and finally take the difference $F(b) - F(a)$. Note that the value is always the value at the upper limit minus the value at the lower limit. Furthermore, the constant of integration which was present in the indefinite integral automatically disappears in the definite integral.

### Properties

1. $\int_a^b f(x)\,dx = -\int_b^a f(x)\,dx$

2. $\int_a^b f(x)\,dx + \int_b^c f(x)\,dx$

$$= \int_a^c f(x)\,dx$$

### Definitions

$$\int_a^{\infty} f(x)\,dx = \lim_{b\to\infty} \int_a^b f(x)\,dx$$

$$\int_{-\infty}^b f(x)\,dx = \lim_{a\to-\infty} \int_a^b f(x)\,dx$$

$$\int_{-\infty}^{\infty} f(x)\,dx = \int_{-\infty}^0 f(x)\,dx + \int_0^{\infty} f(x)\,dx$$

The sum of a sequence of terms is expressed by the symbol

$$\sum_{i=1}^{n} a_i = a_1 + a_2 + \ldots + a_n$$

## ADDITIONAL INFORMATION

**Fundamental Theorem of the Integral Calculus**

Let $f(x)$ be a continuous function over the interval from $x = a$ to $x = b$. Let this interval be divided into $n$ subintervals whose lengths are $\Delta x_i$, $(i = 1,2, \ldots, n)$. Choose an abscissa $x_i$ in each subinterval. Let us now form the sum

$$\sum_{i=1}^{n} f(x_i)\,\Delta x_i = f(x_1)\,\Delta x_1 + f(x_a)\,\Delta x_2 + \ldots + f(x_n)\,\Delta x_n.$$

Then the limiting value of this sum as $n$ increases without bound so that each subinterval approaches zero as a limit equals the value of the definite integral of $f(x)\,dx$ from $a$ to $b$.

$$\lim_{n\to\infty} \sum_{i=1}^{n} f(x_i)\,\Delta x_i = \int_a^b f(x)\,dx = F(b) - F(a)$$

The **area bounded by a curve**, one of the coordinate axes and two straight lines perpendicular to that coordinate axis can be found by a definite integral, if the curve is defined by a continuous function. Consider the curve $C$ defined by $y = f(x)$ and shown in Figure 8–3. The area $A$ bounded by the curve, the $X$-axis and the lines $x = a$ and $x = b$ can be approximated by summing the areas of rectangles whose sides are $y$ and $\Delta x$. This element of area is $\Delta A = y\,\Delta x$. If we cover the interval from $x = a$ to $x = b$ by $n$ increments $\Delta x_i$ and obtain the corresponding $y_i = f(x_i)$ for each $\Delta x_i$, then the area $A$ is approximately equal to the sum of the areas of these rectangles.

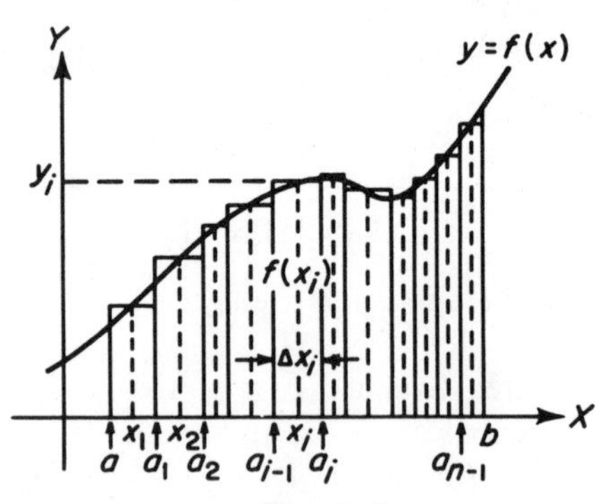


Fig. 8-3

$$A \doteq \sum_{i=1}^{n} \Delta A_i = \sum_{i=1}^{n} y_i\,\Delta x_i = \sum_{i=1}^{n} f(x_i)\,\Delta x_i$$

If we now form more rectangles by taking $\Delta x$ smaller and thus taking $n$ larger, the approximation of the area improves. If we let $n$ increase without bound so that $\Delta x \longrightarrow 0$, we have

$$\lim_{n\to\infty} \sum_{i=1}^{n} f(x_i)\,\Delta x_i = \int_a^b f(x)\,dx = A.$$

The concept of the element of area $\Delta A = y\Delta x$ or the similar $dA = ydx$ is fundamental in integral calculus. It is applied to the element of volume, element of work, element of pressure, etc. The student should have a clear understanding of this important concept.

read "the sum of $a_i$ from $i = 1$ to $i = n$."

**Fundamental Theorem of Integral Calculus**

Let $f(x)$ be continuous in the interval $[a,b]$, then

$$\lim_{n\to\infty} \sum_{i=1}^{n} f(x_i)\,\Delta x_i = \int_a^b f(x)\,dx$$

where the interval $[a,b]$ has been divided into $n$ subintervals of length $\Delta x_i$ and $x_i$ is an abscissa in the corresponding subinterval.

**Area, Work, and Pressure Problems**

Let $y = f(x)$ be a continuous function in $[a,b]$. The **area bounded by the curve** for the values $y = f(x) > 0$, the $X$-axis, and the lines $x = a$ and $x = b$ may be found by $A = \int_a^b y\,dx$.

If $y = f(x) < 0$ (below the $X$-axis), then $A = -\int_a^b y\,dx$. If $y = 0$ at $x = c$ for $a < c < b$, then $A = A_1 + A_2$, where

$$A_1 = \pm \int_a^c y\,dx \text{ and } A_2 = \pm \int_c^b y\,dx$$

the sign depending upon whether $y$ is positive or negative in the corresponding interval.

The area between two curves $y_1 = f_1(x)$ and $y_2 = f_2(x)$ where $y_2 \geq y_1$ can be found directly from $A = \int_a^b (y_2 - y_1)\,dx$ with $a$ and $b$ the limiting values of $x$.

The work done by a force $F$ acting along and in the direction of a straight line is given by $W = \int_{x_1}^{x_2} F\,dx$ providing $F$ is a continuous function of $x$ in the interval $[x_1, x_2]$.

The total force of a fluid acting on a plane area submerged vertically is given by $F = w \int_{h_1}^{h_2} hL\,dh$ where $w$ is the density of the fluid, $h$ is the depth below the surface, $L = f(h)$ is the horizontal length of the plane area, $f(h)$ is a continuous function of $h$ in $[h_1, h_2]$ and $h_2 > h_1 \geq 0$.

Let us now consider the **area bounded by two intersecting curves** as shown in Figure 8–4. The curves are defined by the equations $C_1$: $(y-1)^2 = 4(x-2)$ and $C_2$: $y = 2x - 7$. Since for $C_1$ the variable $y$ has two values for each $x > 2$, we shall take the elements of area to be horizontal. It should be clear from the figure that $dA = (x_2 - x_1)\,dy$ where $x_2$ is on $C_2$ and $x_1$ is on $C_1$. The points of intersection of the curves are $(3,-1)$ and $(6,5)$, so we sum these elements of area as $y$ changes from $-1$ to 5.

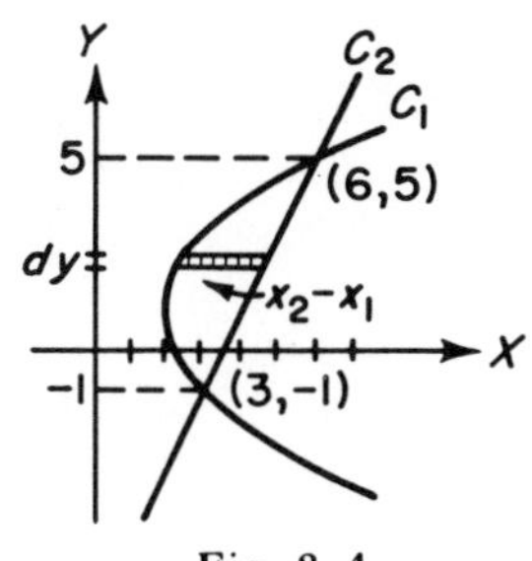


Fig. 8–4

$$A = \int_{-1}^{5} dA = \int_{-1}^{5} (x_2 - x_1)\,dy = \int_{-1}^{5} \left(\frac{y+7}{2} - \frac{y^2 - 2y + 9}{4}\right) dy$$

$$= \frac{1}{4}\int_{-1}^{5} (4y - y^2 + 5)\,dy = \frac{1}{4}\left(2y^2 - \frac{y^3}{3} + 5y\right)\Bigg|_{-1}^{5}$$

$$= \tfrac{1}{4}\left[50 - \tfrac{125}{3} + 25 - \left(2 + \tfrac{1}{3} - 5\right)\right] = \tfrac{1}{4}(36) = 9$$

If we are summing with respect to $x$ and the curve is below the $X$-axis, then $y_i = f(x_i)$ is negative, and the element of area given by $dA = y_i dx$ is negative. This gives rise to *signed* areas although areas ordinarily are considered to be positive quantities. In case the curve crosses the $X$-axis in the interval $[a,b]$ and we desire to find the area bounded by the curve, the $X$-axis and the lines $x = a$ and $x = b$, it is necessary to sum the absolute values of the areas above and below the $X$-axis. Consider the area bounded by the curve $y = x^4 - 4x^3 - 4x^2 + 16x$, the $X$-axis, $x = -2$ and $x = 4$. The curve crosses the $X$-axis at $x = -2, 0, 2$, and 4, since these are the roots of the equation $y = 0$. By calculating $y$ for values of $x$ in each interval, we see that $y < 0$ for $-2 < x < 0$, $y > 0$ for $0 < x < 2$, and $y < 0$ for $2 < x < 4$. Consequently, $\int_{-2}^{0} y\,dx < 0$, $\int_0^2 y\,dx > 0$, and $\int_2^4 y\,dx < 0$. In order to find the area, we sum the absolute values, $A = |\int_{-2}^{0} y\,dx| + |\int_0^2 y\,dx| + |\int_2^4 y\,dx|$. See problem 7 for an example.

The same situation exists if we are summing with respect to $y$ and the curve is to the left of the $Y$-axis. Most books will define the area as $A = -\int_{x_1}^{x_2} y\,dx$ if $y < 0$ in $[x_1, x_2]$, and in order to keep the signs in their proper perspective we take $x_1 < x_2$ when summing with respect to $x$ and $y_1 < y_2$ when summing with respect to $y$. In finding the area between two curves, we take $y_1 > y_2$ in $\int (y_1 - y_2)\,dx$, and if the curves intersect in the open interval $(x_1, x_2)$, then we again break the area into two integrals keeping $y_i - y_j > 0$ in each integral.

## SOLUTIONS TO EXERCISES

I

i. $\int_{-3}^{3} (x^2 - x - 6)\, dx$

$= \left[\frac{x^3}{3} - \frac{x^2}{2} - 6x\right]_{-3}^{3}$

$= \left(9 - \frac{9}{2} - 18\right) - \left(-9 - \frac{9}{2} + 18\right)$

$= -\frac{27}{2} - \frac{9}{2} = -\frac{36}{2} = -18$

ii. $\int_{0}^{r} \sqrt{r^2 - x^2}\; x dx$

$= -\frac{1}{2}\int_{0}^{r} (r^2 - x^2)\,(-2x)\, dx$

$= -\frac{1}{2}\left[\frac{2}{3}(r^2 - x^2)\right]_0^r$

$= -\frac{1}{3}\left[(r^2 - r^2) - (r^2 - 0)\right]$

$= -\frac{1}{3}(0 - r^3) = \frac{r^3}{3}$

iii. $\int_{1}^{\infty} x^{-2}\, dx = \lim_{b\to\infty} \int_{1}^{b} x^{-2}\, dx$

$= \lim_{b\to\infty}\left[-\frac{1}{x}\right]_1^b$

$= \lim_{b\to\infty}\left[-\frac{1}{b} - (-1)\right] = 1$

II

i. The curve $y = x^2 - x - 6$ crosses the $X$-axis at $x = -2$ and $x = 3$, and $y < 0$ for $-2 < x < 3$.

$A = \int_{-3}^{-2} (ydx) - \int_{-2}^{3} ydx$

$= \left[\frac{x^3}{3} - \frac{x^2}{2} - 6x\right]_{-3}^{-2}$

$- \left[\frac{x^3}{3} - \frac{x^2}{2} - 6x\right]_{-2}^{3}$

$= \left(1 - \frac{8}{3} + \frac{9}{2}\right) - \left(-19 - \frac{9}{2} + \frac{8}{3}\right)$

$= 29 - \frac{16}{3} = \frac{71}{3}$

ii. $y = x$ and $y = x^3$. Draw Figure 8-5.

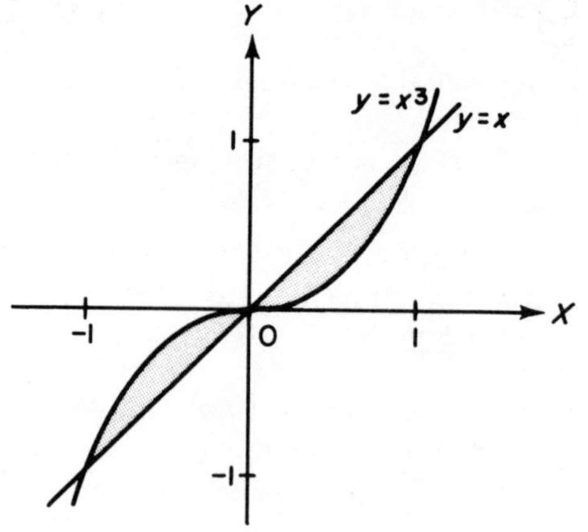


Fig. 8-5

$A = \int_{-1}^{0} (x^3 - x)\, dx + \int_{0}^{1} (x - x^3)\, dx$

$= \left[\frac{x^4}{4} - \frac{x^2}{2}\right]_{-1}^{0} + \left[\frac{x^2}{2} - \frac{x^4}{4}\right]_{0}^{1}$

$= -\left(\frac{1}{4} - \frac{1}{2}\right) + \left(\frac{1}{2} - \frac{1}{4}\right) = \frac{1}{2}$

iii. $y = x - 3$ and $x = y^2 + 1$. Draw Fig. 8-6.

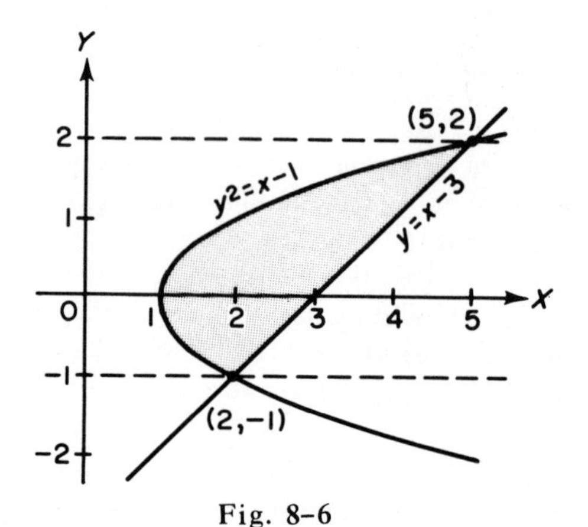


Fig. 8-6

Let $y$ be the independent variable and solve for $x$ as a function of $y$.

$A = \int_{-1}^{2} (y + 3 - y^2 - 1)\, dy$

$= \left[\frac{y^2}{2} - \frac{y^3}{3} + 2y\right]_{-1}^{2}$

$= 2 - \frac{8}{3} + 4 - \left(\frac{1}{2} + \frac{1}{3} - 2\right) = \frac{9}{2}$

# 9 TRIGONOMETRIC FUNCTIONS

## SELF-TEST

DIRECTIONS: Write your answers in the numbered regions to the right. To check your answers, turn the page. Study the solutions to the problems you missed, and do the exercises following any problem or group of problems in which you had errors.

1. Find the exact values of the trigonometric functions.

   a. $\cos 60°$  b. $\tan \frac{\pi}{2}$  c. $\csc(-90°)$

2. Reduce the expressions to functions of a positive acute angle.

   a. $\sin 150°$  b. $\cot 225°$  c. $\csc 330°$
   d. $\cos(-110°)$  e. $\tan 460°$  f. $\sec(-390°)$

3. Find the value of the given expressions.

   a. $\sin \frac{\pi}{3} \tan \frac{\pi}{4} \cos \frac{\pi}{6}$  b. $\dfrac{\sin 30° \tan 60°}{\cot 30°}$

4. Find the following values.

   a. $\cos \text{Arctan } \frac{3}{4}$  b. $\text{Arccos } \frac{\sqrt{2}}{2}$  c. $\text{Arcsin}\left(\sin \frac{\pi}{6}\right)$

5. Given $\cos 2\phi = \frac{3}{5}$. Find $\sin \phi$ and $\cos \phi$.

6. Find $\tan(x - y)$ if $\tan x = \frac{4}{3}$ and $\tan y = \frac{3}{4}$.

7. Find $\cos\left(2 \text{ Arcsin } \frac{12}{13}\right)$.

8. Find the exact value of $\tan 22.5°$.

9. Prove the identity $\dfrac{\sec x}{\cot x + \tan x} = \sin x$.

10. Solve the equation $\dfrac{\tan x + \tan 2x}{1 - \tan x \tan 2x} = 1$.

| |
|---|
| 1 a<br>b<br>c |
| 2 a<br>b<br>c<br>d<br>e<br>f |
| 3 a<br>b |
| 4 a<br>b<br>c |
| 5 |
| 6 |
| 7 |
| 8 |
| 9 |
| 10 |

## SOLUTIONS

1. These values should have been committed to memory.
   a. $\cos 60° = \frac{1}{2}$
   b. $\tan \frac{\pi}{2}$ is undefined.
   c. $\csc(-90°) = -\csc 90° = -1$

### EXERCISES I

Find the exact values of the given expressions.

| | |
|---|---|
| i. $\cos 120°$ | iv. $\sec 330°$ |
| ii. $\tan(-60°)$ | v. $\cot 480°$ |
| iii. $\sin \frac{3\pi}{2}$ | vi. $\csc \frac{5\pi}{4}$ |

2. Use reference angle formulas.
   a. $\sin 150° = \sin 30°$
   b. $\cot 225° = \cot 45°$
   c. $\csc 330° = -\csc 30°$
   d. $\cos(-110°) = \cos 110° = -\cos 70°$
   e. $\tan 460° = \tan 100° = -\tan 80°$
   f. $\sec(-390°) = \sec 390° = \sec 30°$

3. Substitute values and simplify.
   a. $\sin \frac{\pi}{3} \tan \frac{\pi}{4} \cos \frac{\pi}{6}$
   $= \frac{\sqrt{3}}{2}(1)\frac{\sqrt{3}}{2} = \frac{3}{4}$
   b. $\dfrac{\sin 30° \tan 60°}{\cot 30°} = \dfrac{\frac{1}{2}\sqrt{3}}{\sqrt{3}} = \frac{1}{2}$

### EXERCISES II

i. Find the exact value of $\sin 75°$.

ii. Evaluate $\sin \frac{\pi}{3} \cos \frac{\pi}{4}$.

iii. Find $f(n) = \cos \frac{n\pi}{3}$ if $n$ is a nonnegative integer.

4. a. $\cos \text{Arctan} \frac{3}{4}$. Let $y = 3$, and $x = 4$, then $r = \sqrt{x^2 + y^2} = 5$.
   $\cos \text{Arctan} \frac{3}{4} = \frac{4}{5}$
   b. $\text{Arccos} \frac{\sqrt{2}}{2} = \frac{\pi}{4}$
   c. $\text{Arcsin}\left(\sin \frac{\pi}{6}\right) = \frac{\pi}{6}$

### EXERCISES III

i. Find $\arcsin \frac{1}{2}$.

ii. Find $\cos \arctan \frac{3}{4}$.

iii. Find $\text{Arccos}(\tan 45°)$.

5. Given $\cos 2\phi = \frac{3}{5}$.

$$\sin \phi = \sqrt{\frac{1 - \cos 2\phi}{2}} = \sqrt{\frac{1 - \frac{3}{5}}{2}} = \frac{\sqrt{5}}{5}$$

$$\cos \phi = \sqrt{\frac{1 + \cos 2\phi}{2}} = \sqrt{\frac{1 + \frac{3}{5}}{2}} = \sqrt{\frac{4}{5}} = \frac{2\sqrt{5}}{5}$$

6. $$\tan(x - y) = \frac{\tan x - \tan y}{1 + \tan x \tan y} = \frac{\frac{4}{3} - \frac{3}{4}}{1 + \left(\frac{4}{3}\right)\left(\frac{3}{4}\right)} = \frac{7}{24}$$

### EXERCISES IV

If $\tan 2\phi = \dfrac{B}{A - C}$, find

i. $\sin \phi$ if $A = 5$, $B = 4$, $C = 2$.

ii. $\cos \phi$ if $A = 4$, $B = 5$, $C = 4$.

7. $\cos\left(2 \text{Arcsin} \frac{12}{13}\right)$
   Let $\phi = \text{Arcsin} \frac{12}{13}$, then
   $$\cos 2\phi = 1 - 2\sin^2 \phi = 1 - 2\left(\frac{12}{13}\right)^2 = -\frac{119}{169}.$$

8. $$\tan 22.5° = \tan \tfrac{1}{2}(45°) = \frac{1 - \cos 45°}{\sin 45°} = \frac{1 - \frac{\sqrt{2}}{2}}{\frac{\sqrt{2}}{2}} = \sqrt{2} - 1$$

9. $$\frac{\sec x}{\cot x + \tan x} = \sin x$$

$$\frac{\frac{1}{\cos x}}{\frac{\cos x}{\sin x} + \frac{\sin x}{\cos x}} =$$

$$\frac{1}{\cos x} \cdot \frac{\sin x \cos x}{\cos^2 x + \sin^2 x} =$$

$$\sin x = \sin x$$

10. $$\frac{\tan x + \tan 2x}{1 - \tan x \tan 2x} = 1$$

$$\tan 3x = 1$$

$$3x = \frac{\pi}{4} + n\pi; \quad x = \frac{\pi}{12} + \frac{n\pi}{3}$$

EXERCISES V

Solve the following equations.

i. $\sec^2 x - \tan x = 1$

ii. $\cos 2x - \cos x = \sin \frac{x}{2}$

Solutions to the exercises are on page 54.

## ANSWERS

|  | |
|---|---|
| 1. a. $\frac{1}{2}$<br>b. Undefined<br>c. $-1$ | 1 a<br>b<br>c |
| 2. a. $\sin 30°$<br>b. $\cot 45°$<br>c. $-\csc 30°$<br>d. $-\cos 70°$<br>e. $-\tan 80°$<br>f. $\sec 30°$ | 2 a<br>b<br>c<br>d<br>e<br>f |
| 3. a. $\frac{3}{4}$<br>b. $\frac{1}{2}$ | 3 a<br>b |
| 4. a. $\frac{4}{5}$<br>b. $\frac{\pi}{4}$<br>c. $\frac{\pi}{6}$ | 4 a<br>b<br>c |
| 5. $\sin \phi = \frac{\sqrt{5}}{5}$<br>$\cos \phi = \frac{2\sqrt{5}}{5}$ | 5 |
| 6. $\frac{7}{24}$ | 6 |
| 7. $-\frac{119}{169}$ | 7 |
| 8. $\sqrt{2} - 1$ | 8 |
| 9. Identity | 9 |
| 10. $\frac{\pi}{12} + \frac{n\pi}{3}$ | 10 |

## BASIC FACTS

Let any line segment, $OA$, be revolved about one of its end-points, say $O$, until it takes the position $OB$. This revolution generates the **general angle** $AOB$. There is no limit to the number of times the line may revolve about $O$.

In the situation described, the point $O$ is called the **origin**, the revolving line segment is called the **radius vector**, $OA$ is called the **initial side**, and $OB$ the **terminal side**.

If the radius vector moves counterclockwise (clockwise), the angle generated is said to be a **positive (negative) angle**.

If a general angle is placed so that the vertex is at the origin of a rectangular coordinate system and its initial side coincides with the positive $X$-axis, then the angle is said to be in its **standard position.**

Let $\theta$ be any angle in its standard position. Let $P$ be any point (not the origin) having coordinates $(x, y)$ and lying on the terminal side of $\theta$. Let $r$ be the radius vector of $P$. Then the **trigonometric functions** are defined by

$$\sin \theta = \frac{\text{ordinate}}{\text{radius vector}} = \frac{y}{r}$$

$$\cos \theta = \frac{\text{abscissa}}{\text{radius vector}} = \frac{x}{r}$$

$$\tan \theta = \frac{\text{ordinate}}{\text{abscissa}} = \frac{y}{x}$$

$$\cot \theta = \frac{\text{abscissa}}{\text{ordinate}} = \frac{x}{y}$$

$$\sec \theta = \frac{\text{radius vector}}{\text{abscissa}} = \frac{r}{x}$$

$$\csc \theta = \frac{\text{radius vector}}{\text{ordinate}} = \frac{r}{x}$$

The trigonometric functions are not defined if $x$ or $y$ is zero in a denominator.

If $\theta$ is any positive number, then

$$\sin(-\theta) = -\sin \theta$$
$$\cos(-\theta) = \cos \theta$$
$$\tan(-\theta) = -\tan \theta$$
$$\cot(-\theta) = -\cot \theta$$
$$\sec(-\theta) = \sec \theta$$
$$\csc(-\theta) = -\csc \theta$$

## ADDITIONAL INFORMATION

The **general angle** may be measured in the sexagesimal measure or the circular (radian) measure.

| | |
|---|---|
| one **degree** | $= \frac{1}{360}$ of one revolution $= 1^\circ$ |
| one **minute** | $= \frac{1}{60}$ of one degree $= 1'$ |
| one **second** | $= \frac{1}{60}$ of one minute $= 1''$ |

A **radian** is defined to be the central angle subtended by an arc of a circle equal in length to the radius of the circle. Since a radian is an arc length equal to a radius, and since there are $2\pi$ radii on the circumference of a circle ($C = 2\pi R$), there are $2\pi$ radians in one revolution. Thus, we may formulate degree-radian conversion factors.

| | |
|---|---|
| 360 degrees $= 2\pi$ radians | 2 radians $= \frac{360}{\pi}$ degrees |
| 1 degree $= \frac{\pi}{180}$ radians | 1 radian $= \frac{180}{\pi}$ degrees |
| $= 0.017453$ radians | $= 57.2958^\circ$ |

Since there are certain values for which the trigonometric functions do not exist, we should have a clear concept of the **domain for the trigonometric functions.** Let $R$ denote the real numbers and let $n$ denote an integer. Then we have the following summary.

| Function | Domain | Function | Domain | Function | Domain |
|---|---|---|---|---|---|
| sin | $R$ | tan | $R$, except $\frac{\pi}{2} + n\pi$ | sec | $R$, except $\frac{\pi}{2} + n\pi$ |
| cos | $R$ | cot | $R$, except $n\pi$ | csc | $R$, except $n\pi$ |

The inadmissible values of the argument which cause the exceptions in the summary arise from the fact that the division by zero is not defined. Consider, for example, $\frac{\pi}{2}$ for which $x = 0$. Tan $\frac{\pi}{2} = \frac{y}{x} = \frac{y}{0}$ is not defined, and $\frac{\pi}{2}$ is inadmissible for the tangent function.

If we restrict the domain of the **trigonometric functions** to $0 < \theta < 90^\circ$ (the acute angles), then the functions can be defined in terms of the sides of a right triangle.

| | | |
|---|---|---|
| $\sin \theta = \frac{\text{opposite side}}{\text{hypotenuse}}$ | $\tan \theta = \frac{\text{opposite side}}{\text{adjacent side}}$ | $\sec \theta = \frac{\text{hypotenuse}}{\text{adjacent side}}$ |
| $\cos \theta = \frac{\text{adjacent side}}{\text{hypotenuse}}$ | $\cot \theta = \frac{\text{adjacent side}}{\text{opposite side}}$ | $\csc \theta = \frac{\text{hypotenuse}}{\text{opposite side}}$ |

| angle | sin | cos | tan | cot | sec | csc |
|---|---|---|---|---|---|---|
| $30^\circ$ | $\frac{1}{2}$ | $\frac{1}{2}\sqrt{3}$ | $\frac{1}{3}\sqrt{3}$ | $\sqrt{3}$ | $\frac{2}{3}\sqrt{3}$ | 2 |
| $45^\circ$ | $\frac{1}{2}\sqrt{2}$ | $\frac{1}{2}\sqrt{2}$ | 1 | 1 | $\sqrt{2}$ | $\sqrt{2}$ |
| $60^\circ$ | $\frac{1}{2}\sqrt{3}$ | $\frac{1}{2}$ | $\sqrt{3}$ | $\frac{1}{3}\sqrt{3}$ | 2 | $\frac{2}{3}\sqrt{3}$ |

The **reference angle**, $\alpha$, for a given value of $\theta \leq 360^\circ$ is defined in the following table.

| $\theta$ | $\alpha$ |
|---|---|
| $0 < \theta < 90^\circ$ | $\alpha = \theta$ |
| $90^\circ < \theta < 180^\circ$ | $\alpha = 180^\circ - \theta$ |
| $180^\circ < \theta < 270^\circ$ | $\alpha = \theta - 180^\circ$ |
| $270^\circ < \theta < 360^\circ$ | $\alpha = 360^\circ - \theta$ |

Any trigonometric function of an angle $\theta$, in any quadrant, is numerically equal to the same function of the reference angle for $\theta$. The sign is determined by the quadrant in which the terminal side of $\theta$ falls.

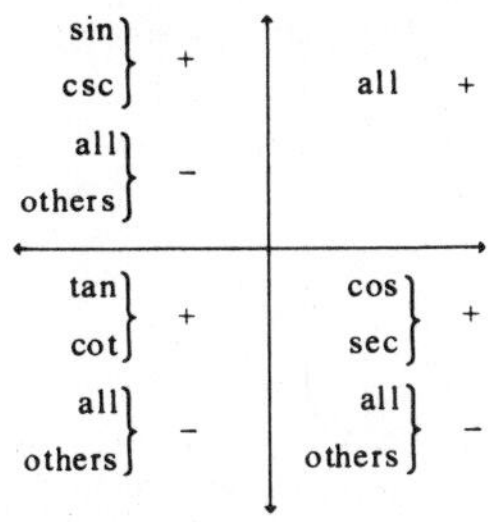

sin, csc +; all others −
all +
tan, cot +; all others −
cos, sec +; all others −

Fig. 9-1

If $\theta > 2\pi$, any trigonometric function of $\theta$ is equal to the same function of $(\theta - 2n\pi)$, where $n$ is any integer.

The word "arc" preceding the name of a trigonometric function indicates its **inverse relation.** Thus, the expression arc sin $N$ represents the angle whose sine is $N$. The inverse trigonometric functions are the **principal values** of the inverse relations. The principal value is usually indicated by capitalizing the first letter of "arc." See Additional Information for the intervals of these values.

All trigonometric functions are periodic with period $2\pi = 360^\circ$. The tangent and cotangent also have a period of $\pi = 180^\circ$. A trigonometric equation has an infinite number of solutions obtained by adding $2n\pi$ or $n\,360^\circ$, where $n$ is an integer, to the basic solutions, $0 \leq x < 2\pi$.

The **quadrantal angles** are those which fall on the axes. The trigonometric functions of these angles are given in the following table.

| angle | sin | cos | tan | cot | sec | csc |
|---|---|---|---|---|---|---|
| 0° | 0 | 1 | 0 | ... | 1 | ... |
| 90° | 1 | 0 | ... | 0 | ... | 1 |
| 180° | 0 | −1 | 0 | ... | −1 | ... |
| 270° | −1 | 0 | ... | 0 | ... | −1 |

The **principal values** of the inverse relations (the values of the inverse functions) are in the following table.

| | | |
|---|---|---|
| $-\frac{\pi}{2} \leq \text{Arcsin } x \leq \frac{\pi}{2}$ | $-\frac{\pi}{2} \leq \text{Arctan } x \leq \frac{\pi}{2}$ | $0 \leq \text{Arcsec } x \leq \pi$ |
| $0 \leq \text{Arccos } x \leq \pi$ | $-\frac{\pi}{2} \leq \text{Arccot } x \leq \frac{\pi}{2}$ | $-\frac{\pi}{2} \leq \text{Arccsc } x \leq \frac{\pi}{2}$ |

See page 155 for identities and multiple angles formulas.

**Suggestions for Proving Identities**

1. Have the fundamental identities well in mind. As one becomes accustomed to using these identities, he should familiarize himself with variations such as $\sec x \cos x = 1$ and $1 - \sin^2 x = \cos^2 x$.
2. Remember that any function can be expressed in terms of any other function; in particular, all functions can be expressed in terms of sines and cosines.
3. If one member of an equation involves only one trigonometric function, it may be best to express the other member entirely in terms of this function.
4. Look for the possibility of factoring expressions. Look particularly for the possibility of removing a common factor.
5. Avoid the use of radicals whenever possible.
6. A preliminary simplification of both sides may suggest a method of attack.
7. Reduce multiple arguments by changing to functions of single arguments.

The **solutions of trigonometric equations** may be expressed in terms of pure numbers with $2n\pi$ as the period or in terms of degrees using $n360^\circ$ as the period. For example, $\sin x = \frac{1}{2}$ has solutions $x = \frac{\pi}{6} + 2n\pi$, $\frac{5\pi}{6} + 2n\pi$, or we may write the solutions as $x = 30^\circ + n360^\circ$, $150^\circ + n360^\circ$. Equations of the form $\sin x = \pm \sin y$ and $\cos x = \pm \cos y$ are analyzed as follows.

| equation | implies | or |
|---|---|---|
| $\sin x = \sin y$ | $y = x + 2n\pi$ | $y = (\pi - x) + 2n\pi$ |
| $\sin x = -\sin y$ | $y = (x + \pi) + 2n\pi$ | $y = -x + 2n\pi$ |
| $\cos x = \cos y$ | $y = x + 2n\pi$ | $y = -x + 2n\pi$ |
| $\cos x = -\cos y$ | $y = (\pi - x) + 2n\pi$ | $y = (\pi + x) + 2n\pi$ |

## SOLUTIONS TO EXERCISES

### I

i. $\cos 120° = -\cos 60° = -\frac{1}{2}$

ii. $\tan(-60°) = -\tan 60° = -\sqrt{3}$

iii. $\sin \frac{3\pi}{2} = \sin 270° = -1$

iv. $\sec 330° = \sec 30° = \frac{2}{3}\sqrt{3}$

v. $\cot 480° = \cot 120° = -\frac{\sqrt{3}}{3}$

vi. $\csc \frac{5\pi}{4} = -\csc 45° = -\sqrt{2}$

### II

i. $\sin 75° = \sin(45° + 30°)$

$= \sin 45° \cos 30° + \cos 45° \sin 30°$

$= \frac{\sqrt{2}}{2} \cdot \frac{\sqrt{3}}{2} + \frac{\sqrt{2}}{2} \cdot \frac{1}{2} = \frac{\sqrt{6} + \sqrt{2}}{4}$

ii. $\sin \frac{\pi}{3} \cos \frac{\pi}{4} = \sin 60° \cos 45°$

$= \frac{\sqrt{3}}{2} \cdot \frac{\sqrt{2}}{2} = \frac{\sqrt{6}}{4}$

iii. $f(n) = \cos \frac{n\pi}{3}$

$f(0) = \cos 0 = 1$

$f(1) = \cos \frac{\pi}{3} = \cos 60° = \frac{1}{2}$

$f(2) = \cos \frac{2\pi}{3} = \cos 120° = -\frac{1}{2}$

$f(3) = \cos \pi = \cos 180° = -1$

$f(4) = \cos \frac{4\pi}{3} = \cos 240° = -\frac{1}{2}$

$f(5) = \cos \frac{5\pi}{3} = \cos 300° = \frac{1}{2}$

$f(6) = \cos 2\pi = \cos 0 = 1$

$f(n) = 1, \frac{1}{2}, -\frac{1}{2}, -1$

### III

i. Since $\sin \frac{\pi}{6} = \frac{1}{2}$, we have

$\arcsin \frac{1}{2} = \frac{\pi}{6} + 2n\pi, \frac{5\pi}{6} + 2n\pi$

where $n$ is an integer.

ii. $\cos \arctan \frac{3}{4} = \cos \theta$, where $\tan \theta = \frac{3}{4}$

$\tan \theta = \frac{y}{x}$ and $r = \pm\sqrt{x^2 + y^2}$

$= \sqrt{4^2 + 3^2} = \pm 5$

$\cos \theta = \frac{x}{r} = \pm \frac{4}{5}$

iii. $\operatorname{Arccos}(\tan 45°) = \operatorname{Arccos} 1 = 0°$

### IV

i. $\tan 2\phi = \frac{B}{A - C} = \frac{4}{5 - 2} = \frac{4}{3}$

$r = \sqrt{16 + 9} = 5$ and $\cos 2\phi = \frac{3}{5}$

$\sin \phi = \sqrt{\frac{1 - \cos 2\phi}{2}}$

$= \sqrt{\frac{2}{5} \cdot \frac{1}{2}} = \frac{1}{\sqrt{5}}$

ii. $\tan 2\phi = \frac{B}{A - C} = \frac{5}{4 - 4} = \frac{5}{0}$

Then $2\phi = 90°$ and $\phi = 45°$;

$\cos \phi = \cos 45° = \frac{\sqrt{2}}{2}$.

### V

i. $\sec^2 x - \tan x = 1$

Use the identities to write

$\tan^2 x + 1 - \tan x = 1.$

$\tan x(\tan x - 1) = 0$

| $\tan x = 0$ | $\tan x = 1$ |
|---|---|
| $x = n\pi$ | $x = \frac{\pi}{4} + n\pi$ |

ii. $\cos 2x - \cos x = \sin \frac{x}{2}$

Use the difference formula on the left member.

$-2 \sin \frac{2x + x}{2} \sin \frac{2x - x}{2} = \sin \frac{x}{2}$

$\sin \frac{x}{2}\left(1 + 2 \sin \frac{3x}{2}\right) = 0$

| $\sin \frac{x}{2} = 0$ | $\sin \frac{3x}{2} = -\frac{1}{2}$ |
|---|---|
| $\frac{x}{2} = 0 + n\pi$ | $\frac{3x}{2} = 210° + n360°$ |
| $x = 0 + 2n\pi$ | $= 330° + n360°$ |
| | $x = 140° + n240°$ |
| | $= 220° + n240°$ |

# 10 LINES AND CIRCLES

## SELF-TEST

DIRECTIONS: Write your answers in the numbered regions to the right. To check your answers, turn the page. Study the solutions to the problems you missed, and do the exercises following any problem or group of problems in which you had errors.

1. Given the equation $2x - 3y + 6 = 0$
   a. Find the two intercepts.
   b. Find the slope of the line defined by the equation.
   c. Find the equation of a line through (3,2) and perpendicular to the given line.

2. Do the three points $(-1, 5)$, $(2, 3)$, and $(5, 1)$ lie on a straight line?

3. Find the distance from the midpoint of the line segment joining $(4, 2)$ and $(2, 6)$ to the line defined by $3x - 4y = 18$.

4. Find the two points on the line $9x - 2y - 3 = 0$ which are at a distance 3 from the line defined by $3x - 4y - 6 = 0$.

5. Find the equation of a line through the point $(2, 5)$ that is perpendicular to the line $y = 3$.

6. Find the center and the radius of the circle whose equation is $x^2 + y^2 - 12x + 8y + 3 = 0$.

7. What is the center and the radius of the circle which passes through $P_1(1, 1)$, $P_2(2, -1)$, and $P_3(2, 3)$?

8. Find the equation of the tangent to the circle

$$x^2 + y^2 + 3x - 4y - 7 = 0 \text{ at } P_1(2, 3).$$

9. Find the equation of the common chord, the points intersection, and the length of the common chord of the circles $x^2 + y^2 - 3x + 3y = 10$ and $x^2 + y^2 + 4x - 4y = 17$.

10. Find the distance from the point $P(6, 4)$ to the point of tangency to the circle $x^2 + y^2 - 2x - 4y = 4$.

| | |
|---|---|
| 1 | a |
| | b |
| | c |
| 2 | |
| 3 | |
| 4 | |
| 5 | |
| 6 | |
| 7 | |
| 8 | |
| 9 | |
| 10 | |

## SOLUTIONS

1. $2x - 3y + 6 = 0$
   at $x = 0$: $-3y = -6$, $y = 2$
   at $y = 0$: $2x = -6$, $x = -3$
   Slope-intercept form:
   $y = \frac{2}{3}x + 2$; $m = \frac{2}{3}$.
   Line ⊥ given line has slope $-\frac{3}{2}$.
   $y - 2 = -\frac{3}{2}(x - 3)$ or
   $3x + 2y - 13 = 0$

2. The line through $(-1, 5)$ and $(2, 3)$:
   $y - 5 = \frac{3-5}{2+1}(x + 1)$
   or $2x + 3y - 13 = 0$.
   Substitute $(5, 1)$ into this equation: $2(5) + 3(1) - 13 = 0$. Since the equation is satisfied the answer is yes.

3. Given $P_1(4, 2)$ and $P_2(2, 6)$. Midpoint of $\overline{P_1P_2}$ has
   $x = \frac{1}{2}(4 + 2) = 3$, $y = \frac{1}{2}(2 + 6) = 4$.
   Distance from $(3, 4)$ to line:
   $$d = \left|\frac{3(3) - 4(4) - 18}{\sqrt{9 + 16}}\right| = \frac{25}{5} = 5.$$

4. Given: $9x - 2y = 3$, $3x - 4y = 6$.
   Let $P(x_1, y_1)$ be a point with $9x_1 - 2y_1 = 3$ and
   $$d = \left|\frac{3x_1 - 4y_1 - 6}{\sqrt{9 + 16}}\right| = 3 \text{ or}$$
   $3x_1 - 4y_1 - 6 = \pm 5(3)$.
   Solve the two systems.
   $9x_1 - 2y_1 = 3$
   $3x_1 - 4y_1 = 6 \pm 15 = (21 \text{ or } -9)$.
   We obtain $(1, 3)$ and $(-1, -6)$.

**EXERCISES I**

i. Write the equation of a line through the point $(0, 0)$ and perpendicular to the line through $(2, -3)$ and $(-3, -4)$.

ii. Find the tangent of the angle between the lines $3y = x - 2$ and $3x + 2y = 4$.

5. The line $y = 3$ is parallel to the $X$-axis and a line ⊥ it would be || the $Y$-axis and of the form $x = k$. If $P(2, 5)$ is on the line, then $2 = k$ and $x = 2$.

6. Complete the square on both the $x$ and $y$ terms.
   $(x^2 - 12x + 36) + (y^2 + 8y + 16) = 49$
   $(x - 6)^2 + (y + 4)^2 = 49$
   $C(6, -4)$, $r = 7$

**EXERCISE II**

i. Two towns $A$ and $B$ are 200 miles apart. A car leaves $A$ at noon to drive $B$ at an average of 40 mi/hr. Another car travels in the opposite direction, leaving $B$ at 1:00 pm and driving at 40 mi/hr. At 2:00 pm a second car leaves $B$ and drives toward $A$ at 50 mi/hr. Determine the interval of time during which the first car will be between the other two.

7. If $(x - h)^2 + (y - k)^2 = r^2$ is satisfied by $P_1(1, 1)$, $P_2(2, -1)$, and $P_3(2, 3)$ then the system
   $(1 - h)^2 + (1 - k)^2 = r^2$
   $(2 - h)^2 + (-1 - k)^2 = r^2$
   $(2 - h)^2 + (3 - k)^2 = r^2$

   can be solved for $h$, $k$, and $r$ to obtain the center $C\left(\frac{7}{2}, 1\right)$, and $r = \frac{5}{2}$. See Additional Information for a more detailed discussion.

8. The slope of the tangent to the circle $x^2 + y^2 + 3x - 4y = 7$ is $y'$ evaluated at $P(2, 3)$. Differentiate implicitly to obtain $2x + 2yy' + 3 - 4y' = 0$ and find $y'(2, 3) = -\frac{7}{2}$. The equation of the tangent at $P(2, 3)$ is
   $y - 3 = \frac{7}{2}(x - 2)$ or $2y + 7x = 20$.

9. $x^2 + y^2 - 3x + 3y = 10$ (1)
   $x^2 + y^2 + 4x - 4y = 17$ (2)
   Subtract: $-7x + 7y = -7$

Common chord: $x - y = 1$

Solve for $x$: $x = y + 1$ (3)

Substitute (3) into (1):

$(y+1)^2 + y^2 - 3(y+1) + 3y = 10$,

$y^2 + y - 6 = 0$ and $y = 2, -3$.

From (3): $x = 3, -2$. The points of intersection are $P_1(3,2)$ and $P_2(-2,-3)$ and the distance is

$$\overline{P_1P_2} = \sqrt{(x_1 - x_2)^2 + (y_1 - y_2)^2}$$
$$= \sqrt{5^2 - 5^2} = 5\sqrt{2}$$

EXERCISES III

i. Find the equation of a circle passing through $(-1,-3)$ and $(-5,3)$ if its center is on the line $2y = x + 2$.

ii. Find the centers of the two circles which are tangent to each of the lines

$$2x = 3y \text{ and } 3x + 2y = 5$$

with their centers on the line $x + y = 7$.

iii. The equation

$$\begin{vmatrix} x^2 + y^2 & x & y & 1 \\ x_1^2 + y_1^2 & x_1 & y_1 & 1 \\ x_2^2 + y_2^2 & x_2 & y_2 & 1 \\ x_3^2 + y_3^2 & x_3 & y_3 & 1 \end{vmatrix} = 0$$

represents the circle through $(x_1, y_1)$, $(x_2, y_2)$, and $(x_3, y_3)$. Use it to find the equation of a circle through $(4,4)$, $(-1,3)$, and $(2,1)$.

10. The distance is found by

$$d = \sqrt{6^2 + 4^2 - 2(6) - 4(4) - 4}$$
$$= \sqrt{36 + 16 - 12 - 16 - 4}$$
$$= \sqrt{20} = 2\sqrt{5}.$$

Solutions to the exercises are on page 60.

## ANSWERS

| | |
|---|---|
| 1. a. $x = -3$, $y = 2$<br>b. $m = \frac{2}{3}$<br>c. $3x + 2y - 13 = 0$ | 1 a<br>b<br>c |
| 2. Yes | 2 |
| 3. 5 | 3 |
| 4. (1,3) and $(-1,-6)$ | 4 |
| 5. $x = 2$ | 5 |
| 6. $C(6,-4)$, $r = 7$ | 6 |
| 7. $C\left(\frac{7}{2}, 1\right)$, $r = \frac{5}{2}$ | 7 |
| 8. $7x + 2y = 20$ | 8 |
| 9. $x - y = 1$, (3,2), $(-2,-3)$, $5\sqrt{2}$ | 9 |
| 10. $2\sqrt{5}$ | 10 |

## BASIC FACTS

The graph of the general linear equation in two variables is a straight line.

The tangent of the angle of inclination of the line with the positive $X$-axis is called the slope of the line. It can be found by the formula,

$$m = \frac{y_2 - y_1}{x_2 - x_1} = \frac{dy}{dx} = y'$$

where $(x_1,y_1)$ and $(x_2,y_2)$ are the coordinates of two points on the line.

The slope of a line parallel to the $X$-axis is zero. The slope of a line parallel to the $Y$-axis is undefined.

If two lines with slopes $m_1$ and $m_2$ are parallel, $m_1 = m_2$.

If two lines are perpendicular, the slopes are negative reciprocals of each other, $m_1 m_2 = -1$.

There are many forms of the linear equation. In the following forms, $k$, $A$, $B$, and $C$ are constants; $m$ is the slope; $a$ is the $x$-intercept; $b$ is the $y$-intercept; and the variables with subscripts are constants representing the coordinates of specific points.

Parallel to $Y$-axis: $x = k$.

Parallel to $X$-axis: $y = k$.

The Point-slope form

$$y - y_1 = m(x - x_1).$$

The Two-point form

$$y - y_1 = \frac{y_2 - y_1}{x_2 - x_1}(x - x_1).$$

The Slope-intercept form

$$y = mx + b.$$

The General form

$$Ax + By + C = 0.$$

The Intercept form

$$\frac{x}{a} + \frac{y}{b} = 1.$$

The midpoint of a line segment

$$x = \tfrac{1}{2}(x_1 + x_2), \quad y = \tfrac{1}{2}(y_1 + y_2).$$

## ADDITIONAL INFORMATION

**The Coordinates of a Point on a Line Segment**

In Figure 10-1, the point $P$ divides the line segment $P_1P_2$ in a given ratio, $k$, so that we have $P_1P = k(P_1P_2)$. From the similar triangles $P_1P_2R_2$ and $P_1PR$,

$$\frac{N_1N}{N_1N_2} = \frac{P_1P}{P_1P_2} = k$$

so that $N_1N = kN_1N_2$ and since $N_1N = x - x_1$ and $N_1N_2 = x_2 - x_1$ we have $x - x_1 = k(x_2 - x_1)$ which, when solved for $x$, gives the abscissa of $P$. The $y$-coordinate can be found in a similar manner by constructing the perpendiculars to the $Y$-axis.

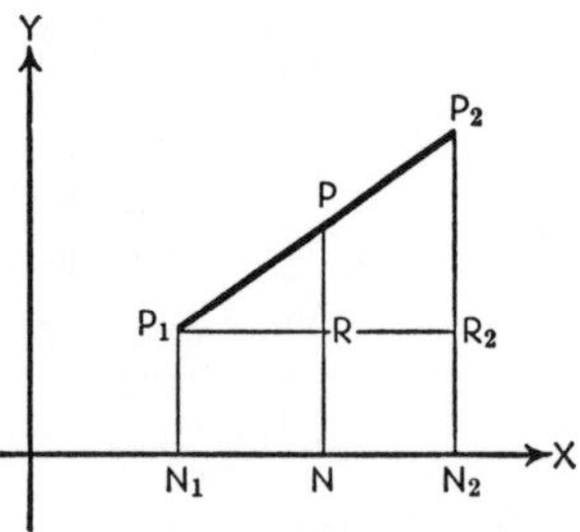


Fig. 10-1

**Distance from a Point to a Line**

Let the line $l_1$ be defined by the equation $Ax + By + C = 0$, and let $P_1(x_1,y_1)$ be a point not on the line. Figure 10-2 shows $l_1$ with an angle of inclination $\alpha$, $P_1R \perp l_1$, and $PN \perp X$-axis. Since the sides of angle $NP_1R$ are respectively perpendicular to the sides of the angle $\alpha$, we have $\angle NP_1R = \alpha$. The (perpendicular) distance from $P_1$ to $l_1$, then, can be expressed in terms of $\cos\alpha$.

$$\cos\alpha = \frac{P_1R}{P_1M}$$

$$P_1R = d = P_1M\cos\alpha$$

$$= (P_1N - MN)\cos\alpha$$

$$= (y_1 - MN)\cos\alpha$$

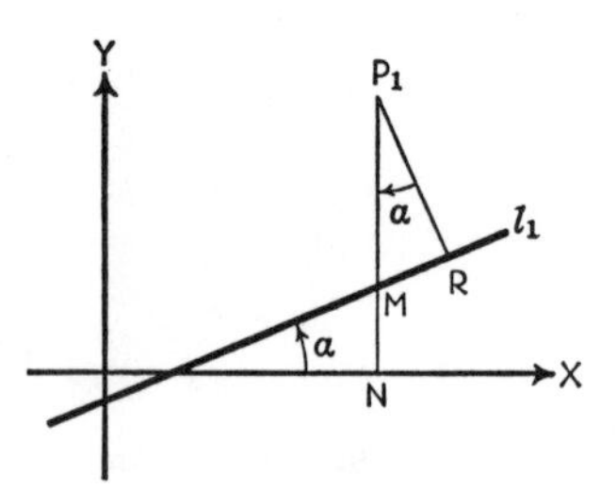


Fig. 10-2

Values must now be found for $(y_1 - MN)$ and $\cos\alpha$ such that when these values are substituted in the above equation, the formula for the distance from a point to a line will have been obtained.

Let us begin by evaluating $(y_1 - MN)$. The coordinates of $M$ are $(x_1, y)$ and since $M$ is on the line $l$ we have $Ax_1 + By + C = 0$. Solving for $y$, $y = \frac{Ax_1 + C}{B}$, if $B \neq 0$. From the figure it can be seen that $MN = y$. Therefore,

$$y_1 - MN = y_1 + \frac{Ax_1 + C}{B} = \frac{1}{B}(By_1 + Ax_1 + C).$$

Now evaluate $\cos\alpha$. Since $\alpha$ is the angle of inclination, we have, by the definition of slope, $\tan\alpha = m = -\frac{A}{B}$. Then by the theorem of Pythagoras, $\cos\alpha = \pm\frac{B}{\sqrt{A^2 + B^2}}$. Substituting these values into the formula for $d$ we have

$$d = \frac{1}{B}(By_1 + Ax_1 + C)\left(\pm\frac{B}{\sqrt{A^2 + B^2}}\right)$$

$$= \pm\frac{Ax_1 + By_1 + C}{\sqrt{A^2 + B^2}}.$$

The angle between two lines can be found from the slopes of the lines.

$$\tan \theta = \frac{m_2 - m_1}{1 + m_1 m_2}$$

The distance from a point $P_1(x_1, y_1)$ to a line $Ax + By + C = 0$ is given by

$$d = \left| \frac{Ax_1 + By_1 + C}{\sqrt{A^2 + B^2}} \right|$$

A **circle** is the locus of a point that moves at a constant distance from a fixed point. The fixed point is called the center, and the constant distance is called the radius.

The **general equation of a circle** is

$$ax^2 + ay^2 + dx + ey + f = 0$$

which can be transformed to the form $x^2 + y^2 + Dx + Ey + F = 0$ by dividing by $a$ and letting $D = \frac{d}{a}$, $E = \frac{e}{a}$, and $F = \frac{f}{a}$. This equation defines a relation, but not a function, between $x$ and $y$.

If the center of a circle is at $(h, k)$ and the fixed distance is $r$, the equation can be written

$$(x - h)^2 + (y - k)^2 = r^2.$$

If the center of the circle is at the origin, $(h, k) = (0, 0)$ and $x^2 + y^2 = r^2$.

Both quadratic terms in the equations of two circles may be eliminated simultaneously since the coefficients of $x^2$ and $y^2$ are the same in the general equation. The result is a linear equation whose graph is a straight line which is called the radical axis of the two circles. If the circles intersect in two points, this line is called the common chord.

The distance from a point $P_1$ on the tangent to the point of contact $Q$ with the circle is given by

$$\overline{P_1Q} = \sqrt{x_1{}^2 + y_1{}^2 + Dx_1 + Ey_1 + F}.$$

The graph of the relation defined by the general quadratic equation

$$ax^2 + bxy + cy^2 + dx + ey + f = 0$$

where $a$, $b$, $c$, $d$, $e$, and $f$ are constants and $a$, $b$, and $c$ are not all zero, is a curve. If $a = c$ and $b = 0$ we have the equation

$$ax^2 + ay^2 + dx + ey + f = 0$$

**and the graph of this relation is a circle.** The equation of a circle can be put into a different form by completing the square on each of the variables and collecting the terms. Thus we obtain

$$(x - h)^2 + (y - k)^2 = r^2.$$

If we let the center of a circle be denoted by the point $(h,k)$ and the radius by $r$, then, according to the definition of a circle, the distance from any point $P(x,y)$ on the circumference of the circle to the center is equal to the radius. Thus we have the relation

$$\sqrt{(x - h)^2 + (y - k)^2} = r.$$

Once we have obtained the equation of a circle in this form, we can read off the coordinates of the center and the value of the radius. The equation $2x^2 + 2y^2 - 8x + 12y + 10 = 0$, for example, may be transformed into

$$2(x^2 - 4x + 4) + 2(y^2 + 6y + 9) = 8 + 18 - 10 = 16$$

$$(x - 2)^2 + (y + 3)^2 = 8$$

from which the coordinates of the center, $C(2, -3)$, and the value of the radius, $r = 2\sqrt{2}$, can be read off.

**The number of conditions required to determine a curve** is equal to the number of independent constants in the equation of the curve, for then the relation between the variables is defined. Since in the equation, $x^2 + y^2 + Dx + Ey + F = 0$, there are three independent constants, $D$, $E$, and $F$, three conditions will determine the equation of a circle. Suppose, for example, that the three conditions are that each of the points (1,3), (−1,1), and (0,4) lie on the circle. The coordinates of these points then satisfy the equation of a circle, and upon substituting them into the equation we obtain three linear equations, $1 + 9 + D + 3E + F = 0$, $1 + 1 - D + E + F = 0$, and $16 + 4E + F = 0$ in the three unknowns $D$, $E$, and $F$. We solve this system of linear equations to obtain $D = 1$, $E = -5$, and $F = 4$. Substituting these values into the equation of a circle, we obtain the equation $x^2 + y^2 + x - 5y + 4 = 0$ which defines the circle.

If the **distance from a point to a line** is one of the given conditions of a problem, then we have an equation which involves the absolute value. In order to solve equations of this type, we must consider both the positive and negative values of the expression inside the absolute value sign. For example, the equation $|x| = k$ has two values: $x = k$ and $x = -k$.

## SOLUTIONS TO EXERCISES

### I

i. Slope of line through (2, 3) and (−3, −4): $m = \frac{-3+4}{2+3} = \frac{1}{5}$. Slope of ⊥ line is −5.

Use point-slope form of line.

$$y - 0 = -5(x - 0) \quad \text{or} \quad y + 5x = 0$$

ii. Slope of $3y = x - 2$: $m_1 = \frac{1}{3}$.

Slope of $3x + 2y = 4$: $m_2 = -\frac{3}{2}$.

$$\tan\theta = \frac{m_2 - m_1}{1 + m_1 m_2} = \frac{-\frac{3}{2} - \frac{1}{3}}{1 + (\frac{1}{3})(-\frac{3}{2})}$$

$$= -\frac{\frac{11}{6}}{1 - \frac{1}{2}} = -\frac{11}{3}$$

### II

i. Let $t$ be the time from noon and $C_i (i = 1,2,3)$ represent the cars. Distance of $C_1$ from $A$: $x_1 = 40t$.

Distance of $C_2$ from $A$:

$$x_2 = 200 - 40(t - 1).$$

Distance of $C_3$ from $A$:

$$x_3 = 200 - 50(t - 2).$$

$$x_1 = x_2 \text{ at } 40t = 200 - 40(t - 1)$$

Solve for $t$ to obtain $t = 3$.

$$x_1 = x_3 \text{ at } 40t = 200 - 50(t - 2)$$

Solve for $t$ to obtain $t = 3\frac{1}{3}$. $C_1$ will be between $C_2$ and $C_3$ during $3{:}00 < t < 3{:}20$ pm.

### III

i. Let the equation of the circle be $(x - h)^2 + (y - k)^2 = r^2$.
At (−1,−3): $(-1 - h)^2 + (-3 - k)^2 = r^2$
At (−5,3): $(-5 - h)^2 + (3 - k)^2 = r^2$
Subtract: $3k = 2h + 6$. Since $(h,k)$ is on $2y = x + 2$, we have $2k = h + 2$. Solution of this linear system is (−6,−2). Substitute this point to get $(-1 + 6)^2 + (-3 + 2)^2 = r^2 = 26$. Equation of the circle is

$$(x + 6)^2 + (y + 2)^2 = 26.$$

ii. The distance from the center to each of the given tangents is equal to the radius.

Use $d = \pm\frac{Ax_1 + By_1 + C}{\sqrt{A^2 + B^2}}$

with $h = x_1$ and $k = y_1$.

$$\left|\frac{2h - 3k}{\sqrt{4 + 9}}\right| = r = \left|\frac{3h + 2k - 5}{\sqrt{9 + 4}}\right|$$

Considering both cases we have

$2h - 3k = 3h + 2k - 5$ and
$2h - 3k = -3h - 2k + 5$, or
$h + 5k = 5$ and $5h - k = 5$.

Combine each of these with $h + k = 7$ and solve to get $h = \frac{15}{2}$, $k = -\frac{1}{2}$ and $h = 2$, $k = 5$.

iii.
$$\begin{vmatrix} x^2 + y^2 & x & y & 1 \\ 32 & 4 & 4 & 1 \\ 10 & -1 & 3 & 1 \\ 5 & 2 & 1 & 1 \end{vmatrix} = 0$$

$$\begin{vmatrix} x^2 + y^2 - 5 & x - 2 & y - 1 & 0 \\ 22 & 5 & 1 & 0 \\ 5 & -3 & 2 & 0 \\ 5 & 2 & 1 & 1 \end{vmatrix} = 0$$

$$\begin{vmatrix} x^2 + y^2 - 5 & x - 2 & y - 1 \\ 22 & 5 & 1 \\ 5 & -3 & 2 \end{vmatrix} = 0$$

$$10(x^2 + y^2 - 5) + 5(x - 2) - 66(y - 1)$$
$$-25(y - 1) - 44(x - 2)$$
$$+ 3(x^2 + y^2 - 5) = 0$$
$$x^2 + y^2 - 3x - 7y + 8 = 0$$

# 11 THE CONICS

## SELF-TEST

DIRECTIONS: Write your answers in the numbered regions to the right. To check your answers, turn the page. Study the solutions to the problems you missed, and do the exercises following any problem or group of problems in which you had errors.

1. Find (a) the vertex, (b) the focus, (c) the equation of the directrix, and (d) the length of the latus rectum of the parabola defined by the equation $y^2 - 4y = 12x + 32$.

2. Find the equation of the parabola with vertex $(2, -1)$ and focus $(2, -3)$.

3. Given $4x^2 + 9y^2 - 16x + 54y + 61 = 0$, find (a) center, (b) semiaxes, (c) vertices, (d) foci, (e) eccentricity, (f) directrices, and (g) the length of the latus rectum.

4. Given $9x^2 - 16y^2 - 18x + 64y = 199$, find (a) center, (b) semiaxes, (c) vertices, (d) foci, (e) eccentricity, (f) directrices, (g) length of latus rectum, and (h) asymptotes.

5. Find the equation of the hyperbola with center at (0,0), the transverse axis along X-axis, the eccentricity equal to $\sqrt{3}$, and the length of the latus rectum equal to 10.

6. Find the equation of the locus of a point which moves so that the sum of its distance from (3,4) and $(3, -2)$ is 8.

7. Use translation and rotation to transform $5x^2 + 4xy + 2y^2 - 2x + 4y = 19$ into a standard form.

1a
b
c
d

2

3a
b
c
d
e
f
g

4a
b
c
d
e
f
g
h

5

6

7

## SOLUTIONS

**1.** Complete the square on $y$:
$y^2 - 4y + 4 = 12x + 32 + 4$
$(y - 2)^2 = 12(x + 3)$
Compare: $(y - k)^2 = 4a(x - h)$
Vertex: $V(h,k) = (-3, 2)$
Since $4a = 12$ we have $a = 3$
Focus: $F(h + a, k) = (0,2)$
Directrix: $x = h - a = -3 - 3 = -6$
Latus rectum: L.R. $= 4a = 12$.

**2.** Given $V(2, -1)$ and $F(2, -3)$. Since the $x$ coordinates ($x = 2$) are equal, the axis $\parallel$ $Y$-axis and the form of the parabola is $(x - h)^2 = 4a(y - k)$ with $h = 2$ and $k = -1$. Since $F(h, k + a)$ we have $-1 + a = -3$ and $a = -2$. The equation of the parabola is
$(x - 2)^2 = -8(y + 1)$.

> **EXERCISES I**
>
> **i.** Find the vertex of
> $2x^2 + 6x - y + 3 = 0$
>
> **ii.** Find the equation of a parabola with axis $\parallel$ $Y$-axis and passing through $(-2, 11), (0, 5), (2, 3)$.
>
> **iii.** Find the equation of the tangent to the curve
> $(x - 2)^2 = 2(y - 3)$ at $(2, 3)$.

**3.** Change to standard form by completing the square on both $x$ and $y$.
$4(x^2 - 4x + 4) + 9(y^2 + 6y + 9) = 36$
$$\frac{(x-2)^2}{9} + \frac{(y+3)^2}{4} = 1$$
Center: $C(h, k) = (2, -3)$
Semiaxes: $a = 3$ and $b = 2$
$V_i(h \pm a, k)$: $V_1(5, -3), V_2(-1, -3)$
$ae = \sqrt{a^2 - b^2} = \sqrt{9 - 4} = \sqrt{5}$
$F_i(h \pm ae, k)$: $F_i(2 \pm \sqrt{5}, -3)$
Eccentricity: $e = \frac{ae}{a} = \frac{\sqrt{5}}{3}$

Directricies: $y = k \pm \frac{a}{e} = 2 \pm \frac{9\sqrt{5}}{5}$

L.R. $= \frac{2b^2}{a} = \frac{2(4)}{3} = \frac{8}{3}$

**4.** Change to standard form.
$9(x^2 - 2x + 1) - 16(y^2 - 4y + 4) = 144$
$$\frac{(x-1)^2}{16} - \frac{(y-2)^2}{9} = 1$$
Center: $C(h, k) = C(1, 2)$
Semiaxes: $a = 4$ and $b = 3$
$ae = \sqrt{a^2 + b^2} = \sqrt{16 + 9} = 5$
$V_i(h \pm a, k)$: $V_1(5, 2), V_2(-3, 2)$
$F_i(h \pm ae, k)$: $F_1(6, 2), F_2(-4, 2)$
Eccentricity: $e = \frac{ae}{a} = \frac{5}{4}$

Directrices: $x = h \pm \frac{a}{e} = 1 \pm \frac{16}{5}$

L.R. $= \frac{2b^2}{a} = \frac{2(9)}{4} = \frac{9}{2}$

Asymptotes: $y - k = \pm \frac{b}{a}(x - h)$

$y - 2 = \pm \frac{3}{4}(x - 1)$

$4y - 8 = \pm 3(x - 1)$

> **EXERCISES II**
>
> **i.** Find the eccentricity of
> $12x^2 + 9y^2 = 108$.
>
> **ii.** Find the foci of
> $7x^2 + 2y^2 + 4y + 16 = 28x$.
>
> **iii.** Find the equation of the tangent to the curve
> $9x^2 + 4y^2 + 8y + 31 = 36x$ at $(3, -1)$.

**5.** Given $e = \sqrt{3}$ and L.R. $= 10$
$ae = \sqrt{a^2 + b^2} = \sqrt{3}a$ or $b^2 = 2a^2$
L.R. $= \frac{2b^2}{a} = 10$ or $b^2 = 5a$

Then $5a = 2a^2$ and $a = \frac{5}{2}$

(The value $a = 0$ is extraneous.)

Then $b^2 = 5a = \frac{25}{2}$ and $a^2 = \frac{25}{4}$

Since the center is at $(0, 0)$ and the transverse axis $\parallel$ $X$-axis,
$$\frac{x^2}{a^2} - \frac{y^2}{b^2} = 1 \text{ or } 4x^2 - 2y^2 = 25.$$

**6.** Let the moving point be $P(x, y)$. Use the distance formula to write
$$\sqrt{(x-3)^2 + (y-4)^2} + \sqrt{(x-3)^2 + (y+2)^2} = 8.$$

Transpose one radical and square:
$(x - 3)^2 + (y - 4)^2$
$= 64 - 16\sqrt{(x-3)^2 + (y+2)^2} + (x-3)^2 + (y+2)^2$

Isolate the radical:
$16\sqrt{(x-3)^2 + (y+2)^2} = 64 + (y+2)^2 - (y-4)^2 = 12y + 52$

Divide by 4 and square:
$16[(x-3)^2 + (y+2)^2] = 9y^2 + 78y + 169$
$16x^2 + 7y^2 - 96x - 14y + 39 = 0$

**EXERCISES III**

**i.** Find the eccentricity of $9x^2 - 16y^2 = 36x - 96y - 36$.

**ii.** Find the equations of the asymptotes of the hyperbola $16x^2 - 81y^2 - 16x + 270y = 257$.

**7.** $5x^2 + 4xy + 2y^2 - 2x + 4y = 19$
$B^2 - 4AC = 16 - 40 < 0$; ellipse
Substitute $x = x' + h$, $y = y' + k$. Setting the resulting coefficients of $x'$ equal to zero yields $10h + 4k = 2$. Then set the coefficients of $y'$ equal to zero $(4h + 4k = -4)$ and solve the two equations simultaneously to obtain $h = 1$ and $k = -2$.
The translation gives us
$5x'^2 + 4x'y' + 2y'^2 = 24$

$$\tan 2\phi = \frac{B}{A - C} = \frac{4}{5-2} = \frac{4}{3}$$

Then $\cos 2\phi = \frac{3}{5}$ and we have

$$\sin\phi = \sqrt{\frac{1 - \cos 2\phi}{2}} = \frac{1}{\sqrt{5}}$$

$$\cos\phi = \sqrt{\frac{1 + \cos 2\phi}{2}} = \frac{2}{\sqrt{5}}$$

Rotate by substituting
$x' = \frac{1}{\sqrt{5}}(2x'' - y'')$; $y' = \frac{1}{\sqrt{5}}(x'' + 2y'')$.

The result is
$6x''^2 + y''^2 = 24$ or $\frac{x''^2}{4} + \frac{y''^2}{24} = 1$.

Solutions to the exercises are on page 66.

## ANSWERS

|  | |
|---|---|
| **1. a.** $V(-3, 2)$ | 1a |
| **b.** $F(0, 2)$ | b |
| **c.** $x = -6$ | c |
| **d.** L.R. $= 12$ | d |
| **2.** $(x - 2)^2 = -8(y + 1)$ | 2 |
| **3.** $C(2, -3)$ | 3a |
| $a = 3, b = 2$ | b |
| $V_1(5, -3)$, $V_2(-1, -3)$ | c |
| $F_i(2 \pm \sqrt{5}, -3)$ | d |
| $e = \frac{\sqrt{5}}{3}$ | e |
| $y = 2 \pm \frac{9\sqrt{5}}{5}$ | f |
| L.R. $= \frac{8}{3}$ | g |
| **4.** $C(1, 2)$ | 4a |
| $a = 4$ and $b = 3$ | b |
| $V_1(5, 2)$ and $V_2(-3, 2)$ | c |
| $F_1(6, 2)$ and $F_2(-4, 2)$ | d |
| $e = \frac{5}{4}$ | e |
| $5x = 21$ and $5x + 11 = 0$ | f |
| L.R. $= \frac{9}{2}$ | g |
| $4y - 8 = \pm(3x - 3)$ | h |
| **5.** $4x^2 - 2y^2 = 25$ | 5 |
| **6.** $16x^2 + 7y^2 = 96x + 14y - 39$ | 6 |
| **7.** $\frac{x''^2}{4} + \frac{y''^2}{24} = 1$ | 7 |

## BASIC FACTS

A conic section is the path of a point which moves so that the ratio of its distance from a fixed point to its distance from a fixed line is constant. The constant ratio is called the **eccentricity** ($e$); the fixed line, the **directrix**; and the fixed point, the **focus**.

The conic sections are divided into three classes.

| if | the conic is |
|---|---|
| $e < 1$ | an ellipse |
| $e = 1$ | a parabola |
| $e > 1$ | a hyperbola |

The line through the focus perpendicular to the directrix is the axis of the parabola, major axis for the ellipse, transverse axis for the hyperbola. The points where the curve intersects the axis are called the **vertices**. The line through the focus parallel to the directrix intersects the curve in two points, $R_1$ and $R_2$; the line segment, $R_1R_2$, is called the latus rectum.

The hyperbola and the ellipse each have a second definition.

A **hyperbola** is the locus of a point which moves so that the difference of its distance from two fixed points is a constant.

An **ellipse** is the locus of a point which moves so that the sum of its distances from two fixed points is a constant.

The general second-degree equation in two variables, $Ax^2 + Bxy + Cy^2 + Dx + Ey + F = 0$, represents a conic. The discriminant of this equation is

$$\Delta = \begin{vmatrix} 2A & B & D \\ B & 2C & E \\ D & E & 2F \end{vmatrix}.$$

If $\Delta = 0$, the conic is degenerate, and the locus may be parallel lines, intersecting lines, or a point. If $\Delta \neq 0$, we have a proper conic.

## ADDITIONAL INFORMATION

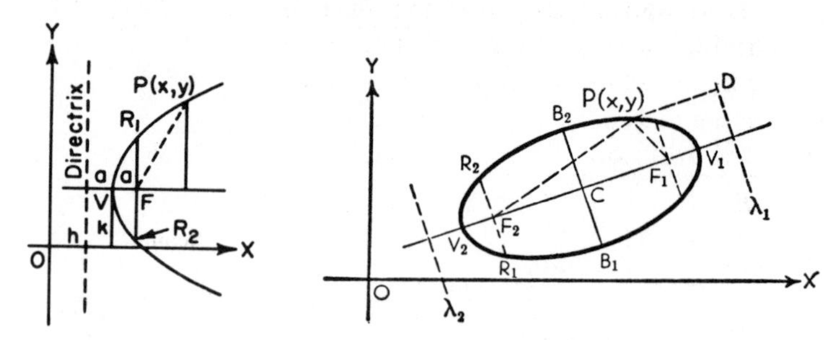


**Fig. 11-1: Parabola** **Fig. 11-2: Ellipse**

There are eight standard forms of parabolas which have their axes parallel to the coordinate axes. These are summarized in the following table. (See Figure 11-1.)

| No. | Equation | Focus | Directrix | Length of Latus Rectum | Curve Opening |
|---|---|---|---|---|---|
| Vertex at (0,0); Axis Parallel to $X$-axis | | | | | |
| 1 | $y^2 = 4ax$ | $(a,0)$ | $x = -a$ | $4a$ | to the right |
| 2 | $y^2 = -4ax$ | $(-a,0)$ | $x = a$ | $4a$ | to the left |
| Vertex at (0,0); Axis Parallel to $Y$-axis | | | | | |
| 3 | $x^2 = 4ay$ | $(0,a)$ | $y = -a$ | $4a$ | upward |
| 4 | $x^2 = -4ay$ | $(0,-a)$ | $y = a$ | $4a$ | downward |
| Vertex at $(h,k)$; Axis Parallel to $X$-axis | | | | | |
| 5 | $(y-k)^2 = 4a(x-h)$ | $(h+a,k)$ | $x = h - a$ | $4a$ | to the right |
| 6 | $(y-k)^2 = -4a(x-h)$ | $(h-a,k)$ | $x = h + a$ | $4a$ | to the left |
| Vertex at $(h,k)$; Axis Parallel to $Y$-axis | | | | | |
| 7 | $(x-h)^2 = 4a(y-k)$ | $(h,k+a)$ | $y = k - a$ | $4a$ | upward |
| 8 | $(x-h)^2 = -4a(y-k)$ | $(h,k-a)$ | $y = k + a$ | $4a$ | downward |

The standard forms and the characteristic elements for the ellipse are summarized in the following table. The letters in parentheses refer to Figure 11-2. In the table the arbitrary constants $a$ and $b$ have been chosen so that $a^2 > b^2$.

| Equation | Center ($C$) | Vertices ($V_1$ and $V_2$) | Foci ($F_1$ and $F_2$) | Directrices ($\lambda_1$ and $\lambda_2$) |
|---|---|---|---|---|
| Major Axis ($V_1V_2$) Parallel to $X$-axis | | | | |
| $\frac{x^2}{a^2} + \frac{y^2}{b^2} = 1$ | $(0,0)$ | $(a,0)$<br>$(-a,0)$ | $(ae,0)$<br>$(-ae,0)$ | $x = \pm\frac{a}{e}$ |
| $\frac{(x-h)^2}{a^2} + \frac{(y-k)^2}{b^2} = 1$ | $(h,k)$ | $(h+a,k)$<br>$(h-a,k)$ | $(h+ae,k)$<br>$(h-ae,k)$ | $x = h \pm \frac{a}{e}$ |

The coefficient of $x^2$, $xy$, and $y^2$ in a second-degree equation in two variables may be used to determine the locus defined by the equation.

| Condition | Locus |
|---|---|
| $B^2 - 4AC = 0$ | a parabola |
| $B^2 - 4AC < 0$ | an ellipse |
| $B^2 - 4AC > 0$ | a hyperbola |

The linear terms of a second-degree equation in two variables may be removed by a **translation**. To translate, let $x = x' + h$ and let $y = y' + k$.

The $xy$ term of a second-degree equation in two variables may be removed by a **rotation**. To rotate, let

$$x = x' \cos\phi - y' \sin\phi$$
$$y = x' \sin\phi + y' \cos\phi$$

where $\tan 2\phi = \dfrac{B}{A - C}$,

$$\sin\phi = \sqrt{\frac{1 - \cos 2\phi}{2}},$$

and $\cos\phi = \sqrt{\dfrac{1 + \cos 2\phi}{2}}$.

The equation $xy = k$ is a special kind of rectangular hyperbola. By rotating, it can be transformed into $\dfrac{x^2}{k} - \dfrac{y^2}{k} = 1$.

**To obtain the locus** from a given equation of the second degree in two variables, first test the discriminant to see if it is a proper conic, secondly test $B^2 - 4AC$ to determine the type of conic. If $B^2 - 4AC \neq 0$, it is a central conic. Remove the linear terms by a translation, then remove the $xy$ term by a rotation. If it is not a central conic, remove the $xy$ term. Reduce the transformed equation to a standard form. Then calculate the elements and sketch the curve.

Translation locates the curve on a new set of axes. Rotation locates the curve in a new quadrant. Rotation is limited by $2\phi < 90°$.

| Major Axis ($\overline{V_1 V_2}$) Parallel to $Y$-axis | | | | |
|---|---|---|---|---|
| $\dfrac{x^2}{b^2} + \dfrac{y^2}{a^2} = 1$ | $(0,0)$ | $(0,a)$<br>$(0,-a)$ | $(0,ae)$<br>$(0,-ae)$ | $y = \pm\dfrac{a}{e}$ |
| $\dfrac{(x-h)^2}{b^2} + \dfrac{(y-k)^2}{a^2} = 1$ | $(h,k)$ | $(h,k+a)$<br>$(h,k-a)$ | $(h,k+ae)$<br>$(h,k-ae)$ | $y = k \pm \dfrac{a}{e}$ |
| Semimajor axis ($\overline{V_2 C}$): $a$ | | Eccentricity $\left(\dfrac{\overline{PF_1}}{\overline{PD}}\right)$: $e = \dfrac{\sqrt{a^2 - b^2}}{a} < 1$ | | |
| Semiminor axis ($\overline{B_1 C}$): $b$ | | Length of latus rectum ($\overline{R_1 R_2}$): $\dfrac{2b^2}{a}$ | | |

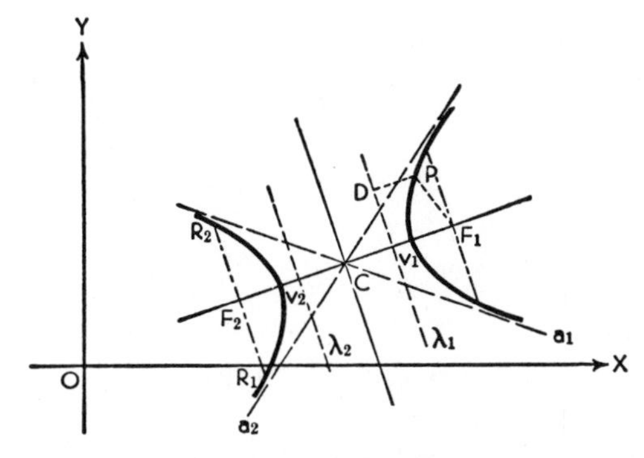


**Fig. 11-3: Hyperbola**

In the following table, the letters in parentheses refer to Figure 11-3.

| Equation | Center ($C$) | Vertices ($V_1$ and $V_2$) | Foci ($F_1$ and $F_2$) | Directrices ($\lambda_1$ and $\lambda_2$) | Asymptotes ($a_1$ and $a_2$) |
|---|---|---|---|---|---|
| Transverse Axis ($\overline{V_1 V_2}$) Parallel to the $X$-axis | | | | | |
| $\dfrac{x^2}{a^2} - \dfrac{y^2}{b^2} = 1$ | $(0,0)$ | $(a,0)$<br>$(-a,0)$ | $(ae,0)$<br>$(-ae,0)$ | $x = \pm\dfrac{a}{e}$ | $y = \pm\dfrac{b}{a}x$ |
| $\dfrac{(x-h)^2}{a^2} - \dfrac{(y-k)^2}{b^2} = 1$ | $(h,k)$ | $(h+a,k)$<br>$(h-a,k)$ | $(h+ae,k)$<br>$(h'-ae,k)$ | $x = h \pm \dfrac{a}{e}$ | $y - k = \pm\dfrac{b}{a}(x-h)$ |
| Transverse Axis ($\overline{V_1 V_2}$) Parallel to the $Y$-axis | | | | | |
| $-\dfrac{x^2}{b^2} + \dfrac{y^2}{a^2} = 1$ | $(0,0)$ | $(0,a)$<br>$(0,-a)$ | $(0,ae)$<br>$(0,-ae)$ | $y = \pm\dfrac{a}{e}$ | $y = \pm\dfrac{a}{b}x$ |
| $-\dfrac{(x-h)^2}{b^2} - \dfrac{(y-k)^2}{a^2} = 1$ | $(h,k)$ | $(h,k+a)$<br>$(h,k-a)$ | $(h,k+ae)$<br>$(h,k-ae)$ | $y = k \pm \dfrac{a}{e}$ | $y - k = \pm\dfrac{a}{b}(x-h)$ |
| Semitransverse axis ($\overline{V_2 C}$): $a$ | | | Eccentricity $\left(\dfrac{\overline{PF_1}}{\overline{PD}}\right)$: $e = \dfrac{\sqrt{a^2 + b^2}}{a} > 1$ | | |
| Semiconjugate axis: $b$ | | | Length of latus rectum ($\overline{R_1 R_2}$): $\dfrac{2b^2}{a}$ | | |

## SOLUTIONS TO EXERCISES

I

**i.** Complete the square.

$2\left(x^2 + 3x + \frac{9}{4}\right) = y + \frac{9}{2} - 3$

$\left(x + \frac{3}{2}\right)^2 = \frac{1}{2}\left(y + \frac{3}{2}\right)$

$V\left(-\frac{3}{2}, -\frac{3}{2}\right)$

**ii.** Substitute the given points into the standard form.

$(x - h)^2 = 4a(y - k)$:

$4 + 4h + h^2 = 4a(11 - k)$ (1)

$h^2 = 4a(5 - k)$ (2)

$4 - 4h + h^2 = 4a(3 - k)$ (3)

(1) − (2): $4 + 4h = 4a(6)$

(1) − (3): $8h = 4a(8)$

$1 + h = 6a$ and $h = 4a$

Then $a = \frac{1}{2}$ and $h = 2$

Substitute into (2):

$4 = 2(5 - k)$ or $k = 3$

Then the equation is

$(x - 2)^2 = 2(y - 3)$ or

$x^2 - 4x - 2y + 10 = 0$

**iii.** Differentiate $(x - 2)^2 = 2y - 6$

$2y' = 2(x - 2)$, $y' = x - 2$

$y'(2, 3) = 2 - 2 = 0$

Equation of tangent

$y - 3 = 0(x - 2)$ or $y = 3$

II

**i.** $12x^2 + 9y^2 = 108$

Divide by 108:

$\frac{x^2}{9} + \frac{y^2}{12} = 1$

The graph is an ellipse with $a^2 = 12$ and $b^2 = 9$

$ae = \sqrt{a^2 - b^2}$

$2\sqrt{3}e = \sqrt{12 - 9} = \sqrt{3}$

$e = \frac{\sqrt{3}}{2\sqrt{3}} = \frac{1}{2}$

**ii.** $7x^2 + 2y^2 + 4y + 16 = 28x$

Put into standard form:

$7(x^2 - 4x + 4) + 2(y^2 + 2y + 1)$

$= 28 + 2 - 16 = 14$

$\frac{(x - 2)^2}{2} + \frac{(y + 1)^2}{7} = 1$

$h = 2$, $k = -1$, $a = \sqrt{7}$, $b = \sqrt{2}$

$ae = \sqrt{7 - 2} = \sqrt{5}$; $F_i(h, k \pm ae)$:

$F_1(2, -1 + \sqrt{5})$

$F_2(2, -1 - \sqrt{5})$

**iii.** $9x^2 + 4y^2 + 8y + 31 = 36x$

Find $y'$ at the point $(3, -1)$:

$18x + 8yy' + 8y' = 36$

$y' = \frac{36 - 18x}{8y + 8}$

$y'(3, -1) = \frac{36 - 54}{-8 + 8} = -\frac{18}{0}$

Since $y'$ is undefined, the tangent is parallel to the $Y$-axis $(x = k)$. Since the tangent passes through $(3, -1)$, $k = 3$ or $x = 3$.

III

**i.** Put into standard form.

$9(x^2 - 4x + 4) - 16(y^2 - 6y + 9)$

$= -36 + 36 - 144 = -144$

$-\frac{(x - 2)^2}{16} + \frac{(y - 3)^2}{9} = 1$

This is a hyperbola with $a^2 = 9$ and $b^2 = 16$, so that $ae = \sqrt{9 + 16} = 5$ and $e = \frac{5}{3}$.

**ii.** Put into standard form.

$16\left(x^2 - x - \frac{1}{4}\right) - 81\left(y^2 - \frac{10}{3}y + \frac{25}{9}\right)$

$= 257 + 4 - 225 = 36$

$\frac{\left(x - \frac{1}{2}\right)^2}{\frac{9}{4}} - \frac{\left(y - \frac{5}{3}\right)^2}{\frac{4}{9}} = 1$

$h = \frac{1}{2}$, $k = \frac{5}{3}$, $a = \frac{3}{2}$, $b = \frac{2}{3}$

Equations of asymptotes

$y - k = \pm\frac{b}{a}(x - k)$

$y - \frac{5}{3} = \pm\frac{4}{9}\left(x - \frac{1}{2}\right)$

$9y - 15 = \pm(4x - 2)$

# 12 LOGARITHMIC, EXPONENTIAL, AND HYPERBOLIC FUNCTIONS

## SELF-TEST

DIRECTIONS: Write your answers in the numbered regions to the right. To check your answers, turn the page. Study the solutions to the problems you missed, and do the exercises following any problem or group of problems in which you had errors.

1. Express as the algebraic sum of logarithms

   $\log N$ if $N = \dfrac{(gv^2)}{t^3\sqrt{h}}$

2. Find the value of $x$ if $\log_3 27 = x$.

3. Find the value of $N$ if $\log_2 N = 3$.

4. Find the value of $a$ if $\log_a 16 = 2$.

5. Find $\log_b N$ if $\log_a N = 4$ and $\log_a b = 3$.

6. Find log 45 if
   $\log 2 = 0.30103$, $\log 3 = 0.47712$ and $\log 5 = 0.69897$.

7. Find the value of $e^{\ln x}$.

8. Given $(2^{2x+2})(6^x) = 3^{x+1}$. Express $x$ as the quotient of two logarithms.

9. Find $x$ if $\log x + \log(x + 3) = 1$.

10. Prove that $\cosh^2 x - \sinh^2 x = 1$.

11. Prove that $\sinh 2x = 2 \sinh x \cosh x$.

12. Derive the logarithmic expression for $\sinh^{-1} x$.

1

2

3

4

5

6

7

8

9

10

11

12

## SOLUTIONS

1. $\log N = \log \dfrac{gv^2}{t^3\sqrt{h}}$

   $= \log (gv^2) - \log (t^3\sqrt{h})$
   $= \log g + 2 \log v$
   $- 3 \log t - \frac{1}{2} \log h$

2. If $\log_3 27 = x$, then $3^x = 27$.
   Since $27 = 3^3$, we have
   $3^x = 3^3$ and $x = 3$.

3. If $\log_2 N = 3$, then $N = 2^3 = 8$.

### EXERCISES I

Write $\ln N$ as the algebraic sum of logarithms.

i. $N = \dfrac{\sqrt{y}}{x}$

ii. $N = \dfrac{\sin^2 x}{\cos y}$

iii. $N = (3)\, 2^e$

4. If $\log_a 16 = 2$, then $16 = a^2$.
   Since $16 = 4^2$, we have $a = 4$.
   Note: By definition $a > 0$, so $a \neq -4$.

5. If $\log_a N = 4$ and $\log_a b = 3$, then

   $\log_b N = \dfrac{\log_a N}{\log_a b} = \dfrac{4}{3}.$

6. $\log 45 = \log (9)(5) = \log 3^2 (5)$
   $= 2 \log 3 + \log 5$
   $2 \log 3 = 2(0.47712) = 0.95424$
   $\log 5 = 0.69897$

   $\log 45 = 1.65321$

7. Let $N = e^{\ln x}$ and take the natural logarithm of both sides:

   $\ln N = \ln (e^{\ln x}) = \ln x \ln e.$

   Since $\ln e = 1$, $\ln N = \ln x$ and

   $N = x$ or $e^{\ln x} = x$.

8. $(2^{2x+2})(6^x) = 3^{x+1}$
   $(2x + 2) \log 2 + x \log 6 = (x + 1) \log 3$
   $x (2 \log 2 + \log 6 - \log 3)$
   $= \log 3 - 2 \log 2$

   $$x = \frac{\log 3 - \log 2^2}{\log 2^2 + \log 6 - \log 3}$$

   $$= \frac{\log \frac{3}{4}}{\log \left(\frac{4 \cdot 6}{3}\right)} = \frac{\log \frac{3}{4}}{\log 8}$$

### EXERCISES II

i. Find colog $N$ if $\log N = 0.8326$.

ii. Find log 72 from the fact that $\log 2 = 0.30103$ and $\log 3 = 0.47712$.

iii. Find ln 1000 given that $\log 1000 = 3$ and that $(\log e)^{-1} = 2.3026$.

iv. Find log tan 10° given that
   $\log \sin 10° = 9.2397 - 10$,
   $\log \cos 10° = 9.9934 - 10$.

9. $\text{lox } x + \log (x + 3) = 1$
   $\log x(x + 3) = 1$
   $x(x + 3) = 10$ or $x^2 + 3x - 10 = 0$
   $(x + 5)(x - 2) = 0$; $x = -5$ and $2$
   Since logarithms are not defined for negative numbers, $x = -5$ is not a solution.

### EXERCISES III

i. Write in exponential form
   $\ln 3 - \ln 2 = x$.

ii. Solve $2^{x^2-2x} = \frac{1}{2}$ for $x$.

iii. Find $x$ if
   $\log (x^2 - 4) = \log (x + 2) + 2 \log 3$.

10. $\cosh^2 x - \sinh^2 x = 1$

   $\frac{1}{4}(e^x + e^{-x})^2 - \frac{1}{4}(e^x - e^{-x})^2$

   $= \frac{1}{4}(e^{2x} + 2 + e^{-2x} - e^{2x} + 2 - e^{-2x})$

   $= \frac{1}{4}(2 + 2) = \frac{4}{4} = 1$

**11.** $\sinh 2x = 2 \sinh x \cosh x$
Use the definitions to write

$$\frac{1}{2}(e^{2x} - e^{-2x})$$
$$= 2\left[\frac{1}{2}(e^{x} - e^{-x})\,\frac{1}{2}(e^{x} + e^{-x})\right]$$
$$= \frac{1}{2}(e^{2x} - e^{-2x}).$$

**12.** If $y = \sinh^{-1} x$, then $x = \sinh y$ and by definition $x = \dfrac{e^{y} - e^{-y}}{2}$. Multiply both sides by $2e^{y}$: $2xe^{y} = e^{2y} - 1$. Transpose: $e^{2y} - 2xe^{y} - 1 = 0$. This is a quadratic equation in $e^{y}$ and $e^{y} = x \pm \sqrt{x^2 + 1}$. However, $e^{y} > 0$ for all values of $y$ and $\sqrt{x^2+1} > x$ for all $x$, so that $x - \sqrt{x^2+1} < 0$ for all $x$ and this solution is extraneous to the original problem. We now solve for $y$ by taking the natural logarithm of both sides.

$$\ln e^{y} = y = \ln(x + \sqrt{x^2 + 1})$$

**EXERCISES IV**

i. If $\sinh x = -\frac{5}{12}$, find the remaining hyperbolic functions.

ii. Prove $\sinh(x - y) = \sinh x \cosh y - \cosh x \sinh y$.

iii. Find the logarithmic expression for $\cosh^{-1} x$ with $x > 1$ and $y \geq 0$.

Solutions to the exercises are on page 72.

## ANSWERS

|  | |
|---|---|
| 1. $\log g + 2 \log v - 3 \log t - \frac{1}{2} \log h$ | 1 |
| 2. 3 | 2 |
| 3. 8 | 3 |
| 4. 4 | 4 |
| 5. $\frac{4}{3}$ | 5 |
| 6. 1.65321 | 6 |
| 7. $x$ | 7 |
| 8. $\dfrac{\log \frac{3}{4}}{\log 8}$ | 8 |
| 9. 2 | 9 |
| 10. Identity | 10 |
| 11. Identity | 11 |
| 12. $\ln(x + \sqrt{x^2 + 1})$ | 12 |

## BASIC FACTS

The function defined by $y = a^x$, $a > 0$, where $a$ is a constant is called an **exponential function.** The inverse of an exponential function is a **logarithmic function.**

The **logarithm** of a number $N$ to the base $a$, $\log_a N$, $a > 0$, $a \neq 1$, is the exponent of the power to which the base must be raised to equal the number $N$.

If $N = a^x$, then $x = \log_a N$. If $a = 10$, the notation is $\log N$. If $a = e$, the notation is $\ln N$.

### Properties of Logarithms

The logarithm of a product is equal to the sum of the logarithms of its factors.

$$\log_a MN = \log_a M + \log_a N$$

The logarithm of a quotient is equal to the logarithm of the numerator minus the logarithm of the denominator.

$$\log_a \frac{M}{N} = \log_a M - \log_a N$$

The logarithm of the $k$th power of a number equals $k$ times the logarithm of the number.

$$\log_a N^k = k \log_a N$$

The logarithm of the $q$th root of a number is equal to the logarithm of the number divided by $q$.

$$\log_a \sqrt[q]{N} = \frac{1}{q} \log_a N$$

The number corresponding to a given logarithm is called the **antilogarithm.**

If $\log N = x$, then $N = \text{antilog } x$.

The **cologarithm** of a number is defined as the logarithm of the reciprocal of the number.

$$\text{colog } N = \log \frac{1}{N} = -\log N$$

## ADDITIONAL INFORMATION

The **exponential and logarithmic functions** are excellent examples of elementary transcendental functions. The domain of the exponential function is the set of real numbers, and the range is $0 < y < \infty$. Since the logarithmic function is the inverse of the exponential function, the graph of $y = \log_a x$ can be obtained by reflecting the graph of $y = a^x$ on the line $y = x$.

We can check these facts and infer other interesting properties by studying the graphs shown in Figures 12–1 and 12–2.

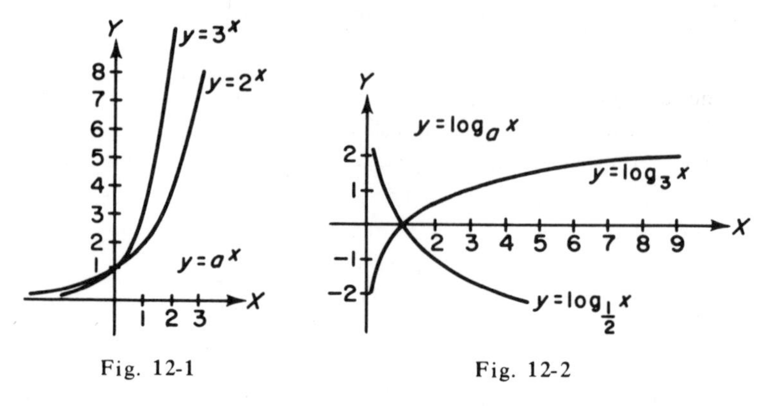


Fig. 12-1 Fig. 12-2

From Figure 12–1 we note that: (a) the function is positive for all values of $x$; (b) if $a > 1$, the function is **montone increasing,** that is as $x$ increases so does $y$; (c) as $x$ decreases algebraically, the function approaches but never becomes zero, consequently, the range is $0 < y < \infty$; (d) when $x = 0$ then $y = 1$ for all values of $a$; (e) since $y > 0$ there are no zeros of the function. Draw the graph for $a = \frac{1}{2}$ (i.e., $y = \frac{1}{2} x$) and check that this function is monotone decreasing.

From Figure 12-2 we note that the domain is the set of all positive real numbers $(x > 0)$ and the range is the set of all real numbers $(-\infty < y < \infty)$. This is true for all values of $a > 0$ and $a \neq 1$.

Although the base of a logarithm can be any positive number except unity, the two most widely employed values of $a$ are $a = 10$, in which case the logarithms are called **common logarithms** ($\log N$), and $a = e = 2.71828\ldots$, an irrational number. This number $e$ occurs frequently in applied mathematics and is defined as a limit $\lim_{v \to 0} (1 + v)^{1/v}$. The logarithms to this base are called **natural logarithms** ($\ln N$). The corresponding exponential function $y = e^x$ is often referred to as *the* exponential function.

The **properties for logarithms** can be proved by using the definition. Let us consider Property I; $\log_a MN = \log_a M + \log_a N$. Let $m = \log_a M$ and $n = \log_a N$, then by definition we have $M = a^m$ and $N = a^n$. Using the properties of exponents we can write $MN = a^m a^n = a^{m+n}$. Now apply the definition of logarithms to this product and write $\log_a MN = m + n = \log_a M + \log_a N$. The other properties are proved in the same manner. The properties can be extended to any number of factors.

The logarithm of a number $N$ to the base $b$ is equal to the quotient of its logarithm to the base $a$ divided by the logarithm of $b$ to the base $a$.

$$\log_b N = \frac{\log_a N}{\log_a b}$$

$$\log N = \frac{\ln N}{\ln 10} = 0.4343 \ln N$$

$$\ln N = \frac{\log N}{\log e} = 2.3026 \log N$$

The exponential function $y = e^x$ is used to define the set of functions known as the **hyperbolic functions**.

$$\sinh x = \frac{e^x - e^{-x}}{2}$$

$$\cosh x = \frac{e^x + e^{-x}}{2}$$

$$\tanh x = \frac{\sinh x}{\cosh x}$$

$$\coth x = \frac{\cosh x}{\sinh x},\ x \neq 0$$

$$\text{sech}\ x = \frac{2}{e^x + e^{-x}}$$

$$\text{csch}\ x = \frac{2}{e^x - e^{-x}},\ x \neq 0$$

If the unknown in an equation is the exponent, the equation is called an **exponential equation.** If the unknown in an equation is involved in an expression whose logarithm is indicated, the equation is called a **logarithmic equation.** These equations are solved by using the properties of logarithms and the antilogarithms.

Expressions for the **inverse hyperbolic functions** are found by solving exponential equations. Thus to find an expression for $\sinh^{-1} x$, solve the exponential equation $2x = e^y - e^{-y}$ for $y$.

The formula for changing from one base to another is proved in the following manner. Let $x = \log_b N$, then by definition $N = b^x$. Take the logarithms of both members of this equation to the base $a$. We have $\log_a N = \log_a b^x = x \log_a b$. Solve for $x$ to obtain $x = \dfrac{\log_a N}{\log_a b}$.

If $x$ is replaced by its defined value, we have the formula for changing the base of a logarithm.

The identities and addition formulas involving the **hyperbolic functions** are derived and proved in the same manner as the trigonometric functions. Consider the identity $\text{csch}^2 x = \coth^2 x - 1$. By definition

$$\left(\frac{2}{e^x - e^{-x}}\right)^2 = \left(\frac{e^x + e^{-x}}{e^x - e^{-x}}\right)^2 - 1 = \frac{(e^x + e^{-x})^2 - (e^x - e^{-x})^2}{(e^x - e^{-x})^2}.$$

Simplifying the numerator of the right member we obtain

$$(e^x + e^{-x})^2 - (e^x - e^{-x})^2 = (2e^x)(2e^{-x}) = 4.$$

The **inverse hyperbolic functions** are obtained by solving the exponential functions obtained from the definitions. Consider for example the expression for $\tanh^{-1} x$. We wish to solve the equation $x = \tanh y$ for $y$. Using the definition for $\tanh y$, we obtain the equation $x(e^y + e^{-y}) = e^y - e^{-y}$, and multiplying by $e^y$ we have $xe^{2y} + x = e^{2y} - 1$ or $(1 - x)e^{2y} = 1 + x$. Theoretically this equation has two solutions: $e^y = \pm\sqrt{\dfrac{1+x}{1-x}}$. However, $e^y > 0$ for all $y$, and the negative radical is extraneous. We now take the natural logarithm of both sides to obtain $y = \frac{1}{2} \ln \dfrac{1+x}{1-x}$. Since the domain of the logarithmic function is the set of positive real numbers, the argument of this expression must be positive. Let us establish the table:

| $x$ | $1+x$ | $1-x$ | Quotient | Remarks |
|---|---|---|---|---|
| $< -1$ | $< 0$ | $> 0$ | $< 0$ | not defined |
| $-1$ | 0 | 2 | 0 | not defined |
| $-1 < x < 1$ | $> 0$ | $> 0$ | $> 0$ | defined |
| 1 | 2 | 0 | not defined | not defined |
| $> 1$ | $> 0$ | $< 0$ | $< 0$ | not defined |

Consequently, the domain of $\tanh^{-1} x$ is $-1 < x < 1$.

The six inverse hyperbolic functions and their domains are:

$$\sinh^{-1} x = \ln(x + \sqrt{x^2 + 1}) \quad (\text{all } x)$$

$$\text{Cosh}^{-1} x = \ln(x + \sqrt{x^2 - 1}) \quad (x \geq 1)$$

$$\tanh^{-1} x = \frac{1}{2} \ln \frac{1+x}{1-x} \quad (x^2 < 1)$$

$$\coth^{-1} x = \frac{1}{2} \ln \left(\frac{x+1}{x-1}\right) \quad (x^2 > 1)$$

$$\text{Sech}^{-1} x = \ln\left(\frac{1}{x} + \sqrt{\frac{1}{x^2} - 1}\right) \quad (0 < x \leq 1)$$

$$\text{csch}^{-1} x = \ln\left(\frac{1}{x} + \sqrt{\frac{1}{x^2} + 1}\right) \quad (\text{all } x \neq 0)$$

## SOLUTIONS TO EXERCISES

I

i. $\ln N = \frac{1}{2} \ln y - \ln x$

ii. $\ln N = 2 \ln \sin x - \ln \cos y$

iii. $\ln N = \ln 3 + e \ln 2$

II

i. $\log N = 0.8326$
$\text{colog } N = 0 - 0.8326$
$= 9.1674 - 10$

ii. $\log 72 = \log 8\,(9)$
$= 3 \log 2 + 2 \log 3$
$3 \log 2 = 3[0.30103] = 0.90309$
$2 \log 3 = 2[0.47712] = 0.95424$
$\log 72 = 1.85733$

iii. $\ln 1000 = \dfrac{\log 1000}{\log e}$
$= (2.3026)\,(3)$
$= 6.9078$

iv. $\log \tan 10° = \log \dfrac{\sin 10°}{\cos 10°}$
$= \log \sin 10° - \log \cos 10°$
$= 9.2397 - 10 - (9.9934 - 10)$
$= 9.2463 - 10$

III

i. $\ln 3 - \ln 2 = x$
$\ln \frac{3}{2} = x,\ e^x = \frac{3}{2}$

ii. $2^{x^2-2x} = \frac{1}{2} = 2^{-1}$
$x^2 - 2x = -1$
$x^2 - 2x + 1 = 0,\ x = 1$

iii. $\log(x^2 - 4) = \log(x + 2) + 2 \log 3$
$\log \dfrac{x^2 - 4}{x + 2} = \log 9$
$x - 2 = 9,\ x = 11$

IV

i. $\sinh x = -\frac{5}{12}$
$\cosh^2 x = 1 + \sinh^2 x$
$= 1 + \frac{25}{144} = \frac{169}{144}$
$\cosh x = \frac{13}{12}$ (Note: $\cosh x \nless 0$)
$\tanh x = \dfrac{\sinh x}{\cosh x} = -\frac{5}{13}$
$\coth x = \dfrac{1}{\tanh x} = -\frac{13}{5}$
$\text{sech } x = \dfrac{1}{\cosh x} = \frac{12}{13}$
$\text{csch } x = \dfrac{1}{\sinh x} = -\frac{12}{5}$

ii. $\sinh(x - y)$
$= \sinh x \cosh y - \cosh x \sinh y$
Left side $= \frac{1}{2}(e^{x-y} - e^{-x+y})$
Right side
$= \frac{1}{4}[(e^x - e^{-x})(e^y + e^{-y})$
$- (e^x + e^{-x})(e^y - e^{-y})]$
$= \frac{1}{4}[e^{x+y} - e^{-x+y} + e^{x-y} - e^{-x-y}$
$- e^{x+y} - e^{-x+y} + e^{x-y} + e^{-x-y}]$
$= \frac{1}{4}[2\ e^{x-y} - 2\ e^{-x+y}]$
$= \frac{1}{2}(e^{x-y} - e^{-x+y})$

iii. Let $x = \cosh y,\ y \geq 0$.
$2x = e^y + e^{-y},\ y \geq 0$
$e^{2y} - 2\,xe^y + 1 = 0,\ y \geq 0$
$e^y = x \pm \sqrt{x^2 - 1},\ y \geq 0$
If $y \geq 0,\ e^y \geq 1$ and since
$(x + \sqrt{x^2 - 1})(x - \sqrt{x^2 - 1}) = 1$,
we have for $x > 1,\ x + \sqrt{x^2 - 1} > 1$
and $x - \sqrt{x^2 - 1} < 1$. Thus we take the plus sign. Now take the natural logarithms of both sides.
$y = \ln(x + \sqrt{x^2 - 1}),\ x \geq 1$

# 13 DIFFERENTIATION AND INTEGRATION OF TRANSCENDENTAL FUNCTIONS

**SELF-TEST**

DIRECTIONS: Write your answers in the numbered regions to the right. To check your answers, turn the page. Study the solutions to the problems you missed, and do the exercises following any problem or group of problems in which you had errors.

In **1–5** find $\frac{dy}{dx}$.

1. $y = e^{2x} \cos \frac{x}{2}$

2. $y = \ln \sqrt{\frac{1 + \sin x}{1 - \sin x}}$

3. $y = \left(\frac{a}{x}\right)^x$

4. $y = \text{arc tan } 2x$

5. $y = x^2 \ln \sinh x^2$

In **6–10** find $\int y\,dx$.

6. $y = \sin 2x \cos 2x$

7. $y = \frac{x + 2}{x - 1}$

8. $y = x \sec 2x^2$

9. $y = \sqrt{4 - x^2}$

10. $y = (4x^2 + 9)^{-\frac{1}{2}}$

11. Find the maximum, minimum, and inflection points of $y = \frac{x}{\ln x}$.

12. Find the equation of the tangent to $y = \cos x$ at the point where $x = \frac{\pi}{2}$.

13. Find the area bounded by $y = \sin x$, $X$-axis, $x = 0$, and $x = \pi$.

14. Find the area under the catenary $y = \frac{a}{2}\left(e^{\frac{x}{a}} + e^{-\frac{x}{a}}\right)$ from $x = -a$ to $x = a$.

1

2

3

4

5

6

7

8

9

10

11

12

13

14

## SOLUTIONS

MEMORIZE the formulas given in the Basic Facts.

1. $y = e^{2x} \cos \frac{x}{2}$

$y' = e^{2x}(2) \cos \frac{x}{2} + e^{2x} \left(-\sin \frac{x}{2}\right)\left(\frac{1}{2}\right)$

$= e^{2x}\left(2 \cos \frac{x}{2} - \frac{1}{2} \sin \frac{x}{2}\right)$

2. $y = \ln \sqrt{\dfrac{1 + \sin x}{1 - \sin x}}$

Let $u = \sqrt{\dfrac{(1 + \sin x)(1 - \sin x)}{(1 - \sin x)(1 - \sin x)}}$

$= \dfrac{\cos x}{1 - \sin x}$

$y' = \dfrac{1}{u} D_x u$

$D_x u = \dfrac{(1 - \sin x)(-\sin x) - \cos x(-\cos x)}{(1 - \sin x)^2}$

$= \dfrac{-\sin x + \sin^2 x + \cos^2 x}{(1 - \sin x)^2}$

$= \dfrac{1 - \sin x}{(1 - \sin x)^2} = \dfrac{1}{1 - \sin x}$

$y' = \dfrac{1 - \sin x}{\cos x} \cdot \dfrac{1}{1 - \sin x} = \sec x$

3. $y = \left(\dfrac{a}{x}\right)^x$ and $\ln y = x \ln \left(\dfrac{a}{x}\right)$

$\dfrac{y'}{y} = \ln \left(\dfrac{a}{x}\right) + x\left(\dfrac{a}{x}\right)^{-1}\left(-\dfrac{a}{x^2}\right)$

$y' = y\left[\ln \left(\frac{a}{x}\right) - \frac{x^2}{a}\left(\frac{a}{x^2}\right)\right]$

$= \left(\frac{a}{x}\right)^x \left[\ln \left(\frac{a}{x}\right) - 1\right]$

4. $y = \arctan 2x$

$y' = \dfrac{D_x(2x)}{1 + (2x)^2} = \dfrac{2}{1 + 4x^2}$

### EXERCISES I

Find $y''$.

i. $y = e^x \sin 2x$

ii. $y = \ln \dfrac{x}{x + 1}$

iii. $y = \tan 2x$

iv. $y = \arcsin \sqrt{x}$

5. $y = x^2 \ln \sinh x^2$

$y' = 2x \ln \sinh x^2 + x^2 \dfrac{2x \cosh x^2}{\sinh x^2}$

$= 2x \ln \sinh x^2 + 2x^3 \coth x^2$

6. $\int \sin 2x \cos 2x\, dx$
Let $u = \sin 2x$, $du = 2 \cos 2x\, dx$.

$\dfrac{1}{2} \displaystyle\int u\, du = \dfrac{1}{4} u^2 + C = \dfrac{1}{4} \sin^2 2x + C$

7. $y = \dfrac{x + 2}{x - 1} = 1 + \dfrac{3}{x - 1}$

$\int y dx = \displaystyle\int \left(1 + \dfrac{3}{x - 1}\right) dx$

$= x + 3 \ln (x - 1) + C$

8. $\int y dx = \int x \sec 2x^2 dx$
Let $u = 2x^2$, $du = 4x\, dx$

$\int y dx = \frac{1}{4} \int \sec 2x^2 (4x dx)$

$= \frac{1}{4} \ln (\sec 2x^2 + \tan 2x^2) + C$

9. $y = \sqrt{4 - x^2}$, $a = 2$, $u = x$

$\int y dx = \frac{x}{2} \sqrt{4 - x^2} + 2 \operatorname{Arcsin} \frac{x}{2} + C$

10. $y = (4x^2 + 9)^{-\frac{1}{2}}$, $u = 2x$, $a = 3$.

$\int y dx = \frac{1}{2} \displaystyle\int \dfrac{2dx}{\sqrt{4x^2 + 9}}$

$= \frac{1}{2} \ln (2x + \sqrt{4x^2 + 9}) + C$

### EXERCISES II

Find $\int y dx$.

i. $y = \sec^2 3x$

ii. $y = \csc 2x \cot 2x$

iii. $y = (4 + 9x^2)^{-1}$

iv. $y = e^{-2x}$

11. $y = \dfrac{x}{\ln x}$ $\quad y' = \dfrac{\ln x - x(x^{-1})}{(\ln x)^2}$

$y' = 0$: $\ln x - 1 = 0$, $x = e$

$y(e) = \frac{e}{\ln e} = \frac{e}{1} = e$

$y'' = D_x[(\ln x)^{-1} - (\ln x)^{-2}]$

$= -(\ln x)^{-2}\left(\frac{1}{x}\right) + 2(\ln x)^{-3}\left(\frac{1}{x}\right)$

$= \frac{1}{x}(\ln x)^{-2}[-1 + 2(\ln x)^{-1}]$

$y''(e) = \frac{1}{e}(1)^{-2}[-1 + 2(1)^{-1}] = \frac{1}{e} > 0$

There is a minimum at $(e, e)$.

$y'' = 0$: $-1 + \frac{2}{\ln x} = 0$, $\ln x = 2$

and $x = e^2$, $y(e^2) = \frac{e^2}{\ln e^2} = \frac{e^2}{2}$

Point of inflection at $\left(e^2, \frac{e^2}{2}\right)$.

12. $y = \cos x$, $y' = -\sin x$

$y\left(\frac{\pi}{2}\right) = 0$, $y'\left(\frac{\pi}{2}\right) = -1$

Equation of the tangent is

$y - 0 = -1\left(x - \frac{\pi}{2}\right)$ or $y + x = \frac{\pi}{2}$.

EXERCISES III

Find the equations of the tangents to $y = f(x)$ at the indicated points.

i. $y = \tan x$ at $\left(\frac{\pi}{4}, 1\right)$

ii. $y = \text{Arcsin}\, 2x$ at $x = \frac{1}{4}$

iii. $y = e^{3x}$ at $x = \frac{1}{3}$

iv. $y = \sin x + \cos x$ at $x = \frac{3\pi}{4}$

13. $A = \int_a^b y\,dx = \int_0^\pi \sin x\,dx$

$= -\cos x\Big|_0^\pi = -(\cos\pi - \cos 0)$

$= -(-1 - 1) = 2$

14. $A = \frac{a}{2}\int_{-a}^{a}\left(e^{\frac{x}{a}} + e^{-\frac{x}{a}}\right)dx$

$= \frac{a}{2}\left[a\int_{-a}^{a} e^{\frac{x}{a}}\left(\frac{dx}{a}\right) - a\int_{-a}^{a} e^{-\frac{x}{a}}\left(-\frac{dx}{a}\right)\right]$

$= \frac{a^2}{2}\left(e^{\frac{x}{a}}\Big|_{-a}^{a} - e^{-\frac{x}{a}}\Big|_{-a}^{a}\right)$

$= \frac{a^2}{2}[e - e^{-1} - (e^{-1} - e)]$

$= \frac{a^2}{2}[2e - 2e^{-1}] = a^2\left(e - \frac{1}{e}\right)$

## ANSWERS

| | |
|---|---|
| 1. $e^{2x}\left(2\cos\frac{x}{2} - \frac{1}{2}\sin\frac{x}{2}\right)$ | 1 |
| 2. $\sec x$ | 2 |
| 3. $\left(\frac{a}{x}\right)^x\left[\ln\left(\frac{a}{x}\right) - 1\right]$ | 3 |
| 4. $\frac{2}{1 + 4x^2}$ | 4 |
| 5. $2x \ln\sinh x^2 + 2x^3\coth x^2$ | 5 |
| 6. $\frac{1}{4}\sin^2 2x + C$ | 6 |
| 7. $x + 3\ln(x - 1) + C$ | 7 |
| 8. $\frac{1}{4}\ln(\sec 2x^2 + \tan 2x^2) + C$ | 8 |
| 9. $\frac{x}{2}\sqrt{4 - x^2} + 2\,\text{Arcsin}\,\frac{x}{2} + C$ | 9 |
| 10. $\frac{1}{2}\ln(2x + \sqrt{4x^2 + 9}) + C$ | 10 |
| 11. Minimum $(e, e)$; Point of inflection $\left(e^2, \frac{e^2}{2}\right)$ | 11 |
| 12. $y + x = \frac{\pi}{2}$ | 12 |
| 13. 2 | 13 |
| 14. $a^2\left(e - \frac{1}{e}\right)$ | 14 |

EXERCISE IV

i. Find the area between $y = \cos x$ and $y = \sin x$ from $x = 0$ to $x = \frac{\pi}{2}$.

## BASIC FACTS

### Derivatives of Transcendental Functions

$$\frac{d}{dx}(\log_e v) = \frac{1}{v}\frac{dv}{dx}$$

$$\frac{d}{dx}(\log_{10} v) = \frac{\log_{10} e}{v}\frac{dv}{dx}$$

$$\frac{d}{dx}(a^v) = a^v \log_e a \frac{dv}{dx}$$

$$\frac{d}{dx}(e^v) = e^v \frac{dv}{dx}$$

$$\frac{d}{dx}(u^v) = vu^{v-1}\frac{du}{dx} + u^v \log_e u \frac{dv}{dx}$$

$$\frac{d}{dx}(\sin v) = \cos v \frac{dv}{dx}$$

$$\frac{d}{dx}(\cos v) = -\sin v \frac{dv}{dx}$$

$$\frac{d}{dx}(\tan v) = \sec^2 v \frac{dv}{dx}$$

$$\frac{d}{dx}(\cot v) = -\csc^2 v \frac{dv}{dx}$$

$$\frac{d}{dx}(\sec v) = \sec v \tan v \frac{dv}{dx}$$

$$\frac{d}{dx}(\csc v) = -\csc v \cot v \frac{dv}{dx}$$

$$\frac{d}{dx}(\arcsin v) = \frac{1}{\sqrt{1-v^2}}\left(\frac{dv}{dx}\right)$$

$$\frac{d}{dx}(\arccos v) = -\frac{1}{\sqrt{1-v^2}}\left(\frac{dv}{dx}\right)$$

$$\frac{d}{dx}(\arctan v) = \frac{1}{1+v^2}\left(\frac{dv}{dx}\right)$$

$$\frac{d}{dx}(\text{arccot } v) = -\frac{1}{1+v^2}\left(\frac{dv}{dx}\right)$$

$$\frac{d}{dx}(\text{arcsec } v) = \frac{1}{x\sqrt{x^2-1}}\left(\frac{dv}{dx}\right)$$

$$\frac{d}{dx}(\text{arccsc } v) = \frac{-1}{x\sqrt{x^2-1}}\left(\frac{dv}{dx}\right)$$

$$\frac{d}{dx}(\sinh v) = \cosh v\left(\frac{dv}{dx}\right)$$

$$\frac{d}{dx}(\cosh v) = \sinh v\left(\frac{dv}{dx}\right)$$

## ADDITIONAL INFORMATION

In order to obtain the derivatives of the trigonometric functions, we need the following theorem.

**Theorem.** If $\theta$ is measured in radians then

$$\lim_{\theta\to 0}\frac{\sin\theta}{\theta} = 1 \quad\text{and}\quad \lim_{\theta\to 0}\frac{1-\cos\theta}{\theta} = 0$$

*Proof.* Let $\theta$ approach zero from the positive side $\left(0 < \theta < \frac{\pi}{2}\right)$ so that $\sin\theta > 0$. Construct a portion of the unit circle and the related lines shown in Figure 13-1. From the figure we see that $OA = r = 1$, $y = \sin\theta$, $z = \tan\theta$ and area $\Delta CAP <$ area sector $OAP <$ area $\Delta OAR$. Area $\Delta OAP = \frac{1}{2}\overline{OA}\, y = \frac{1}{2}\sin\theta$, $(OA = 1)$. Area sector $OAP = \frac{1}{2}r^2\theta = \frac{1}{2}\theta$. Area $\Delta OAR = \frac{1}{2}\overline{OA}\, z = \frac{1}{2}\tan\theta$. Thus we have the inequalities

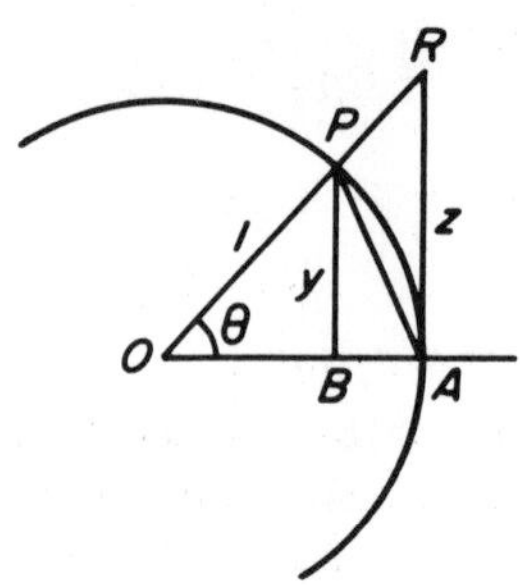


Fig. 13-1

$$\tfrac{1}{2}\sin\theta < \tfrac{1}{2}\theta < \tfrac{1}{2}\tan\theta \quad\text{or}$$

$$1 < \frac{\theta}{\sin\theta} < \frac{1}{\cos\theta}$$

as we multiply by 2 and divide by $\sin\theta > 0$. If we take the reciprocals of each expression, the inequalities are reversed and we have $1 > \frac{\sin\theta}{\theta} > \cos\theta$. As $\theta$ tends to zero $\cos\theta \longrightarrow 1$, and the quotient of $\sin\theta$ and $\theta$ is "squeezed" between 1 and 1 indicating that this fraction approaches 1. Consider now

$$\frac{1-\cos\theta}{\theta} = \frac{(1-\cos\theta)(1+\cos\theta)}{\theta(1+\cos\theta)} = \frac{\sin\theta}{\theta}\cdot\frac{\sin\theta}{1+\cos\theta}$$

Since $\lim_{\theta\to 0}\frac{\sin\theta}{\theta} = 1$ and $\lim_{\theta\to 0}\frac{\sin\theta}{1+\cos\theta} = \frac{0}{2} = 0$, we have

$$\lim_{\theta\to 0}\frac{1-\cos\theta}{\theta} = (1)(0) = 0.$$

If $\theta$ tends to zero from the negative side, we have $f(-\theta) = \frac{\sin(-\theta)}{-\theta} = \frac{-\sin\theta}{-\theta} = \frac{\sin\theta}{\theta} = f(\theta)$, and the limit would be the same. A function for which $f(-x) = f(x)$ is called an even function. If $f(-x) = -f(x)$, the function is called an odd function.

Let us apply the five step rule to obtain a derivative of the function $f(x) = \sin x$.

STEP 1: $f(x) = \sin x$ STEP 2: $f(x+h) = \sin(x+h)$

STEP 3: $\Delta f = \sin(x+h) - \sin x = \sin x\cos h + \cos x\sin h - \sin x$

STEP 4: $\frac{\Delta f}{h} = \cos x\frac{\sin h}{h} + \sin x\frac{\cos h - 1}{h}$

STEP 5: Take the limit as $h \longrightarrow 0$. The left member is $f'(x)$. By the above theorem the right member is $(\cos x)(1) + (\sin x)(0) = \cos x$.

$$\frac{d}{dx}(\tanh v) = \text{sech}^2 v \left(\frac{dv}{dx}\right)$$

$$\frac{d}{dx}(\coth v) = -\text{csch}^2 v \left(\frac{dv}{dx}\right)$$

$$\frac{d}{dx}(\text{sech } v) = -\text{sech } v \tanh v \left(\frac{dv}{dx}\right)$$

$$\frac{d}{dx}(\text{csch } v) = -\text{csch } v \coth v \left(\frac{dv}{dx}\right)$$

**Integrals**

$$\int \frac{du}{u} = \log_e u + c$$

$$\int a^u\, du = \frac{a^u}{\log_e a} + c$$

$\int e^u\, du = e^u + c$
$\int \sin u\, du = -\cos u + c$
$\int \cos u\, du = \sin u + c$
$\int \sec^2 u\, du = \tan u + c$
$\int \csc^2 u\, du = -\cot u + c$
$\int \sec u \tan u\, du = \sec u + c$
$\int \csc u \cot u\, du = -\csc u + c$
$\int \tan u\, du = \log_e \sec u + c$
$\int \cot u\, du = \log_e \sin u + c$
$\int \sec u\, du = \log_e (\sec u + \tan u) + c$
$\int \csc u\, du = \log_e (\csc u - \cot u) + c$

$$\int \frac{du}{u^2 + a^2} = \frac{1}{a} \text{arc tan } \frac{u}{a} + c$$

$$\int \frac{du}{u^2 - a^2} = \frac{1}{2a} \log_e \frac{u - a}{u + a} + c$$

$$\int \frac{du}{a^2 - u^2} = \frac{1}{2a} \log_e \frac{a + u}{a - u} + c$$

$$\int \frac{du}{\sqrt{a^2 - u^2}} = \text{arc sin } \frac{u}{a} + c$$

$$\int \sqrt{a^2 - u^2}\, du = \frac{u}{2}\sqrt{a^2 - u^2} + \frac{a^2}{2} \text{arc sin } \frac{u}{a} + c$$

$$\int \frac{du}{u\sqrt{u^2 - 1}} = \text{arcsec } u + C$$

$$\int \frac{du}{\sqrt{u^2 \pm a^2}} = \log_e (u + \sqrt{u^2 \pm a^2}) + C$$

For the integrals of the hyperbolic functions take the antiderivatives of the formulas for the derivatives.

If $y = \sin u$ where $u$ is a function of $x$, then we use the chain rule and write $\frac{dy}{dx} = \cos u \frac{du}{dx}$, and in terms of differentials we have $d \sin u = \cos u\, du$. This last expression leads to the formula for the integral of $\cos u$: $\int \cos u\, du = \int d \sin u = \sin u + C$.

The **natural logarithm of $x$** can be defined in terms of a definite integral with a variable limit.

**Definition.** For $x > 0$ define $\ln x = \int_1^x \frac{1}{t}\, dt$.

The use of this definition and the theorem which states that $\frac{d}{dx}\int_a^x f(t)\, dt = f(x)$ yields the derivative of $\ln x$ directly. However, it can also be obtained by the five step rule and the definition of $e$. In the fifth step we multiply by $\frac{x}{x}$ and then use the exponential property of logarithms to write

$$\frac{\Delta f}{h} = \frac{1}{h} \ln \left(1 + \frac{h}{x}\right) = \frac{1}{x}\frac{x}{h} \ln \left(1 + \frac{h}{x}\right) = \frac{1}{x} \ln \left(1 + \frac{h}{x}\right)^{\frac{x}{h}}$$

In taking the limit as $h \longrightarrow 0$, we have $\frac{h}{x} \longrightarrow 0$ and by definition

$$\lim_{h \to 0} \left(1 + \frac{h}{x}\right)^{\frac{x}{h}} = e \text{ so that } f'(z) = \frac{1}{x} \ln e = \frac{1}{x}.$$

The **derivative of an exponential function** may be found by the use of logarithms. Given $y = a^x$, then $\ln y = x \ln a$. Implicit differentiation yields $\frac{1}{y} y' = \ln a$, and solving for $y'$ we have $y' = y \ln a = a^x \ln a$. The exponential function $y = e^x$ is also expressed as $y = \exp x$. The derivative of $y = e^x$ is $y' = e^x$, and this function is the only function (except the trivial function $y = 0$) which is its own derivative.

Logarithmic differentiation is used to find the derivative of functions of the form, $y = u(x)^{v(x)}$. Consider for example the interesting function $y = x^x$. Take the logarithm of both sides to obtain $\ln y = x \ln x$. Implicit differentiation yields

$$\frac{1}{y} y' = x\left(\frac{1}{x}\right) + \ln x(1) = 1 + \ln x \text{ or } y' = x^x(1 + \ln x).$$

Logarithmic differentiation is also used if the function is a complex product of algebraic expressions. Consider, for example, $y = \frac{\sqrt{x + 1}}{\sqrt{x + 2}\sqrt{x + 3}}$. Take the logarithm of both sides: $\ln y = \frac{1}{2} \ln (x + 1) - \frac{1}{2} \ln (x + 2) - \frac{1}{2} \ln (x + 3)$. Then

$$\frac{y'}{y} = \frac{1}{2}\left(\frac{1}{x + 1} - \frac{1}{x + 2} - \frac{1}{x + 3}\right) = \frac{1}{2}\left(\frac{1 - 2x - x^2}{(x + 1)(x + 2)(x + 3)}\right)$$

$$y' = \frac{1 - 2x - x^2}{2\sqrt{x + 1}\,(x + 2)^{\frac{3}{2}}(x + 3)^{\frac{3}{2}}}.$$

Since $\ln x$ is defined only for $x > 0$, we consider the absolute value in logarithmic differentiation. Thus for negative values of $x$ and $y = \ln |x|$, we have $y' = \frac{1}{x}$, $x \neq 0$.

## SOLUTIONS TO EXERCISES

### I

i. $y = e^x \sin 2x$

$y' = e^x \sin 2x + (e^x \cos 2x)(2)$
$= e^x (\sin 2x + 2 \cos 2x)$

$y'' = e^x (\sin 2x + 2 \cos 2x) + e^x (2 \cos 2x - 4 \sin 2x)$
$= e^x (4 \cos 2x - 3 \sin 2x)$

ii. $y = \ln \dfrac{x}{x+1}$

$$y' = \left(\frac{x}{x+1}\right)^{-1} \left[\frac{x+1-x}{(x+1)^2}\right]$$

$$= \frac{x+1}{x} \cdot \frac{1}{(x+1)^2} = \frac{1}{x^2+x}$$

$$y'' = -\frac{(2x+1)}{(x^2+x)^2}$$

iii. $y = \tan 2x$

$y' = (\sec^2 2x)(2)$

$y'' = 2(2) \sec 2x (\sec 2x \tan 2x)(2)$
$= 8 \sec^2 2x \tan 2x$

iv. $y = \arcsin \sqrt{x}$

$$y' = \frac{1}{\sqrt{1-x}}\left(\frac{1}{2} x^{-\frac{1}{2}}\right) = \frac{1}{2}(x - x^2)^{-\frac{1}{2}}$$

$$y'' = \frac{1}{2}\left(-\frac{1}{2}\right)(x - x^2)^{-\frac{3}{2}}(1 - 2x)$$

$$= -\frac{1-2x}{4(x-x^2)^{\frac{3}{2}}}$$

$$= \frac{2x-1}{4\sqrt{(x-x^2)^3}}$$

### II

i. $\int \sec^2 3x\, dx = \frac{1}{3} \tan 3x + C$

ii. $\int \csc 2x \cot 2x = -\frac{1}{2} \csc 2x + C$

iii. $\displaystyle\int \frac{dx}{4+9x^2} = \frac{1}{6} \text{Arctan} \frac{3x}{2} + C$

$(u = 3x,\ du = 3dx,\ a = 2)$

iv. $\int e^{-2x}\, dx = -\frac{1}{2} e^{-2x} + C$

### III

i. $y = \tan x,\ y' = \sec^2 x$

$y'\left(\frac{\pi}{4}\right) = (\sqrt{2})^2 = 2$

Equation of tangent

$y - 1 = 2\left(x - \frac{\pi}{4}\right)$

$y - 2x = 1 - \frac{\pi}{2}$

ii. $y = \text{Arcsin } 2x,\ y' = \dfrac{2}{\sqrt{1-4x^2}}$

$$y'\left(\tfrac{1}{4}\right) = \frac{2}{\sqrt{1 - \frac{4}{16}}} = \frac{4}{\sqrt{3}}$$

$y\left(\frac{1}{4}\right) = \text{Arcsin } \frac{1}{2} = \frac{\pi}{6}$

$y - \frac{\pi}{6} = \frac{4}{\sqrt{3}}\left(x - \frac{1}{4}\right)$

$\sqrt{3}\, y - 4x = \frac{\pi\sqrt{3}}{6} - 1$

iii. $y = e^{3x},\ y' = 3e^{3x}$

$y\left(\frac{1}{3}\right) = e,\ y'\left(\frac{1}{3}\right) = 3e$

$y - e = 3e\left(x - \frac{1}{3}\right)$

$y = 3ex$

iv. $y = \sin x + \cos x$

$y' = \cos x - \sin x$

$y\left(\frac{3\pi}{4}\right) = \frac{\sqrt{2}}{2} - \frac{\sqrt{2}}{2} = 0$

$y'\left(\frac{3\pi}{4}\right) = -\frac{\sqrt{2}}{2} - \frac{\sqrt{2}}{2} = -\sqrt{2}$

$y = -\sqrt{2}\left(x - \frac{3\pi}{4}\right)$

### IV

i. The curves $y = \cos x$ and $y = \sin x$ intersect in $0 \le x \le \frac{\pi}{2}$ at $x = \frac{\pi}{4}$.

$A = A_1 + A_2$

$$A_1 = \int_0^{\frac{\pi}{4}} (\cos x - \sin x)\, dx$$

$$= \sin x + \cos x \Big|_0^{\frac{\pi}{4}}$$

$$= \frac{\sqrt{2}}{2} + \frac{\sqrt{2}}{2} - (0 - 1) = \sqrt{2} - 1$$

$$A_2 = \int_{\frac{\pi}{4}}^{\frac{\pi}{2}} (\sin x - \cos x)\, dx$$

$$= -\cos x - \sin x \Big|_{\frac{\pi}{4}}^{\frac{\pi}{2}}$$

$$= 0 - 1 + \frac{\sqrt{2}}{2} + \frac{\sqrt{2}}{2} = \sqrt{2} - 1$$

$A = \sqrt{2} - 1 + \sqrt{2} - 1 = 2\sqrt{2} - 2$

# 14 PARAMETRIC EQUATIONS, CURVATURE AND ARC LENGTH

## SELF-TEST

DIRECTIONS: Write your answers in the numbered regions to the right. To check your answers, turn the page. Study the solutions to the problems you missed, and do the exercises following any problem or group of problems in which you had errors.

1

2

3

4

5

6

7

8

In **1–3** find the equations of the lines tangent and normal to the curve at the point corresponding to the given value of $t$.

**1.** $x = t^3$, $y = 2t - 1$; $t = 2$

**2.** $x = \sin^2 t$, $y = -\cos 2t$; $t = \frac{\pi}{4}$

**3.** $x = e^{2t}$, $y = e^{-2t}$; $t = 0$

In **4–6** find the arc length.

**4.** $y = x^{\frac{3}{2}} + 1$; $0 \leq x \leq 2$

**5.** $x = a(t - \sin t)$, $y = a(1 - \cos t)$; $0 \leq t \leq 2\pi$

**6.** $y = \ln \sec x$; $0 \leq x \leq \frac{\pi}{4}$

In **7–8** find the radius and center of curvature.

**7.** $x = t^2 - 2t$, $y = 2 - 4t$ at $t = 2$

**8.** $y^2 = x - 2y$ at $(3, 1)$

## SOLUTIONS

1. $x = t^3$, $y = 2t - 1$, at $t = 2$
$D_tx = 3t^2$, $D_ty = 2$
$D_xy = \dfrac{2}{3t^2} = m$ and at $t = 2$
$m_1 = \frac{1}{6}$, $x_1 = 8$, and $y_1 = 3$
Tangent: $y - 3 = \frac{1}{6}(x - 8)$
$6y = x + 10$
Normal: $y - 3 = -6(x - 8)$
$y + 6x = 51$

2. $x = \sin^2 t$, $y = -\cos 2t$, $t = \frac{\pi}{4}$
$D_tx = 2 \sin t \cos t$, $D_ty = 2 \sin 2t$
At $t = \frac{\pi}{4}$: $D_tx = 2\left(\frac{\sqrt{2}}{2}\right)\left(\frac{\sqrt{2}}{2}\right) = 1$
$D_ty = 2 \sin \frac{\pi}{2} = 2$
$m_1 = D_xy\left(t = \frac{\pi}{4}\right) = \frac{2}{1} = 2$
At $t = \frac{\pi}{4}$: $x_1 = \left(\frac{\sqrt{2}}{2}\right)^2 = \frac{1}{2}$, $y_1 = 0$
Tangent: $y - 0 = 2\left(x - \frac{1}{2}\right)$; $y = 2x - 1$
Normal: $y - 0 = -\frac{1}{2}\left(x - \frac{1}{2}\right)$
$4y + 2x = 1$

3. $x = e^{2t}$, $y = e^{-2t}$, at $t = 0$
$D_tx = 2e^{2t}$, $D_ty = -2e^{-2t}$
At $t = 0$: $D_tx = 2$, $D_ty = -2$
$m_1 = D_xy\,(t = 0) = \frac{-2}{2} = -1$
At $t = 0$: $x_1 = 1$ and $y_1 = 1$
Tangent: $y - 1 = -(x - 1)$
$y + x = 2$
Normal: $y - 1 = x - 1$; $y = x$

### EXERCISES I

Find equations of the tangent and normal at indicated point.

i. $x = \ln(t - 3)$, $y = (t - 2)^{-3}$ at $t = 4$

ii. $x = 2t - \sin t$, $y = 2t - \cos t$ at $t = \frac{\pi}{2}$

iii. $x = e^t \cos t$, $y = e^t \sin t$, $t = \pi$

4. $y = x^{\frac{3}{2}} + 1$, $0 \le x \le 2$
$y' = \frac{3}{2}\sqrt{x}$, $(y')^2 = \frac{9}{4}x$
$$s = \int_0^2 \tfrac{1}{2}\sqrt{4 + 9x}\,dx$$
$$= \tfrac{1}{18}(4 + 9x)^{\frac{3}{2}}\left(\tfrac{2}{3}\right)\Big|_0^2$$
$$= \tfrac{1}{27}[22\sqrt{22} - 8]$$

5. $x = a(t - \sin t)$, $y = a(1 - \cos t)$
$D_tx = a(1 - \cos t)$
$D_ty = a \sin t$
$$ds = [a^2(1 - 2\cos t + 1]^{\frac{1}{2}}\,dt$$
$$= a\sqrt{2 - 2\cos t}\,dt$$
$$= a\sqrt{2}\,\sqrt{2}\sin \tfrac{t}{2}\,dt$$
Since $\sqrt{\dfrac{1 - \cos t}{2}} = \sin \frac{1}{2} t$
$$s = \int_0^{2\pi} 2a \sin \tfrac{t}{2}\,dt$$
$$= 2 \cdot 2a\left[-\cos \tfrac{t}{2}\right]_0^{2\pi} = 8a$$

6. $y = \ln \sec x$, $0 \le x \le \frac{\pi}{4}$
$y' = \dfrac{1}{\sec x}(\sec x \tan x) = \tan x$
$\sqrt{1 + (y')^2} = \sqrt{1 + \tan^2 x} = \sec x$
$$s = \int_0^{\frac{\pi}{4}} \sec x\,dx$$
$$= \ln(\sec x + \tan x)\Big|_0^{\frac{\pi}{4}}$$
$$= \ln(\sqrt{2} + 1) - \ln(1 + 0)$$
$$= \ln(\sqrt{2} + 1)$$

### EXERCISES II

Find the arc length.

i. $x = e^{-t} \cos t$, $y = e^{-t} \sin t$, $0 \le t \le \pi$

ii. $y = \sqrt{9 - x^2}$; $0 \le x \le 3$

7. $x = t^2 - 2t$, $y = 2 - 4t$ at $t = 2$
$D_tx = 2t - 2$, $D_ty = -4$, $y' = \dfrac{-2}{t - 1}$
$D_t^2x = 2$, $D_t^2y = 0$, $y'' = \dfrac{D_ty'}{D_tx} = (t - 1)^{-3}$
At $t = 2$: $D_tx = 2$, $D_ty = -4$, $D_t^2x = 2$,
$D_t^2y = 0$, $y' = -2$, $y'' = 1$, $x_1 = 0$, $y_1 = -6$
$$K(t) = \frac{D_tx\,D_t^2y - D_ty\,D_t^2x}{[(D_tx)^2 + D_ty)^2]^{\frac{3}{2}}}$$
$$= \frac{0 - (-4)(2)}{(4 + 16)^{\frac{3}{2}}} = \frac{8}{8(5\sqrt{5})} = \frac{1}{5\sqrt{5}}$$

$$R = \frac{1}{|K|} = 5\sqrt{5}$$

$$X = 0 - \frac{2(1+4)}{1} = 10$$

$$Y = -6 + \frac{1+4}{1} = -1$$

8. $y^2 = x - 2y$, $(3, 1)$
$2yy' = 1 - 2y'$, $y' = (2y + 2)^{-1}$
$y'' = -2y'(2y + 2)^{-2}$
At $(3, 1)$: $y' = (2 + 2)^{-1} = \frac{1}{4}$
$1 + (y')^2 = \frac{17}{16}$
$y'' = -\frac{1}{32}$

$$K = -\frac{1}{32} \frac{1}{\frac{17}{64}\sqrt{17}} = -\frac{2}{17\sqrt{17}}$$

$$R = \frac{1}{|K|} = \frac{17\sqrt{17}}{2}$$

$$X = 3 - \frac{1}{4}\left(\frac{17}{16}\right)(-32) = \frac{23}{2}$$

$$Y = 1 + \frac{17}{16}(-32) = -33$$

EXERCISES III

Find the radius and center of curvature at indicated point.

i. $x = 2 \cos t$, $y = 3 \sin t$ at $t = \frac{\pi}{4}$

ii. $x = e^t \cos t$, $y = e^t \sin t$ at $t = \pi$

iii. $y = e^x + e^{-x}$ at $(0, 2)$

Solutions to the exercises are on page 84.

## ANSWERS

1. Tangent: $6y = x + 10$
   Normal: $y + 6x = 51$
2. Tangent: $y = 2x - 1$
   Normal: $4y + 2x = 1$
3. Tangent: $y + x = 2$
   Normal: $y = x$
4. $\frac{1}{27}(22\sqrt{22} - 8)$
5. $8a$
6. $\ln(\sqrt{2} + 1)$
7. $K = \frac{1}{5\sqrt{5}}$, $R = 5\sqrt{5}$
   $X = 10$, $Y = -1$
8. $K = -\frac{2}{17\sqrt{17}}$, $R = \frac{17\sqrt{17}}{2}$
   $X = \frac{23}{2}$, $Y = -33$

1

2

3

4

5

6

7

8

## BASIC FACTS

If $x$ and $y$ are related by $x = f(t)$ and $y = g(t)$, we say that the relation is defined by a set of **parametric equations** and the independent variable $t$ is called the **parameter.**

### Derivatives

$$\frac{dy}{dx} = \frac{dy}{dt}\frac{dt}{dx} = \frac{D_t y}{D_t x} = y'$$

$$\frac{d^2y}{dx^2} = \frac{dy'}{dx} = \frac{D_t y'}{D_t x}$$

The parametric equations $x = f(t)$ and $y = g(t)$ define a curve $C$. If $f$ and $g$ have continuous derivatives which are not both zero for the same value of $t$, then the arc length $s$ is a function of $t$, and we write the differential of $s$ as

$$ds = \sqrt{(D_t x)^2 + (D_t y)^2}\, dt.$$

If $y = f(x)$, then the differential of arc length is given by

$$ds = \sqrt{1 + (y')^2}\, dx = \sqrt{(D_y x)^2 + 1}\, dy$$
$$= \sqrt{dx^2 + dy^2}.$$

The length of arc $s$ along the curve $C$ from a point $P_1(x_1, y_1)$ to a point $P_2(x_2, y_2)$ may be found by the formulas

$$s = \int_{x_1}^{x_2} \sqrt{1 + (y')^2}\, dx$$

$$= \int_{y_1}^{y_2} \sqrt{1 + \left(\frac{dx}{dy}\right)^2}\, dy$$

$$= \int_{t_1}^{t_2} \sqrt{\left(\frac{dx}{dt}\right)^2 + \left(\frac{dy}{dt}\right)^2}\, dt.$$

### Definitions

Any arc of a curve which has length is said to be **rectifiable.**

The **curvature** $K$ of an arc is the rate of change of the angle of inclination of the tangent line $\tau$ with respect to the arc length $s$, $K = \dfrac{d\tau}{ds}$.

The **radius of curvature** $R$ of an arc at a point is $\dfrac{1}{|K|}$.

## ADDITIONAL INFORMATION

The graphs of the relations defined by **parametric equations** may be found by plotting the points $P(x, y)$ that are obtained by giving arbitrary values to the parameter $t$ and calculating the corresponding values of $x$ and $y$. The parameter $t$ can also be eliminated between the two equations $x = f(t)$ and $y = g(t)$. The result is an equation relating the rectangular coordinates $x$ and $y$. Consider, for example, the parametric equations $x = r \cos t$ and $y = r \sin t$. If we square each of these equations and add the results, we obtain $x^2 + y^2 = r^2(\cos^2 t + \sin^2 t) = r^2$, the familiar equation of a circle with the center at the origin. A similar situation exists with the hyperbolic functions $x = \cosh u$ and $y = \sinh u$ since $x^2 - y^2 = \cosh^2 u - \sinh^2 u = 1$ is the unit hyperbola with the center at the origin. It is from this relation that the hyperbolic functions derived their names.

The parametric equations $x = a(\theta - \sin\theta)$ and $y = a(1 - \cos\theta)$ define a curve called the **cycloid.** The graph is shown in Figure 14-1. It is the path of a point on the circumference of a circle of radius $a$ and rolls along a straight line. The initial position is taken at the origin, and the straight line is chosen along the $X$-axis. Elimination of the parameter $\theta$ results in the rectangular equation $x = a \operatorname{arcos}\left(\frac{a-y}{a}\right) - \sqrt{2ay - y^2}$. This rectangular equation is more difficult to work with than the parametric equations.

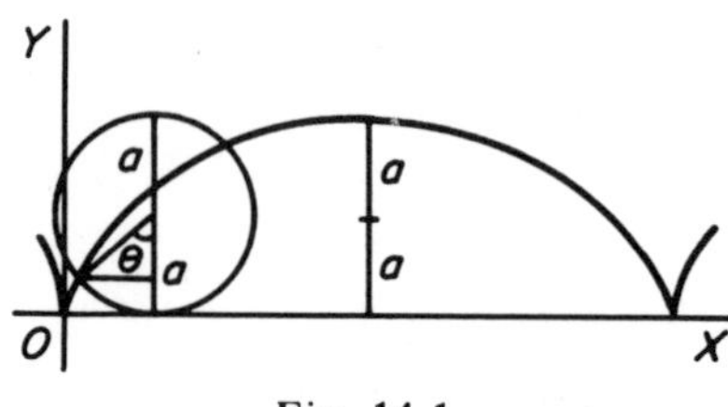


Fig. 14-1

Consider the curve defined by $y = f(x)$, a portion of which is shown in Figure 14-2. Let $P(x, y)$ and $Q(x + \Delta x, y + \Delta y)$ be two points on the curve and let the arc $PQ$ be of length $\Delta s$. We shall say that the curve is rectifiable and an arc between two points has length $s$ if as $Q \longrightarrow P$ the limit of $\dfrac{\text{chord } PQ}{\text{arc } PQ} = 1$. From the right triangle $PRQ$ we have $\overline{PQ}^2 = (\Delta x)^2 + (\Delta y)^2$. This equation may be rewritten in the form $\left(\frac{PQ}{\Delta s}\right)^2 \left(\frac{\Delta s}{\Delta x}\right)^2 = 1 + \left(\frac{\Delta y}{\Delta x}\right)^2$. As $Q \longrightarrow P$, we have $\Delta x \longrightarrow 0$ and the limit of $\left(\frac{PQ}{\Delta s}\right) = 1$, so that $\left(\frac{ds}{dx}\right)^2 = 1 + \left(\frac{dy}{dx}\right)^2$, and we have the forms for $ds$ given in the Basic Facts.

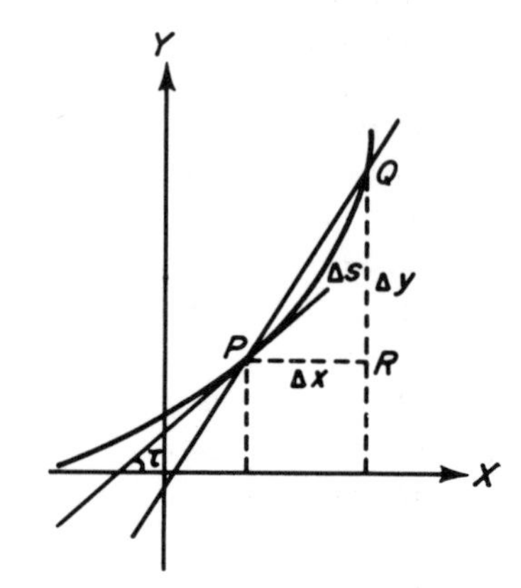


Fig. 14-2

The **curvature of an arc** is defined as the rate of change of the

The **circle of curvature** of an arc at a point $P$ is that circle passing through $P$ whose radius is $R$ and whose center $C$ lies on the concave side of the curve and on the normal to the curve through $P$.

**Formulas.** Curve given by $y = f(x)$.

$$K = \frac{y''}{[1 + (y')^2]^{\frac{3}{2}}}$$

Curve given by $x = f(t)$ and $y = g(t)$.

$$K(t) = \frac{D_t x D_t^2 y - D_t y D_t^2 x}{[(D_t x)^2 + (D_t y)^2]^{\frac{3}{2}}}$$

Center of circle of curvature, $C(X, Y)$

$$X = x_1 - \frac{y_1'[1 + (y_1')^2]}{y_1''}$$

$$Y = y_1 + \frac{1 + (y_1')^2}{y_1''}$$

**Definition.** When a point $P$ moves along a curve defined by $y = f(x)$, the center of curvature $C$ corresponding to $P$ moves along another curve called the **evolute** of the given curve. The original curve is called the **involute** of the evolute.

The equation of the evolute may be found by eliminating $x$ and $y$ from the three equations given by the original curve and the coordinates of the center of curvature at a variable point $P(x, y)$ to yield an equation in $X$ and $Y$.

If the curve $C$ is defined by parametric equations $x = f(t)$ and $y = g(t)$, then $y'$ and $y''$ are expressed in terms of the parameter $t$, and upon substitution of these into the formulas for $X$ and $Y$ we obtain $X = X(t)$ and $Y = Y(t)$ which are the parametric equations of the evolute.

**Theorem.** The normal at $P(x, y)$ to the given curve is tangent to the evolute at the center of curvature $C(X, Y)$ for $P$.

**Theorem.** The curvature of a circle is the same at all points and equals the reciprocal of the radius.

angle of inclination of the tangent line with respect to the arc length. The angle of inclination is given by $\tau = \arctan y'$ (Chapter 4). Differentiate this equation with respect to $s$ and use the above formula for $D_x s$ to obtain the following.

$$K = \frac{d\tau}{ds} = \frac{d\tau}{dx}\frac{dx}{ds} = \frac{d}{dx}(\arctan y')\frac{1}{D_x s} = \frac{y''}{1 + (y')^2}\frac{1}{\sqrt{1 + (y')^2}}$$

Remember $\dfrac{d}{dx}(\arctan u) = \dfrac{D_x u}{1 + u^2}$ and $\dfrac{dx}{ds} = \dfrac{1}{D_x s}$.

$$K = \frac{y''}{[1 + (y')^2]^{\frac{3}{2}}}, \text{ where } y' = \frac{dy}{dx} \text{ and } y'' = \frac{d^2y}{dx^2}$$

The **radius of curvature** is defined as the reciprocal of the curvature and is a positive quantity. Let us use this as the radius of a circle which is tangent to the tangent line to the curve $y = f(x)$ at the point $P$. The center of this circle must lie on the normal to the curve at $P$. Let the coordinates of the center be $(X, Y)$, then the equation of the circle is $(x - X)^2 + (y - Y)^2 = R^2$. (See Chapter 11.) Differentiating this equation implicitly with respect to $x$, we have $2(x - X) + 2(y - Y)y' = 0$ or $y' = -\dfrac{x - X}{y - Y}$. Differentiate again with respect to $x$ to obtain the following form of $y''$.

$$y'' = -\frac{(y - Y) - (x - X)y'}{(y - Y)^2} = -\frac{(y - Y)^2 + (x - X)^2}{(y - Y)^3} = -\frac{R^2}{(y - Y)^3}$$

Note that in the second step we substituted for $y'$ and simplified. Solve this equation for $(y - Y)^3$ and substitute the reciprocal of the curvature for $R$ to obtain

$$(y - Y)^3 = -\frac{[1 + (y')^2]^3}{(y'')^3} \quad \text{or} \quad y - Y = -\frac{1 + (y')^2}{y''}$$

which may be solved for $Y$ to yield the formula for this coordinate. From the above equation for $y'$ we have

$$x - X = -y'(y - Y) = \frac{y'[1 + (y')^2]}{y''}$$

and solving for $X$ we obtain the formula for this coordinate.

Throughout this chapter we have exhibited formulas in which the roles of the **independent and dependent variables** have been interchanged; for example, $ds = \sqrt{1 + (D_x y)^2}\,dx = \sqrt{(D_y x)^2 + 1}\,dy$. The student should have a clear understanding of this operation. Let us use the notation $D_u f$ in which the subscript indicates the variable with respect to which the differentiation is made $\left(D_u f = \dfrac{df}{du}\right)$. The fundamental relation is $D_x y = \dfrac{1}{D_y x}$. Using this relation we have

$$\sqrt{1 + (D_x y)^2}\,dx = \sqrt{1 + \left(\frac{1}{D_y x}\right)^2}\,dx = \sqrt{(D_y x)^2 + 1}\,\frac{dx}{D_y x}$$
$$= \sqrt{(D_y x)^2 + 1}\,D_x y\,dx = \sqrt{(D_y x)^2 + 1}\,dy$$

since $dy = D_x y\,dx$ (see Chapter 6).

## SOLUTIONS TO EXERCISES

I

i. $x = \ln(t-3)$, $y = (t-2)^{-3}$

$D_t x = \frac{1}{t-3}$, $D_t y = -3(t-2)^{-4}$

At $t = 4$: $D_t x = 1$, $D_t y = -\frac{3}{16}$

$m_1 = y'(t=4) = -\frac{3}{16}$

$x_1 = \ln 1 = 0$, $y_1 = 2^{-3} = \frac{1}{8}$

Tangent: $y - \frac{1}{8} = -\frac{3}{16}x$

$16y + 3x = 2$

Normal: $y - \frac{1}{8} = \frac{16}{3}x$

$24y = 128x + 3$

ii. $x = 2t - \sin t$, $y = 2t - \cos t$

$D_t x = 2 - \cos t$, $D_t y = 2 + \sin t$

$t = \frac{\pi}{2}$: $D_t x = 2 - 0 = 2$

$D_t y = 2 + 1 = 3$

$m_1 = D_x y\left(t = \frac{\pi}{2}\right) = \frac{3}{2}$

$x_1 = \pi - 1$, $y_1 = \pi$.

Tangent: $y - \pi = \frac{3}{2}(x - \pi + 1)$

$2y = 3x - \pi + 3$

Normal: $y - \pi = -\frac{2}{3}(x - \pi + 1)$

$3y + 2x = 5\pi - 2$

iii. $x = e^t \cos t$, $y = e^t \sin t$

$D_t x = e^t(\cos t - \sin t)$

$D_t y = e^t(\sin t + \cos t)$

$D_x y = \frac{\sin t + \cos t}{\cos t - \sin t}$

At $t = \pi$: $x_1 = -e^{\pi}$, $y_1 = 0$

$y' = \frac{0-1}{-1-0} = 1$

Tangent: $y = x + e^{\pi}$

Normal: $y + x + e^{\pi} = 0$

II

i. $x = e^{-t}\cos t$, $y = e^{-t}\sin t$

$D_t x = e^{-t}(-\cos t - \sin t)$

$D_t y = e^{-t}(-\sin t + \cos t)$

$(D_t x)^2 + (D_t y)^2$

$= e^{-2t}(1 + 2\sin t \cos t)$

$+ e^{-2t}(1 - 2\sin t\cos t)$

$= 2e^{-2t}$

$s = \int_0^{\pi} \sqrt{2}\, e^{-t}\, dt = \sqrt{2}\,[-e^{-t}]_0^{\pi}$

$= \sqrt{2}\,(-e^{-\pi} + 1)$

ii. $y = \sqrt{9 - x^2}$, $0 \le x \le 3$

$y' = \frac{1}{2}(9 - x^2)^{-\frac{1}{2}}(-2x)$

$(y')^2 = x^2(9 - x^2)^{-1}$

$\sqrt{1 + (y')^2} = 3(9 - x^2)^{-\frac{1}{2}}$

$s = \int_0^3 \frac{3dx}{\sqrt{9 - x^2}} = 3 \arcsin \frac{x}{3}\Big|_0^3$

$= 3\arcsin 1 - 3\arcsin 0$

$= 3\left(\frac{\pi}{2}\right) - 0 = \frac{3\pi}{2}$

III

i. $x = 2\cos t$, $y = 3\sin t$

$D_t x = -2\sin t$, $D_t y = 3\cos t$

$y' = -\frac{3}{2}\frac{\cos t}{\sin t} = -\frac{3}{2}\cot t$

$D_t y' = \frac{3}{2}\csc^2 t$, $y'' = -\frac{3}{4}\csc^3 t$

At $t = \frac{\pi}{4}$: $x_1 = \sqrt{2}$, $y_1 = \frac{3}{2}\sqrt{2}$

$y' = -\frac{3}{2}$, $y'' = -\frac{3}{2}\sqrt{2}$

$1 + (y')^2 = 1 + \frac{9}{4} = \frac{13}{4}$

$K = \frac{-3\sqrt{2}\,(8)}{2\,(13\sqrt{13})} = -\frac{12\sqrt{2}}{13\sqrt{13}}$

$R = \frac{1}{|K|} = \frac{13\sqrt{13}}{12\sqrt{2}}$

$X = \sqrt{2} - \frac{13}{4\sqrt{2}} = -\frac{5}{8}\sqrt{2}$; $Y = \frac{5}{12}\sqrt{2}$

ii. $D_t y' = \frac{2}{(\cos t - \sin t)^2}$

$y'' = \frac{D_t y'}{D_t x} = \frac{2}{e^t(\cos t - \sin t)^3}$

$1 + (y')^2 = \frac{2}{1 - \sin 2t}$

At $t = \pi$: $y''(\pi) = -2e^{-\pi}$

$1 + y'^2 = 2$

$K = -\frac{\sqrt{2}}{2}e^{-\pi}$, $R = \sqrt{2}\,e^{\pi}$

$X = -e^{\pi} - \frac{1\,(2)}{(-2)\,e^{-\pi}} = 0$

$Y = 0 + \frac{2}{(-2)\,e^{-\pi}} = -e^{\pi}$

iii. $y = e^x + e^{-x}$, $(0, 2)$

$y' = e^x - e^{-x}$, $y'' = e^x + e^{-x}$

At $(0, 2)$: $y' = 0$, $y'' = 2$

$1 + (y')^2 = 1 + 0 = 1$

$K = 2$ and $R = \frac{1}{2}$

$X = 0$ and $Y = 2 + \frac{1}{2} = \frac{5}{2}$

# 15 POLAR COORDINATES

## SELF-TEST

DIRECTIONS: Write your answers in the numbered regions to the right. To check your answers, turn the page. Study the solutions to the problems you missed, and do the exercises following any problem or group of problems in which you had errors.

| |
|---|
| 1 |
| 2 |
| 3 |
| 4 |
| 5 |
| 6 |
| 7 |
| 8 |
| 9 |
| 10 |
| 11 |

In **1–3** transform the given equations into equations involving polar coordinates.

1. $x^2 + y^2 - 6x + 4y = 0$

2. $xy = 3$

3. $x^4 - y^4 = 2xy$

In **4–6** transform the given equations into equations involving rectangular coordinates.

4. $r(3 \sin \theta - 4 \cos \theta) = 25$

5. $r^2 \cos 2\theta = 9$

6. $r(2 - 4 \cos \theta) = 3$

7. Find $\tan \psi$ where $\psi$ is the angle between the tangent line at $P$ and the radius vector through $P$, if $r(1 - 2 \cos \theta) = 2$ at $\theta = \frac{\pi}{4}$.

8. Find the length of arc of $r = 2 \sin \theta$ from $\theta = 0$ to $\theta = \frac{\pi}{4}$

9. Find the area inclosed by the cardioid $r = a(1 - \cos \theta)$.

10. Find the area inside the circle $r = 6 \cos \theta$ and outside $r = 3$.

11. Do $r_1 = a(1 + \cos \theta)$ and $r_2 = a(1 - \cos \theta)$ intersect at right angles?

## SOLUTIONS

1. $x^2 + y^2 - 6x + 4y = 0$
$r^2 - 6r\cos\theta + 4r\sin\theta = 0$
$r = 6\cos\theta - 4\sin\theta$

2. $xy = 3$
$(r\cos\theta)(r\sin\theta) = 3$
$r^2\sin\theta\cos\theta = 3$
$\frac{1}{2}r^2(2\sin\theta\cos\theta) = 3$
$r^2\sin 2\theta = 6$

3. $x^4 - y^4 = 2xy$
$(x^2 + y^2)(x^2 - y^2) = 2xy$
$r^2(r^2\cos^2\theta - r^2\sin^2\theta)$
$= 2r^2\sin\theta\cos\theta$
$r^2\cos 2\theta = \sin 2\theta$
$r^2 = \tan 2\theta$

4. $r(3\sin\theta - 4\cos\theta) = 5$
$3y - 4x = 5$

5. $r^2\cos 2\theta = 9$
$r^2(\cos^2\theta - \sin^2\theta) = 9$
$x^2 - y^2 = 9$

6. $r(2 - 4\cos\theta) = 3$
$2\sqrt{x^2 + y^2} - 4x = 3$
$4(x^2 + y^2) = 16x^2 + 24x + 9$
$4y^2 - 12x^2 - 24x = 9$

> **EXERCISES I**
>
> Convert the following from polar to rectangular coordinates.
>
> i. $r = a(1 - \cos\theta)$
>
> ii. $r = \cos 3\theta$

7. $r(1 - 2\cos\theta) = 2;\ \theta = \frac{\pi}{4}$

$\tan\psi = \frac{r}{r'},\ r' = D_\theta r$

$r'(1 - 2\cos\theta) + r(2\sin\theta) = 0$

At $\theta = \frac{\pi}{4}$: $r = \frac{2}{1 - \sqrt{2}} = -2(1 + \sqrt{2})$

$r'(1 - \sqrt{2}) - (2 + 2\sqrt{2})(\sqrt{2}) = 0$

$r' = \frac{2(\sqrt{2} + 2)}{1 - \sqrt{2}}$

$$\tan\psi = \frac{-2(1 + \sqrt{2})(1 - \sqrt{2})}{2(\sqrt{2} + 2)} = 1 - \frac{\sqrt{2}}{2}$$

> **EXERCISES II**
>
> Find $\tan\psi$.
>
> i. $r = 2\cos\theta - 3\sin 2\theta$
>
> ii. $r = 3\theta^2$

8. $r = 2\sin\theta,\ 0 \le \theta \le \frac{\pi}{4}$

$r' = 2\cos\theta,\ r'^2 = 4\cos^2\theta$
$r^2 + r'^2 = 4\cos^2\theta + 4\sin^2\theta = 4$

$$s = \int_0^{\pi/4} 2\,d\theta = 2\left[\frac{\pi}{4} - 0\right] = \frac{\pi}{2}$$

> **EXERCISES III**
>
> Find $\frac{ds}{d\theta}$.
>
> i. $r = e^{-2\theta}$
>
> ii. $r = 2a\cos 2\theta$

9. $r = a(1 - \cos\theta)$. See Fig. 15-8. The cardioid is symmetric with respect to the polar axis.

$$A = 2\left(\frac{1}{2}\right)\int_0^{\pi} r^2\,d\theta$$

$$= \int_0^{\pi} a^2(1 - \cos\theta)^2\,d\theta$$

$(1 - \cos\theta)^2 = 1 - 2\cos\theta + \cos^2\theta$
$= 1 - 2\cos\theta + \frac{1}{2}\cos 2\theta + \frac{1}{2}$

$$A = a^2\left[\tfrac{3}{2}\theta - 2\sin\theta + \tfrac{1}{4}\sin 2\theta\right]_0^{\pi}$$
$$= a^2\left[\tfrac{3}{2}\pi - 0 + 0 - 0 + 0 - 0\right]$$
$$= \tfrac{3}{2}\pi a^2$$

10.

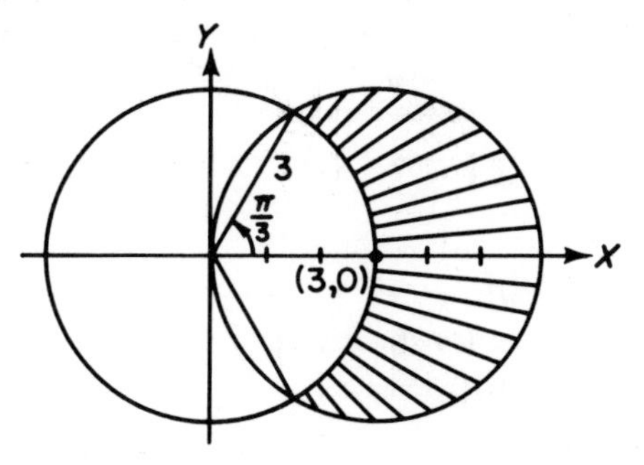


Fig. 15-1

Substitute $r = 3$ into $r = 6 \cos \theta$ to obtain $\cos \theta = \frac{1}{2}$ and the common values $\theta = \frac{\pi}{3}$ and $-\frac{\pi}{3}$. The area is symmetric with respect to the polar axis so that the area is
$A = 2\left(\frac{1}{2}\right)\int_0^{\pi/3} [(6 \cos \theta)^2 - 3^2]\, d\theta$.
$36 \cos^2\theta - 9 = 9(4 \cos^2\theta - 1)$
$= 9(2 + 2 \cos 2\theta - 1)$
Since $2 \cos^2\theta - 1 = \cos 2\theta$

$$A = 9\int_0^{\pi/3} (2 \cos 2\theta + 1)\, d\theta$$
$$= 9[\sin 2\theta + \theta]_0^{\pi/3}$$
$$= 9\left(\frac{\sqrt{3}}{2} + \frac{\pi}{3} - 0 - 0\right)$$
$$= \frac{9}{2}\sqrt{3} + 3\pi$$

**EXERCISES IV**

i. Find the arc length of $r = a \sin^3 \frac{\theta}{3}$ from $\theta = 0$ to $\theta = \frac{\pi}{4}$.

ii. Find the area enclosed by the lemniscate $r^2 = a^2 \cos 2\theta$.

iii. Find the slope of the tangent to the curve $r = a \cos 2\theta$ at $\theta = \frac{\pi}{6}$.

11. $r_1 = a(1 + \cos\theta)$, $r_2 = a(1 - \cos\theta)$
$r_1' = -a \sin\theta$, $r_2' = a \sin\theta$

$$\tan \psi_1 = \frac{a(1 + \cos\theta)}{-a \sin\theta} = -\frac{1 + \cos\theta}{\sin\theta}$$

$$\tan \psi_2 = \frac{a(1 - \cos\theta)}{a \sin\theta} = \frac{1 - \cos\theta}{\sin\theta}$$

$$\tan \phi = \frac{-\frac{1 + \cos\theta}{\sin\theta} - \frac{1 - \cos\theta}{\sin\theta}}{1 - \frac{(1 + \cos\theta)}{\sin\theta}\frac{(1 - \cos\theta)}{\sin\theta}}$$

$$= -\frac{2 \sin\theta}{\sin^2\theta - (1 - \cos^2\theta)}$$

$$= -\frac{2}{0}\sin\theta$$

Since the denominator is zero, $\phi$ is a right angle.

## ANSWERS

| | |
|---|---|
| 1. $r = 6 \cos\theta - 4 \sin\theta$ | 1 |
| 2. $r^2 \sin 2\theta = 6$ | 2 |
| 3. $r^2 = \tan 2\theta$ | 3 |
| 4. $3y - 4x = 5$ | 4 |
| 5. $x^2 - y^2 = 9$ | 5 |
| 6. $4y^2 - 12x^2 - 24x = 9$ | 6 |
| 7. $1 - \frac{\sqrt{2}}{2}$ | 7 |
| 8. $\frac{\pi}{2}$ | 8 |
| 9. $\frac{3}{2}\pi a^2$ | 9 |
| 10. $\frac{9}{2}\sqrt{3} + 3\pi$ | 10 |
| 11. Yes | 11 |

Solutions to the exercises are on page 90.

## BASIC FACTS

Consider the line $OX$ (usually horizontal) and the point $O$ on this line. The line $OX$ is called the **polar axis**, and the point $O$ is called the **pole.**

The position of any point $P$ may be determined by the length $OP = r$ (frequently represented, $\rho$) and the angle $XOP = \theta$. The length, $r$, is called the **radius vector**; and the angle, $\theta$, the **polar angle.** Together, $r$ and $\theta$ are called the **polar coordinates** of a point, and the point is designated $P(r, \theta)$.

The polar angle is positive when measured counterclockwise. The radius vector is positive if it is measured on the terminal side of $\theta$. A negative radius vector is measured on the terminal side of $\theta$ produced through $O$.

Figure 15-2 shows the graph of three points given by polar coordinates.

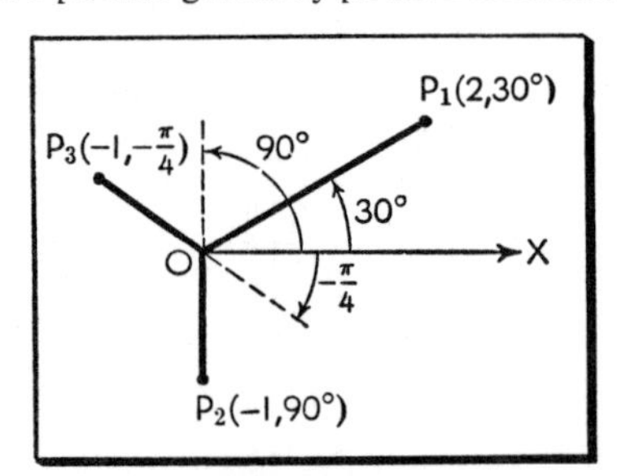


Fig. 15-2

To ease the plotting of points designated by polar coordinates, special graph paper (polar coordinate paper) is used as illustrated in Figure 15-3.

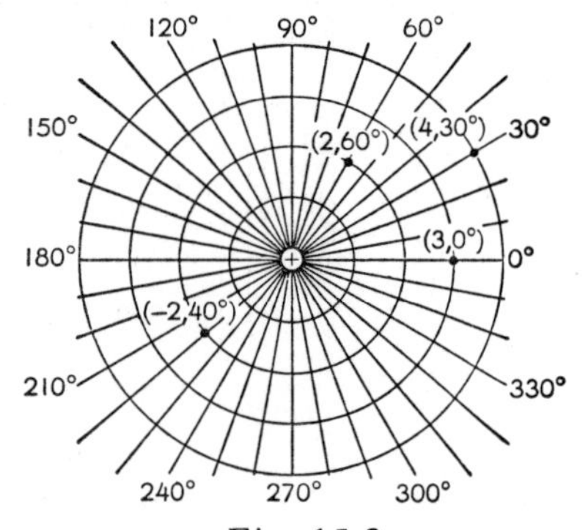


Fig. 15-3

## ADDITIONAL INFORMATION

The **polar coordinates** of a point in a plane are not unique. The same point can be designated by an infinite number of polar coordinates since $P(r, \theta)$, $P(-r, \theta + 180°)$, and $P(r, \theta + n\ 360°)$ all locate the same point.

**To trace the curve** of an equation in polar coordinates, calculate a small table of values and then take advantage of symmetry and the repeating values of the trigonometric functions. The following table is sufficient to trace the curve, $r = 5 \cos 2\theta$.

| $\theta$ | $0°$ | $15°$ | $22\frac{1}{2}°$ | $30°$ | $45°$ | $60°$ |
|---|---|---|---|---|---|---|
| $2\theta$ | $0°$ | $30°$ | $45°$ | $60°$ | $90°$ | $120°$ |
| $r$ | $5$ | $\frac{5}{2}\sqrt{3}$ | $\frac{5}{2}\sqrt{2}$ | $\frac{5}{2}$ | $0$ | $-\frac{5}{2}$ |

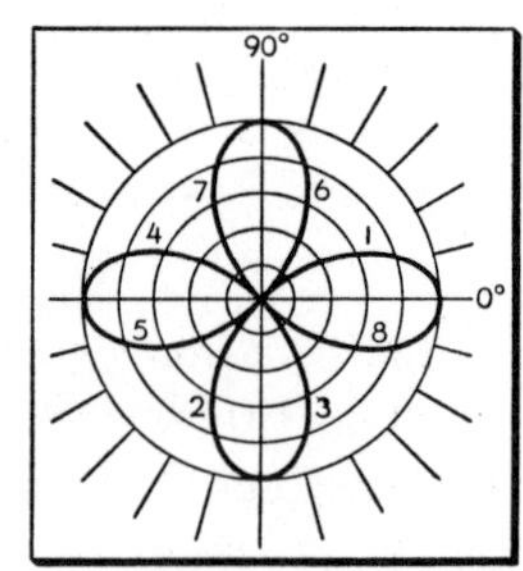


Fig. 15-4

It is easily seen that as $\theta$ varies from 45° to 90°, $2\theta$ varies from 90° to 180°, and we obtain the negative values of $r$ in reverse order from the above table. Plotting the points we obtain the half-loops marked 1 and 2 on Fig. 15-4. We also note that the curve is symmetric with respect to the polar axis. Continuing for all values of $\theta$ up to 360° and using the symmetric property, we obtain the entire curve of Fig. 15-4. Note the order of the half-loops. The polar equations for some well-known loci are given below.

Three-leaved Rose

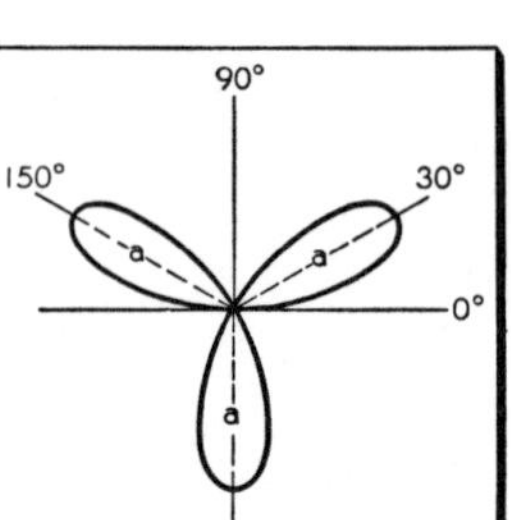


Fig. 15-5: $r = a \sin 3\theta$

Four-leaved Rose

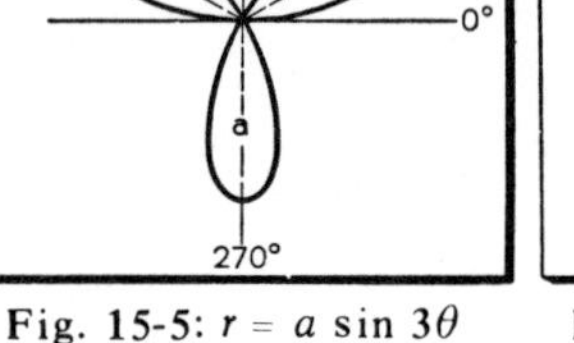


Fig. 15-6: $r = a \sin 2\theta$

Lemniscate of Bernoulli

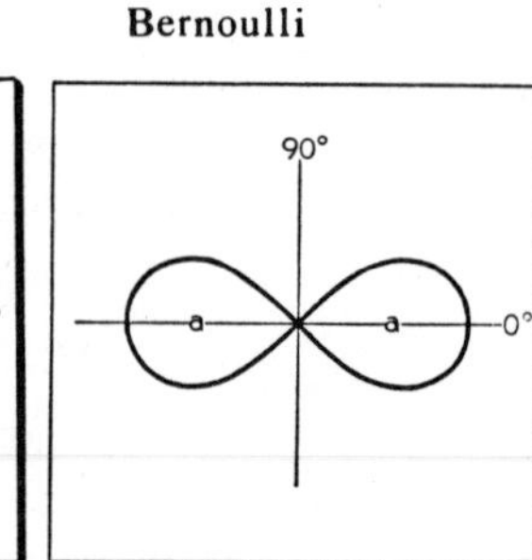


Fig. 15-7: $r^2 = a^2 \cos 2\theta$

Cardioid

Limacon

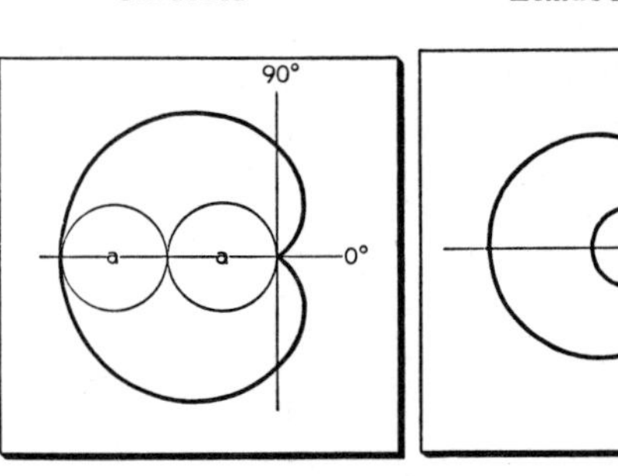


Fig. 15-8: $r = a(1 - \cos \theta)$

Fig. 15-9: $r = b - a \cos \theta$, $(b < a)$

Spiral of Archimedes

Fig. 15-10: $r = a\theta$

The following formulas may be used to transform between polar coordinates and rectangular coordinates if we locate the pole at the origin and let the polar axis coincide with the positive $X$-axis.

$$x = r\cos\theta \qquad \sin\theta = \frac{y}{\pm\sqrt{x^2+y^2}}$$
$$y = r\sin\theta$$
$$r^2 = x^2 + y^2 \qquad \cos\theta = \frac{x}{\pm\sqrt{x^2+y^2}}$$
$$\tan\theta = \frac{y}{x}$$

The relation $r = f(\theta)$ defines a set of points $(r, \theta)$ which when joined by a smooth curve describes the locus of the equation.

**Symmetry.** If the equation $r = f(\theta)$ is unchanged when: (a) $\theta$ is replaced by $-\theta$, the locus is symmetric with respect to the polar axis; (b) $\theta$ is replaced by $\pi - \theta$, the locus is symmetric with respect to the 90°-axis; (c) $\theta$ is replaced by $\pi + \theta$ or $r$ is replaced by $-r$, the locus is symmetric with respect to the pole.

If $r = f(\theta)$, then $\dfrac{dr}{d\theta} = \dfrac{d}{d\theta}[f(\theta)] = r'$.

If $\psi$ is the angle between the **tangent line** to $r = f(\theta)$ at $P$ and the radius vector through $P$, then $\tan\psi = \dfrac{r}{r'}$.

**Arc length** $s$ between $\theta = \alpha$ and $\theta = \beta$:

$$s = \int_\alpha^\beta \sqrt{(r')^2 + r^2}\, d\theta.$$

The **area** $A$ bounded by the line $\theta = \alpha$ and $\theta = \beta$ and the curve $r = f(\theta)$:

$$A = \frac{1}{2}\int_\alpha^\beta r^2 d\theta = \frac{1}{2}\int_\alpha^\beta [f(\theta)]^2 d\theta.$$

Slope of tangent line at $P(r, \theta)$:

$$\tan\tau = \frac{\tan\theta + \tan\psi}{1 - \tan\theta\tan\psi}.$$

Angle between two curves:

$$\tan\phi = \frac{\tan\psi_1 - \tan\psi_2}{1 + \tan\psi_1\tan\psi_2}.$$

If $r = f(\theta)$, the **transformation equations** may be written $x = r\cos\theta = f(\theta)\cos\theta$ and $y = r\sin\theta = f(\theta)\sin\theta$, and these equations may be regarded as the parametric equations of a curve with $\theta$ as the parameter. Differentiating with respect to $\theta$ we have

$$\frac{dx}{d\theta} = f'(\theta)\cos\theta - f(\theta)\sin\theta = r'\cos\theta - r\sin\theta$$

$$\frac{dy}{d\theta} = f'(\theta)\sin\theta + f(\theta)\cos\theta = r'\sin\theta + r\cos\theta.$$

Since $\tau$ is the angle of inclination of the **tangent** line

$$\tan\tau = \frac{dy}{dx} = \frac{r'\sin\theta + r\cos\theta}{r'\cos\theta - r\sin\theta}$$

From Figure 15-11, we see that $\tau = \psi + \theta$, since it is the exterior angle and $\psi$ and $\theta$ are the opposite interior angles of triangle $OPR$. We can then write $\psi = \tau - \theta$ and obtain

$$\tan\psi = \tan(\tau - \theta) = \frac{\tan\tau - \tan\theta}{1 + \tan\tau\tan\theta}.$$

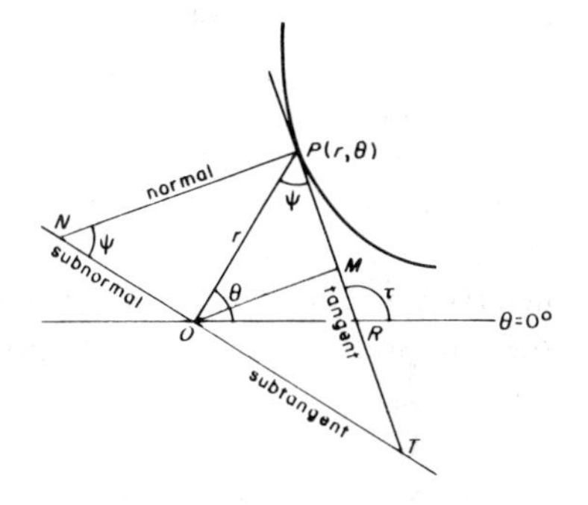


**Fig. 15-11.**

If we now substitute the value for $\tan\tau$ and simplify, we obtain $\tan\psi = \dfrac{r}{r'}$. The differential of **arc length** $ds$ can be obtained from the formula

$$ds^2 = dx^2 + dy^2 = (r'\cos\theta - r\sin\theta)^2 d\theta^2 + (r'\sin\theta + r\cos\theta)^2 d\theta^2.$$

As the right member is multiplied out and $\sin^2\theta + \cos^2\theta$ is replaced by 1, we obtain $ds^2 = (r'^2 + r^2)d\theta^2$. The arc length $s$ is obtained by integrating the differential $ds$ between given limits.

The usual difficulty in finding the **area** swept out by the radius vector as it moves in describing a curve defined by a polar equation is the determination of the limits of integration. It is recommended that a sketch be made of the curve and that the area be broken into symmetrical loops or parts. The total area is then obtained by summing these parts. The process is illustrated in the problems and exercises.

> The integrals $\int\sin^2 x\, dx$ and $\int\cos^2 x\, dx$ are evaluated by making use of the trigonometric identities of the double angles
>
> $$\sin^2 x = \tfrac{1}{2} - \tfrac{1}{2}\cos 2x \quad\text{and}\quad \cos^2 x = \tfrac{1}{2} + \tfrac{1}{2}\cos 2x$$
>
> to yield $\int\sin^2 x\, dx = \dfrac{x}{2} - \dfrac{\sin 2x}{4}$ and $\int\cos^2 x\, dx = \dfrac{x}{2} + \dfrac{\sin 2x}{4}$.

Note: In the transformation between polar and rectangular coordinates, it is often necessary to square both sides of an equation. This process may introduce extraneous values. Check your results.

## SOLUTIONS TO EXERCISES

I

i. $r = a(1 - \cos\theta)$

$$\sqrt{x^2+y^2} = a\left(1 - \frac{x}{\sqrt{x^2+y^2}}\right)$$

$$x^2+y^2 = a\sqrt{x^2+y^2} - ax$$

$$x^2+y^2+ax = a\sqrt{x^2+y^2}$$

$$(x^2+y^2+ax)^2 = a^2(x^2+y^2)$$

ii. $r = \cos 3\theta = 4\cos^3\theta - 3\cos\theta$

$$\sqrt{x^2+y^2} = \frac{x}{\sqrt{x^2+y^2}}\left(\frac{4x^2}{x^2+y^2} - 3\right)$$

$$(x^2+y^2)^2 = x(4x^2 - 3x^2 - 3y^2)$$

$$= x(x^2 - 3y^2)$$

II

i. $r = 2\cos\theta - 3\sin 2\theta$

$$r' = -2\sin\theta - 6\cos 2\theta$$

$$\tan\psi = -\frac{2\cos\theta - 3\sin 2\theta}{2\sin\theta + 6\cos 2\theta}$$

ii. $r = 3\theta^2$ and $r' = 6\theta$

$$\tan\psi = \frac{3\theta^2}{6\theta} = \frac{1}{2}\theta$$

III

i. $r = e^{-2\theta}$ and $r' = -2e^{-2\theta}$

$$\frac{ds}{d\theta} = \sqrt{(r')^2 + r^2}$$

$$= \sqrt{4e^{-4\theta} + e^{-4\theta}}$$

$$= e^{-2\theta}\sqrt{5}$$

ii. $r = 2a\cos 2\theta$, $r' = -4a\sin 2\theta$

$$\frac{ds}{d\theta} = \sqrt{16a^2\sin^2 2\theta + 4a^2\cos^2 2\theta}$$

$$= 2a\sqrt{4\sin^2 2\theta + \cos^2 2\theta}$$

$$= 2a\sqrt{3\sin^2 2\theta + 1}$$

IV

i. $r = a\sin^3\dfrac{\theta}{3}$

$$r' = 3a\sin^2\frac{\theta}{3}\cos\frac{\theta}{3}\left(\frac{1}{3}\right)$$

$$ds^2 = a^2\sin^6\frac{\theta}{3} + a^2\sin^4\frac{\theta}{3}\cos^2\frac{\theta}{3}$$

$$= a^2\sin^4\frac{\theta}{3}$$

$$s = \int_0^{\pi/4} a\sin^2\frac{\theta}{3}\,d\theta$$

$$= 3a\left[\frac{\theta}{6} - \frac{1}{4}\sin\frac{2\theta}{3}\right]_0^{\pi/4}$$

$$= 3a\left[\frac{\pi}{24} - \frac{1}{4}\left(\frac{1}{2}\right)\right] - 0 + 0$$

$$= \frac{a}{8}(\pi - 3)$$

ii. $r^2 = a^2\cos 2\theta$. See Fig. 15-6. One half of one loop is generated as $\theta$ varies from 0 to $\frac{\pi}{4}$.

$$A = \frac{4}{2}\int_0^{\pi/4} r^2\,d\theta$$

$$= \frac{2}{2}a^2\int_0^{\pi/4}\cos 2\theta\,(2d\theta)$$

$$= a^2[\sin 2\theta]_0^{\pi/4} = a^2$$

iii. $r = a\cos 2\theta$

$$r' = -2a\sin 2\theta$$

At $\theta_1 = \frac{\pi}{6}$ we have $2\theta_1 = \frac{\pi}{3}$.

$$r_1 = a\cos\frac{\theta}{3} = \frac{a}{2}$$

$$r_1' = -2a\sin\frac{\theta}{3} = -a\sqrt{3}$$

$$\tan\psi_1 = \frac{r}{r'} = -\frac{a}{2(a\sqrt{3})} = -\frac{\sqrt{3}}{6}$$

$$\tan\theta_1 = \tan\frac{\pi}{6} = \frac{\sqrt{3}}{3}$$

$$\tan\tau = \frac{\frac{\sqrt{3}}{3} - \frac{\sqrt{3}}{6}}{1 + \frac{\sqrt{3}}{3}\frac{\sqrt{3}}{6}} = \frac{\frac{\sqrt{3}}{6}}{1 + \frac{1}{6}}$$

$$= \frac{\sqrt{3}}{7}$$

# 16 VECTORS AND CURVILINEAR MOTION

## SELF-TEST

DIRECTIONS: Write your answers in the numbered regions to the right. To check your answers, turn the page. Study the solutions to the problems you missed, and do the exercises following any problem or group of problems in which you had errors.

In **1** and **2** express the vector $\mathbf{v}(a, b)$ in terms of $\mathbf{i}$ and $\mathbf{j}$ and find a unit vector $\mathbf{u}$ in the direction of $\mathbf{v}$.

**1.** $\mathbf{v}(a, b) = \mathbf{v}(3, -2)$

**2.** $\mathbf{v}(a, b) = \mathbf{v}(\sqrt{2}, \sqrt{7})$

In **3–5** find $\mathbf{v} + \mathbf{w}$, $\mathbf{v} - \mathbf{w}$, $\mathbf{v} \cdot \mathbf{w}$ and $\cos \theta$.

**3.** $\mathbf{v} = 3\mathbf{i} - 4\mathbf{j}$ and $\mathbf{w} = 5\mathbf{i} - 12\mathbf{j}$

**4.** $\mathbf{v} = 2\mathbf{i} + 2\mathbf{j}$ and $\mathbf{w} = 4\mathbf{i} + 4\mathbf{j}$

**5.** $\mathbf{v} = 7\mathbf{i} + 21\mathbf{j}$ and $\mathbf{w} = -12\mathbf{i} + 4\mathbf{j}$

In **6–8** find $\mathbf{f}'(t)$ and $\mathbf{f}''(t)$.

**6.** $\mathbf{f}(t) = (2 - 3t)\mathbf{i} + (3t^2)\mathbf{j}$

**7.** $\mathbf{f}(t) = (\cos 2t)\mathbf{i} + (\tan t)\mathbf{j}$

**8.** Find $\mathbf{f}(t) \cdot \mathbf{f}'(t)$ if $\mathbf{f}(t) = t^2\mathbf{i} + 2t\mathbf{j}$.

**9.** Find $\mathbf{v}(t)$, $\mathbf{a}(t)$, $\mathbf{v}(t) \cdot \mathbf{a}(t)$, $s'(t)$, $s''(t)$ and $|\mathbf{a}(t)|$ if $\mathbf{f}(t) = (\sin t)\mathbf{i} + (\cos t)\mathbf{j}$.

**10.** Find the tangential and normal components of acceleration of problem **9**.

**11.** Let a curvilinear motion be defined by

$$\mathbf{f}(t) = (3t^2 + t)\mathbf{i} + (2t^3 + t)\mathbf{j}.$$

Find $\mathbf{v}$, $\mathbf{a}$, $\frac{ds}{dt}$, $\mathbf{T}$, $a_T$, $a_N$, and $K$.

**12.** If $P$ moves upward along the parabola $y^2 = 4x$ with constant speed of 6 units, find $\mathbf{v}$ and $\mathbf{T}$ at $P(1, 2)$.

| |
|---|
| 1 |
| 2 |
| 3 |
| 4 |
| 5 |
| 6 |
| 7 |
| 8 |
| 9 |
| 10 |
| 11 |
| 12 |

## SOLUTIONS

1. $\mathbf{v}(3,-2) = 3\mathbf{i} - 2\mathbf{j}$
   $|\mathbf{v}| = \sqrt{9+4} = \sqrt{13}$
   $\mathbf{u} = \dfrac{\mathbf{v}}{|\mathbf{v}|} = \dfrac{3}{\sqrt{13}}\mathbf{i} - \dfrac{2}{\sqrt{13}}\mathbf{j}$

2. $\mathbf{v}(\sqrt{2},\sqrt{7}) = \sqrt{2}\,\mathbf{i} + \sqrt{7}\,\mathbf{j}$
   $|\mathbf{v}| = \sqrt{2+7} = 3$
   $\mathbf{u} = \dfrac{\mathbf{v}}{|\mathbf{v}|} = \dfrac{\sqrt{2}}{3}\mathbf{i} + \dfrac{\sqrt{7}}{3}\mathbf{j}$

### EXERCISES I

Find $\mathbf{v} = v_x\mathbf{i} + v_y\mathbf{j}$ if $\mathbf{v} = \overrightarrow{AB}$.

i. $A(2,3)$ and $B(4,-2)$

ii. $A(3,-4)$ and $B(-2,5)$

3. $\mathbf{v} = 3\mathbf{i} - 4\mathbf{j}$, $\mathbf{w} = 5\mathbf{i} - 12\mathbf{j}$
   $\mathbf{v} + \mathbf{w} = (3+5)\mathbf{i} - (4+12)\mathbf{j} = 8\mathbf{i} - 16\mathbf{j}$
   $\mathbf{v} - \mathbf{w} = (3-5)\mathbf{i} + (-4+12)\mathbf{j} = -2\mathbf{i} + 8\mathbf{j}$
   $\mathbf{v}\cdot\mathbf{w} = (3)(5) + (-4)(-12) = 63$
   $|\mathbf{v}| = \sqrt{9+16} = 5$
   $|\mathbf{w}| = \sqrt{25+144} = 13$
   $\cos\theta = \dfrac{63}{(5)(13)} = \dfrac{63}{65}$

4. $\mathbf{v} = 2\mathbf{i} + 2\mathbf{j}$, $\mathbf{w} = 4\mathbf{i} + 4\mathbf{j}$
   $\mathbf{v} + \mathbf{w} = (2+4)\mathbf{i} + (2+4)\mathbf{j} = 6\mathbf{i} + 6\mathbf{j}$
   $\mathbf{v} - \mathbf{w} = (2-4)\mathbf{i} + (2-4)\mathbf{j} = -2\mathbf{i} - 2\mathbf{j}$
   $\mathbf{v}\cdot\mathbf{w} = (2)(4) + 2(4) = 16$
   $|\mathbf{v}| = \sqrt{4+4} = 2\sqrt{2}$
   $|\mathbf{w}| = \sqrt{16+16} = 4\sqrt{2}$
   $\cos\theta = \dfrac{16}{(2\sqrt{2})(4\sqrt{2})} = \dfrac{16}{16} = 1$

5. $\mathbf{v} = 7\mathbf{i} + 21\mathbf{j}$, $\mathbf{w} = -12\mathbf{i} + 4\mathbf{j}$
   $\mathbf{v} + \mathbf{w} = -5\mathbf{i} + 25\mathbf{j}$, $\mathbf{v} - \mathbf{w} = 19\mathbf{i} + 17\mathbf{j}$
   $\mathbf{v}\cdot\mathbf{w} = 7(-12) + 21(4) = -84 + 84 = 0$
   $\cos\theta = 0$

### EXERCISES II

i. Find $3\mathbf{v} - 4\mathbf{w}$ if $\mathbf{v} = 2\mathbf{i} - 6\mathbf{j}$, $\mathbf{w} = 3\mathbf{i} - 2\mathbf{j}$.

ii. Find $\cos\theta$ if $\theta$ is the angle between $2\mathbf{v} + 3\mathbf{w}$ and $3\mathbf{v} - 2\mathbf{w}$ if $\mathbf{v} = -\mathbf{i} + \mathbf{j}$, $\mathbf{w} = 3\mathbf{i} - 4\mathbf{j}$.

6. $\mathbf{f}(t) = (2 - 3t)\mathbf{i} + (3t^2)\mathbf{j}$
   $\mathbf{f}'(t) = -3\mathbf{i} + 6t\mathbf{j}$
   $\mathbf{f}''(t) = 6\mathbf{j}$

7. $\mathbf{f}(t) = (\cos 2t)\mathbf{i} + (\tan t)\mathbf{j}$
   $\mathbf{f}'(t) = (-2\sin 2t)\mathbf{i} + (\sec^2 t)\mathbf{j}$
   $\mathbf{f}''(t) = (-4\cos 2t)\mathbf{i} + (2\sec^2 t\tan t)\mathbf{j}$

8. $\mathbf{f}(t) = t^2\mathbf{i} + 2t\mathbf{j}$
   $\mathbf{f}'(t) = 2t\mathbf{i} + 2\mathbf{j}$
   $\mathbf{f}\cdot\mathbf{f}' = 2t^3 + 4t$

9. $\mathbf{f}(t) = (\sin t)\mathbf{i} + (\cos t)\mathbf{j}$
   $\mathbf{v}(t) = \mathbf{f}'(t) = \cos t\,\mathbf{i} - \sin t\,\mathbf{j}$
   $\mathbf{a}(t) = \mathbf{v}'(t) = -\sin t\,\mathbf{i} - \cos t\,\mathbf{j}$
   $\mathbf{v}(t)\cdot\mathbf{a}(t) = -\cos t\sin t + \sin t\cos t = 0$
   $s'(t) = |\mathbf{v}(t)| = \sqrt{\cos^2 t + \sin^2 t} = 1$
   $s''(t) = \dfrac{ds'(t)}{dt} = 0$
   $|\mathbf{a}(t)| = \sqrt{\sin^2 t + \cos^2 t} = 1$

### EXERCISES III

i. Find $\mathbf{f}'(t)\cdot\mathbf{f}''(t)$ if $\mathbf{f}(t) = (\tan t)\,\mathbf{i} + (\sec t)\,\mathbf{j}$.

ii. Find $|\mathbf{a}(t)|$ at $t = 2$ if $\mathbf{f} = \sqrt{t-1}\,\mathbf{i} + \sqrt{t+1}\,\mathbf{j}$.

10. $a_T = s'' = 0$, $a_N = |K|v^2$
    $K = \dfrac{x'y'' - x''y'}{(x'^2 + y'^2)^{\frac{3}{2}}}$ where the primes denote differentiation with respect to the variable $t$.
    $x'y'' - x''y'$
    $= \cos t(-\cos t) - (-\sin t)(-\sin t)$
    $= -(\cos^2 t + \sin^2 t) = -1$
    $x'^2 + y'^2 = \cos^2 t + \sin^2 t = 1$
    $|K| = 1$, $v^2 = [s'(t)]^2 = 1$
    Therefore $a_N = 1$

11. $\mathbf{f}(t) = (3t^2 + t)\mathbf{i} + (2t^3 + t)\mathbf{j}$
    $\mathbf{v} = (6t+1)\mathbf{i} + (6t+1)\mathbf{j}$
    $\mathbf{a} = 6\mathbf{i} + 6\mathbf{j}$
    $\dfrac{ds}{dt} = |\mathbf{v}| = \sqrt{(6t+1)^2 + (6t+1)^2}$
    $= (6t+1)\sqrt{2}$
    $\mathbf{T} = \dfrac{\mathbf{v}}{|\mathbf{v}|} = \dfrac{\sqrt{2}}{2}\mathbf{i} + \dfrac{\sqrt{2}}{2}\mathbf{j}$

$$\mathbf{T}' = 0, \ \frac{d\mathbf{T}}{ds} = 0, \ |K| = \left|\frac{d\mathbf{T}}{ds}\right| = 0$$

$$a_T = \frac{d^2s}{dt^2} = 6\sqrt{2}; \ a_N = K(s')^2 = 0$$

**EXERCISES IV**

i. Find $\mathbf{K}$, $\mathbf{a}_T$, and $\mathbf{a}_N$ of Problem **12**.

ii. Find $\mathbf{a}_T$ and $\mathbf{a}_N$ at $t = \frac{2}{3}$ if the motion is defined by the vector $\mathbf{f}(t) = -t^2\mathbf{i} + t^3\mathbf{j}$.

iii. A particle moves in a clockwise direction around the circle $x^2 + y^2 = 25$. Find the points where $v_y$ is twice $v_x$.

iv. A point moves along $y = \frac{1}{3}x^3$ with $v_y = 8$. Find $\mathbf{a}_T$ at $x = 2$.

**12.** $y^2 = 4x$, $P(1,2)$, $|\mathbf{v}| = 6$
Differentiate with respect to $t$.

$$2y\frac{dy}{dt} = 4\frac{dx}{dt} \text{ or } v_x = \frac{1}{2}y\,v_y$$

Since $v^2 = v_x{}^2 + v_y{}^2 = 36$, we have $\frac{1}{4}y^2v_y{}^2 + v_y{}^2 = (x+1)v_y{}^2 = 36$ or

$$v_y = \frac{6}{\sqrt{x+1}} \text{ and } v_x = \frac{3y}{\sqrt{x+1}}.$$

$$\mathbf{v} = \frac{3}{\sqrt{x+1}}(y\mathbf{i} + 2\mathbf{j})$$

$$\mathbf{T} = \frac{\mathbf{v}}{|\mathbf{v}|} = \frac{1}{2\sqrt{x+1}}(y\mathbf{i} + 2\mathbf{j})$$

At $P(1,2)$: $v_y = \frac{6}{\sqrt{2}}$, $v_x = \frac{6}{\sqrt{2}}$

$$\mathbf{v} = 3\sqrt{2}\,\mathbf{i} + 3\sqrt{2}\,\mathbf{j}, \ \mathbf{T} = \frac{\sqrt{2}}{2}\mathbf{i} + \frac{\sqrt{2}}{2}\mathbf{j}.$$

Solutions to the exercises are on page 96.

## ANSWERS

**1.** $3\mathbf{i} - 2\mathbf{j}$, $\frac{3}{\sqrt{13}}\mathbf{i} - \frac{2}{\sqrt{13}}\mathbf{j}$ — 1

**2.** $\sqrt{2}\,\mathbf{i} + \sqrt{7}\,\mathbf{j}$, $\frac{\sqrt{2}}{3}\mathbf{i} + \frac{\sqrt{7}}{3}\mathbf{j}$ — 2

**3.** $8\mathbf{i} - 16\mathbf{j}$, $-2\mathbf{i} + 8\mathbf{j}$, $63$, $\frac{63}{65}$ — 3

**4.** $6\mathbf{i} + 6\mathbf{j}$, $-2\mathbf{i} - 2\mathbf{j}$, $16$, $1$ — 4

**5.** $-5\mathbf{i} + 25\mathbf{j}$, $19\mathbf{i} + 17\mathbf{j}$, $0$, $0$ — 5

**6.** $-3\mathbf{i} + 6t\mathbf{j}$, $6\mathbf{j}$ — 6

**7.** $(-2\sin 2t)\mathbf{i} + (\sec^2 t)\mathbf{j}$
$(-4\cos 2t)\mathbf{i} + (2\sec^2 t\tan t)\mathbf{j}$ — 7

**8.** $2t^3 + 4t$ — 8

**9.** $\cos t\,\mathbf{i} - \sin t\,\mathbf{j}$
$-\sin t\,\mathbf{i} - \cos t\,\mathbf{j}$
$0, 1, 0, 1$ — 9

**10.** $a_T = 0$, $a_N = 1$ — 10

**11.** $(6t+1)\mathbf{i} + (6t+1)\mathbf{j}$
$6\mathbf{i} + 6\mathbf{j}$, $(6t+1)\sqrt{2}$
$\frac{\sqrt{2}}{2}\mathbf{i} + \frac{\sqrt{2}}{2}\mathbf{j}$, $6\sqrt{2}$, $0$, $0$ — 11

**12.** $3\sqrt{2}\,\mathbf{i} + 3\sqrt{2}\,\mathbf{j}$
$\frac{\sqrt{2}}{2}\mathbf{i} + \frac{\sqrt{2}}{2}\mathbf{j}$ — 12

## BASIC FACTS

A two dimensional vector is an ordered pair of numbers $(a, b)$. The numbers $a$ and $b$ are the **components** of the vector. A vector is denoted by a bold face letter, $\mathbf{v}$. A vector is also defined as a directed line segment; $\mathbf{v} = \overrightarrow{AB}$ indicates the vector from $A$ to $B$.

The length of a vector is given by

$$|\mathbf{v}| = v = \sqrt{a^2 + b^2}.$$

Unit vector: $\mathbf{u} = \dfrac{\mathbf{v}}{|\mathbf{v}|}$

$\mathbf{i}$ is the unit vector along the positive $X$-axis.

$\mathbf{j}$ is the unit vector along the positive $Y$-axis.

Components along $\mathbf{i}, \mathbf{j}$: $\mathbf{v} = a\mathbf{i} + b\mathbf{j}$.
Let $\mathbf{v} = a\mathbf{i} + b\mathbf{j}$ and $\mathbf{w} = c\mathbf{i} + d\mathbf{j}$.
Angle between $\mathbf{v}$ and $\mathbf{w}$:

$$\cos\theta = \frac{ac + bd}{|\mathbf{v}||\mathbf{w}|}.$$

Addition and subtraction of $\mathbf{v}$ and $\mathbf{w}$:

$$\mathbf{v} \pm \mathbf{w} = (a \pm c)\mathbf{i} + (b \pm d)\mathbf{j}.$$

Multiplying by a scalar, $k$

$$k\mathbf{v} = ka\mathbf{i} + kb\mathbf{j}.$$

**Dot** (scalar) **product** of $\mathbf{v}$ and $\mathbf{w}$:

$$\mathbf{v} \cdot \mathbf{w} = vw \cos\theta = ac + bd.$$

Associative laws: $k(n\mathbf{v}) = (kn)\mathbf{v}$
$\mathbf{u} + (\mathbf{v} + \mathbf{w}) = (\mathbf{u} + \mathbf{v}) + \mathbf{w}$.

Commutative laws:

$$\mathbf{u} + \mathbf{v} = \mathbf{v} + \mathbf{u} \text{ and } \mathbf{v} \cdot \mathbf{w} = \mathbf{w} \cdot \mathbf{v}.$$

Distributive laws:

$$(k + n)\mathbf{v} = k\mathbf{v} + n\mathbf{v}$$

$$k(\mathbf{v} + \mathbf{w}) = k\mathbf{v} + k\mathbf{w}$$

$$\mathbf{u} \cdot (\mathbf{v} + \mathbf{w}) = \mathbf{u} \cdot \mathbf{v} + \mathbf{u} \cdot \mathbf{w}.$$

Projection of $\mathbf{v} = a\mathbf{i} + b\mathbf{j}$ upon $\mathbf{w} = c\mathbf{i} + d\mathbf{j}$ is a vector $\mathbf{u} = x\mathbf{i} + y\mathbf{i}$ where $x = kc$ and $y = kd$ and where $k = \pm \dfrac{ac + bd}{\sqrt{c^2 + d^2}}$ (plus if $\mathbf{v}$ is in the same direction as $\mathbf{w}$, and minus if $\mathbf{v}$ is in opposite direction of $\mathbf{w}$). The vector $\mathbf{v}$ is perpendicular to vector $\mathbf{w}$

## ADDITIONAL INFORMATION

The sum and difference of two vectors can be obtained geometrically as the diagonals of the parallelogram drawn with the given vectors as sides taking into account their respective directions. See Figure 16-1. The sum vector, $\overrightarrow{OR}$, is called the resultant of the vectors $\mathbf{v}$ and $\mathbf{w}$. We can derive the formula for the cosine of the angle between two vectors by using triangle $OPQ$. From the law of cosines we have

$$(1)\ \cos\theta = \frac{\overline{OP}^2 + \overline{OQ}^2 - \overline{PQ}^2}{2\ \overline{OP}(\overline{OQ})}$$

$$\overline{OP}^2 = v^2 = a^2 + b^2;\ \overline{OP} = |\mathbf{v}|$$

$$\overline{OQ}^2 = w^2 = c^2 + d^2;\ \overline{OQ} = |\mathbf{w}|$$

Fig. 16-1

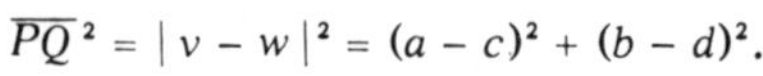

$$\overline{PQ}^2 = |v - w|^2 = (a - c)^2 + (b - d)^2.$$

Substituting into (1) we get

$$\cos\theta = \frac{a^2 + b^2 + c^2 + d^2 - (a^2 - 2ac + c^2) - (b^2 - 2bd + d^2)}{2\ |\mathbf{v}||\mathbf{w}|}$$

$$= \frac{2ac + 2bd}{2\ |\mathbf{v}||\mathbf{w}|} = \frac{ac + bd}{|\mathbf{v}||\mathbf{w}|}.$$

We use the term **scalar** as a quantity that has a magnitude describable by a real number and no direction. The length of a vector is a scalar quantity, $|\mathbf{v}|$. In symbolic notation the bold face $\mathbf{v}$ is a vector and the light face $v$ is a scalar indicating the magnitude of $\mathbf{v}$, then $|\mathbf{v}| = v$. The zero vector $\mathbf{v}(0, 0)$ has zero length and is considered to be orthogonal to all vectors.

The motion of a particle along a curve $C$ in the plane can be discussed in terms of vectors. Let the curve $C$ be defined by parametric equations $x = x(t)$ and $y = y(t)$. The same curve will be traced out by the head of the vector function defined by $\mathbf{f}(t) = x(t)\mathbf{i} + y(t)\mathbf{j}$. If $t$ denotes the time, then the vector $\dfrac{d\mathbf{f}}{dt} = x'(t)\mathbf{i} + y'(t)\mathbf{j}$ is the velocity vector at the time $t$. This vector is always tangent to the curve $C$ describing the motion. We can then construct the velocity diagram shown in Figure 16-2, where

$$\frac{d\mathbf{f}}{dt} = \mathbf{v} = v_x\mathbf{i} + v_y\mathbf{j}$$

$$v_x = x'(t) = \frac{dx}{dt};\ v_y = y'(t) = \frac{dy}{dt}$$

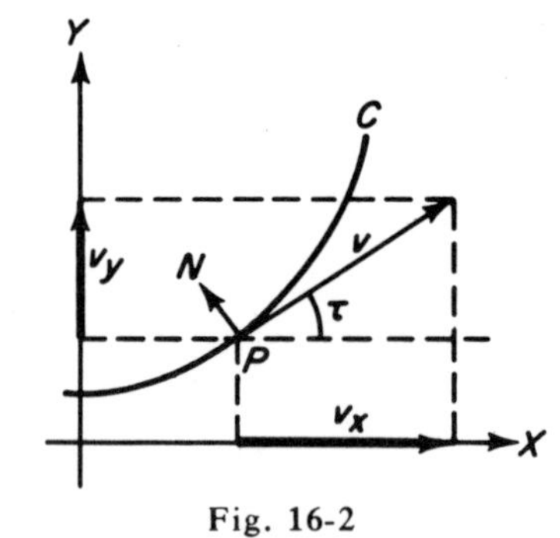


Fig. 16-2

are shown at a variable point $P(x, y)$. The angle $\tau$ that $\mathbf{v}$ makes with the positive $x$-direction is the usual angle of inclination of the

if $\mathbf{v} \cdot \mathbf{w} = 0$, and conversely. If two vectors are perpendicular to each other they are **orthogonal.**

A **vector function** $\mathbf{f}(t)$ is a collection of ordered pairs $(t, \mathbf{v})$ where $t$ is a real number, $\mathbf{v}$ is a vector, and no two pairs have the same first element. The domain is the set of all possible values of $t$, and the range is the set of resulting vectors. Differentiation of vector functions:

$$\mathbf{f}(t) = x(t)\mathbf{i} + y(t)\mathbf{j}$$
$$\mathbf{f}'(t) = x'(t)\mathbf{i} + y'(t)\mathbf{j}$$
$$\mathbf{f}''(t) = x''(t)\mathbf{i} + y''(t)\mathbf{j}$$

where the primes indicate differentiation with respect to $t$.

The head of the vector $\mathbf{f}(t)$ traces a curve coincident with the curve defined by the parametric equations $x = x(t)$ and $y = y(t)$.

The **velocity** vector $\mathbf{v}(t)$:

$$\mathbf{v}(t) = \mathbf{f}'(t) = x'(t)\mathbf{i} + y'(t)\mathbf{j} = v_x\mathbf{i} + v_y\mathbf{j}$$

is tangent to the curve.

Speed: $|\mathbf{v}(t)| = |\mathbf{f}'(t)| = \dfrac{ds}{dt} = \sqrt{v_x^2 + v_y^2} = v.$

The **acceleration** vector

$$\mathbf{a}(t) = \mathbf{v}'(t) = \mathbf{f}''(t).$$

Unit tangent vector:

$$\mathbf{T}(t) = \frac{\mathbf{v}(t)}{|\mathbf{v}(t)|} = \frac{\mathbf{v}(t)}{s'};\ s' = \frac{ds}{dt} = v$$

$$\mathbf{T}(t) = \cos \tau\ \mathbf{i} + \sin \tau\ \mathbf{j};\ |\mathbf{T}| = 1$$

$$\mathbf{T}'(t) = (-\sin \tau\ \mathbf{i} + \cos \tau\ \mathbf{j})\frac{d\tau}{dt}$$

$\mathbf{T}$ and $\mathbf{T}'$ are orthogonal.

$$\frac{d\mathbf{T}}{ds} = \frac{\mathbf{T}'}{v} = \frac{x'}{v}\mathbf{i} + \frac{y'}{v}\mathbf{j};\ \left|\frac{d\mathbf{T}}{ds}\right| = |\mathbf{K}|$$

Unit **normal** vector, $\mathbf{N}(t)$:

$$\mathbf{N}(t) = \frac{1}{|\mathbf{K}(t)|}\frac{d\mathbf{T}}{ds} = \frac{1}{vK}(x''\mathbf{i} + y''\mathbf{j})$$

$$\mathbf{a}(t) = \frac{d^2s}{dt^2}\mathbf{T}(t) + v^2|\mathbf{K}|\mathbf{N}(t)$$

$$a_T = \frac{d^2s}{dt^2} = \frac{dv}{dt},\ a_N = v^2|\mathbf{K}|.$$

tangent line, and we have $\tan \tau = \dfrac{dy}{dx} = \dfrac{y'(t)}{x'(t)} = \dfrac{v_y}{v_x}$. We notice that $v_x = |\mathbf{v}| \cos \tau$ and $v_y = |\mathbf{v}| \sin \tau$, and we can then define the unit vector along $\mathbf{v}$ as $\mathbf{T} = \dfrac{\mathbf{v}}{|\mathbf{v}|} = \cos \tau\ \mathbf{i} + \sin \tau\ \mathbf{j}$. The time derivative of this vector is $\mathbf{T}'(t) = -\sin \tau \dfrac{d\tau}{dt}\mathbf{i} + \cos \tau \dfrac{d\tau}{dt}\mathbf{j}$, and the dot product (also called the **inner product**) is

$$T - T' = \frac{d\tau}{dt}(-\cos \tau \sin \tau + \sin \tau \cos \tau) = \frac{d\tau}{dt}(0) = 0$$

so that $\mathbf{T}$ and $\mathbf{T}'$ are orthogonal. A vector perpendicular to $\mathbf{T}$ would be along the normal to the curve $C$. Let us now take the derivative of $\mathbf{T}$ with respect to $s$, the arc length along $C$.

$$\frac{d\mathbf{T}}{ds} = \frac{D_t\mathbf{T}}{D_ts} = \frac{D_t\tau}{D_ts}(-\sin \tau\ \mathbf{i} + \cos \tau\ \mathbf{j}) = \frac{d\tau}{ds}(-\sin \tau\ \mathbf{i} + \cos \tau\ \mathbf{j})$$

Now recall (see Chapter 14) that $\dfrac{d\tau}{ds}$ is the curvature $K$.

$$\left|\frac{d\mathbf{T}}{ds}\right| = \left|\frac{d\tau}{ds}\right|(\sin^2 \tau + \cos^2 \tau) = \left|\frac{d\tau}{ds}\right| = |K|$$

The unit vector normal to $C$ is the defined as

$$\mathbf{N}(t) = \frac{1}{|K(t)|}\frac{d\mathbf{T}}{ds} = -\sin \tau\ \mathbf{i} + \cos \tau\ \mathbf{j}.$$

The **acceleration** is the derivative of the velocity with respect to time. In vector notation

$$\mathbf{a}(t) = \mathbf{v}'(t) = \mathbf{f}''(t) = x''(t)\mathbf{i} + y''(t)\mathbf{j} = a_x\mathbf{i} + a_y\mathbf{j}.$$

This vector has components along the tangent line and the normal to the curve at $P$. (See Figure 16-3.) We can express $\mathbf{a}$ in terms of these components by differentiating $\mathbf{v} = v\mathbf{T} = D_ts\mathbf{T}$.

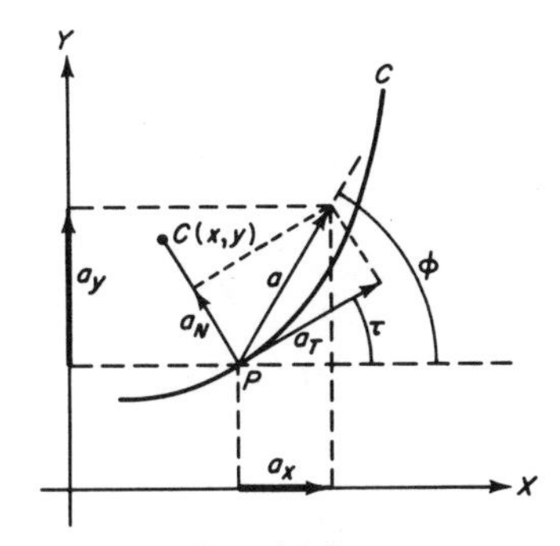

**Fig. 16-3**

$$\mathbf{a}(t) = \frac{d^2s}{dt^2}\mathbf{T} + \left(\frac{ds}{dt}\right)\frac{d\mathbf{T}}{dt} = \frac{d^2s}{dt^2}\mathbf{T} + \left(\frac{ds}{dt}\right)\left(\frac{ds}{dt}\right)\frac{d\mathbf{T}}{ds} = a_T\mathbf{T} + a_N\mathbf{N}$$

where $a_T = \dfrac{d^2s}{dt^2} = \dfrac{dv}{dt}$ and $a_N = \left(\dfrac{ds}{dt}\right)^2|K| = v^2|K|$ since $v^2 = \left(\dfrac{ds}{dt}\right)^2$ and $\mathbf{N} = \dfrac{d\mathbf{T}}{ds}\dfrac{1}{|K|}$.

If the curve $C$ is defined by $y = f(x)$ the **curvilinear motion** of a particle along $C$ is defined in the same manner as above. The function $y = f(x)$ is differentiated with respect to $t$ to obtain $D_ty = D_tf(x) = D_xfD_tx$ or $v_y = f'(x)v_x$ which is used with the relation $v_x^2 + v_y^2 = v^2$ to obtain values of the components. Since these equations involve two unknowns we need a condition on the velocity or one of the components to completely specify the motion. If the condition is that of a constant velocity there may still be a normal component of acceleration. (See Exercises IV–i.)

## SOLUTIONS TO EXERCISES

### I

i. $A(2,3)$ and $B(4,-2)$
$\mathbf{v} = (4-2)\mathbf{i} + (-2-3)\mathbf{j}$
$= 2\mathbf{i} - 5\mathbf{j}$

ii. $A(3,-4)$ and $B(-2,5)$
$\mathbf{v} = (-2-3)\mathbf{i} + (5+4)\mathbf{j}$
$= -5\mathbf{i} + 9\mathbf{j}$

### II

i. $\mathbf{v} = 2\mathbf{i} - 6\mathbf{j}$, $\mathbf{w} = 3\mathbf{i} - 2\mathbf{j}$
$3\mathbf{v} - 4\mathbf{w} = (6-12)\mathbf{i} - (18-8)\mathbf{j}$
$= -6\mathbf{i} - 10\mathbf{j}$

ii. $\mathbf{v} = -\mathbf{i} + \mathbf{j}$, $\mathbf{w} = 3\mathbf{i} - 4\mathbf{j}$
$\mathbf{r} = 2\mathbf{v} + 3\mathbf{w} = 7\mathbf{i} - 10\mathbf{j}$; $r = \sqrt{149}$
$\mathbf{u} = 3\mathbf{v} - 2\mathbf{w} = -9\mathbf{i} + 11\mathbf{j}$; $u = \sqrt{202}$
$$\cos\theta = \frac{-63-110}{\sqrt{149}\sqrt{202}} = \frac{-173}{\sqrt{149}\sqrt{202}}$$

### III

i. $\mathbf{f}(t) = (\tan t)\mathbf{i} + (\sec t)\mathbf{j}$
Let $f_1 = \tan t$ and $f_2 = \sec t$
$f_1' = \sec^2 t$, $f_2' = \sec t \tan t$
$f_1'' = 2\sec^2 t \tan t$,
$f_2'' = \sec t \tan^2 t + \sec^3 t$
$f_1' f_1'' = 2\sec^4 t \tan t$
$f_2' f_2'' = \sec^2 t \tan^3 t + \sec^4 t \tan t$
$\mathbf{f}' \cdot \mathbf{f}'' = f_1' f_1'' + f_2' f_2''$
$= 3\sec^4 t \tan t + \sec^2 t \tan^3 t$

ii. $\mathbf{f} = \sqrt{t-1}\,\mathbf{i} + \sqrt{t+1}\,\mathbf{j}$
$f_1 = \sqrt{t-1}$, $f_1' = \frac{1}{2}(t-1)^{-\frac{1}{2}}$
$f_1'' = -\frac{1}{4}(t-1)^{-\frac{3}{2}}$, $f_1''(2) = -\frac{1}{4}$
$f_2 = \sqrt{t+1}$, $f_2' = \frac{1}{2}(t+1)^{-\frac{1}{2}}$
$f_2'' = -\frac{1}{4}(t+1)^{-\frac{3}{2}}$
$f_2''(2) = -\frac{1}{4}(3)^{-\frac{3}{2}}$
$|\mathbf{a}(2)|^2 = \left(-\frac{1}{4}\right)^2 + \left(-\frac{1}{4}\right)^2 (3)^{-3}$
$= \frac{1}{16}\left(1 + \frac{1}{27}\right) = \frac{1}{4}\left(\frac{7}{27}\right)$
$|\mathbf{a}(2)| = \frac{1}{2}\left(\frac{1}{3}\right)\sqrt{\frac{7}{3}} = \frac{1}{18}\sqrt{21}$

### IV

i. Given $y^2 = 4x$, $v = 6$, $P(1,2)$.
Since $v = 6$, $\frac{dv}{dt} = a_T = 0$
$$\frac{dy}{dx} = y' = \frac{2}{y},\ y'' = -\frac{4}{y^3} = -\frac{1}{xy}$$
$$1 + y'^2 = 1 + \frac{4}{y^2} = 1 + \frac{1}{x}$$
$$K = y''(1 + y'^2)^{-\frac{3}{2}}$$
$$= -\frac{1}{xy}\left(1 + \frac{1}{x}\right)^{-\frac{3}{2}}$$
At $P(1,2)$:
$$K = -\frac{1}{2}(1+1)^{-\frac{3}{2}} = -\frac{\sqrt{2}}{8}$$
$$a_N = v^2|K| = 36\left(\frac{\sqrt{2}}{8}\right) = \frac{9}{2}\sqrt{2}$$

ii. $\mathbf{f}(t) = -t^2\mathbf{i} + t^3\mathbf{j}$
$\mathbf{v} = -2t\mathbf{i} + 3t^2\mathbf{j}$
$\mathbf{a} = -2\mathbf{i} + 6t\mathbf{j}$
$v = (4t^2 + 9t^4)^{\frac{1}{2}}$
$$\frac{dv}{dt} = (4 + 18t^2)(4 + 9t^2)^{-\frac{1}{2}}$$
$|\mathbf{a}| = (4 + 36t^2)^{\frac{1}{2}} = 2\sqrt{1 + 9t^2}$
At $t = \frac{2}{3}$: $|\mathbf{a}| = 2\sqrt{5}$
$a_T = (4+8)(4+4)^{-\frac{1}{2}} = 3\sqrt{2}$
$a_N = \sqrt{a^2 - a_T^2} = \sqrt{20 - 18} = \sqrt{2}$
*See* Fig. 16-3 for relationship.

iii. $x^2 + y^2 = 25$
$2xv_x + 2yv_y = 0$, $xv_x = -yv_y$
If $v_y = 2v_x$, $xv_x = -2yv_x$
or $x = -2y$. Substitute into
$x^2 + y^2 = 25$ and $4y^2 + y^2 = 25$
Then $y = \pm\sqrt{5}$ and $x = \mp 2\sqrt{5}$
$P_1(2\sqrt{5}, -\sqrt{5})$ and $P_2(-2\sqrt{5}, \sqrt{5})$

iv. $y = \frac{x^3}{3}$, $v_y = 8$
$v_y = x^2 v_x = 8$ and $v_x = 8x^{-2}$
$|\mathbf{v}| = (64 + 64x^{-4})^{\frac{1}{2}}$
$= 8(1 + x^{-4})^{\frac{1}{2}}$
$$\frac{dv}{dt} = a_T = \frac{8}{2}(1 + x^{-4})^{-\frac{1}{2}}(-4x^{-5})v_x$$
At $x = 2$: $v_x = 8(2)^{-2} = 2$
$$a_T = 4\left(1 + \frac{1}{16}\right)^{-\frac{1}{2}}\left(-\frac{4}{32}\right)(2)$$
$$= -\left(\frac{17}{16}\right)^{-\frac{1}{2}} = -\frac{4}{\sqrt{17}}$$

# 17 METHODS OF INTEGRATION I

## SELF-TEST

DIRECTIONS: Write your answers in the numbered regions to the right. To check your answers, turn the page. Study the solutions to the problems you missed, and do the exercises following any problem or group of problems in which you had errors.

Find the indefinite integrals.

1. $\int x^2 e^{ax}\, dx$
2. $\int x \ln x\, dx$
3. $\int e^x \sin x\, dx$
4. $\int \sec^3 x\, dx$
5. $\int \sin^4 2x\, dx$
6. $\int \tan^3 x \sec^3 x\, dx$
7. $\int \cot^5 x \csc^4 x\, dx$
8. $\int \frac{\sqrt{x^2 - 4}}{x^2}\, dx$
9. $\int \frac{dx}{x\sqrt{x^2 + 9}}$
10. $\int \sin 3x \cos 5x\, dx$
11. $\int \sqrt{2x - x^2}\, dx$
12. $\int x^2 \sqrt{3 - x^2}\, dx$

1

2

3

4

5

6

7

8

9

10

11

12

## SOLUTIONS

**1.** $\int x^2 e^{ax}\, dx = I$

Let $u = x^2$ and $dv = e^{ax}\, dx$

$du = 2x\, dx;\ v = \int e^{ax} dx = \dfrac{e^{ax}}{a}$

$I = \frac{1}{a} x^2 e^{ax} - \frac{2}{a} \int x e^{ax} dx$

Let $u = x$ and $v = \frac{1}{a} e^{ax}$

$$I = \frac{x^2}{a} e^{ax} - \frac{2}{a}\left(\frac{x}{a} e^{ax} - \frac{1}{a}\int e^{ax} dx\right)$$

$$= e^{ax}\left(\frac{x^2}{a} - \frac{2x}{a^2} + \frac{2}{a^3}\right) + C$$

**2.** $\int x \ln x\, dx = I$

Let $u = \ln x$ and $dv = x\, dx$

$du = \dfrac{dx}{x}$ and $v = \frac{1}{2} x^2$

$I = \frac{1}{2} x^2 \ln x - \frac{1}{2} \int x\, dx$

$= \frac{1}{2} x^2 \ln x - \frac{1}{4} x^2 + C$

**3.** $\int e^x \sin x\, dx = I$

Let $u = e^x$ and $dv = \sin x\, dx$

$du = e^x dx$ and $v = -\cos x$

$I = -e^x \cos x + \int e^x \cos x\, dx$

Let $u = e^x$ and $dv = \cos x\, dx$

$du = e^x dx$ and $v = \sin x$

$I = -e^x \cos x + e^x \sin x - \int e^x \sin x\, dx$

$= e^x (\sin x - \cos x) - I$

$I = \frac{1}{2} e^x (\sin x - \cos x) + C$

### EXERCISES I

Integrate by parts.

i. $\int e^{2x} \sin 3x\, dx$

ii. $\int \arctan x\, dx$

iii. $\int \sin x \cos 3x\, dx$

**4.** $\int \sec^3 x\, dx = I$

Let $u = \sec x$ and $dv = \sec^2 x\, dx$

$du = \sec x \tan x\, dx;\ v = \tan x$

$I = \tan x \sec x - \int \sec x \tan^2 x\, dx$

Substitute $\tan^2 x = \sec^2 x - 1$

$I = \tan x \sec x - \int \sec^3 x\, dx + \int \sec x\, dx$

$= \tan x \sec x - I + \ln(\sec x + \tan x)$

$I = \frac{1}{2} \tan x \sec x$
$\quad + \frac{1}{2} \ln(\sec x + \tan x) + C$

**5.** $\int \sin^4 2x\, dx = I$

$\sin^4 2x = \sin^2 2x\,(1 - \cos^2 2x)$

$= \sin^2 2x - \sin^2 2x \cos^2 2x$

$= \frac{1}{2} - \frac{1}{2} \cos 4x - \frac{1}{4} \sin^2 4x$

$= \frac{1}{2} - \frac{1}{2} \cos 4x - \frac{1}{8} + \frac{1}{8} \cos 8x$

$I = \int \left(\frac{3}{8} - \frac{1}{2} \cos 4x + \frac{1}{8} \cos 8x\right) dx$

$= \frac{3}{8} x - \frac{1}{8} \sin 4x + \frac{1}{64} \sin 8x + C$

**6.** $\int \tan^3 x \sec^3 x\, dx = I$

$\tan^3 x \sec^3 x = \tan x \sec x\,(\tan^2 x \sec^2 x)$

$= \tan x \sec x\,(\sec^4 x - \sec^2 x)$

Recall $d(\sec x) = \tan x \sec x\, dx$.

$I = \frac{1}{5} \sec^5 x - \frac{1}{3} \sec^3 x + C$

**7.** $\int \cot^5 x \csc^4 x\, dx = I$

$\cot^5 x \csc^4 x = (\cot^5 x + \cot^7 x) \csc^2 x$

$I = -\frac{1}{6} \cot^6 x - \frac{1}{8} \cot^8 x + C$

### EXERCISES II

Find the indefinite integrals.

i. $\int \sin^2 x \cos^4 x\, dx$

ii. $\int \cot^2 x\, dx$

**8.** $\displaystyle\int \frac{\sqrt{x^2 - 4}}{x^2}\, dx = I$

Let $x = 2 \sec z;\ x^2 = 4 \sec^2 z$.

$dx = 2 \sec z \tan z\, dz$

$x^2 - 4 = 4 \sec^2 z - 4 = 4 \tan^2 z$

$$\frac{\sqrt{x^2 - 4}}{x^2}\, dx = \frac{2 \tan z}{4 \sec^2 z} (2 \sec z \tan z) dz$$

$$= (\sec z - \cos z) dz$$

$$I = \ln(\sec z + \tan z) - \sin z + C$$

$$= \ln(x + \sqrt{x^2 - 4}) - \frac{\sqrt{x^2 - 4}}{x} + C_1$$

Note: $-\ln 2$ has been included in $C_1$.

9. $\displaystyle\int \frac{dx}{x\sqrt{x^2 + 9}} = I$

Let $x = 3 \tan z$; $dx = 3 \sec^2 z\, dz$

$$I = \int \frac{3 \sec^2 z\, dz}{9 \tan z \sec z} = \frac{1}{3} \int \frac{\sec z}{\tan z}\, dz$$

$$= \frac{1}{3} \ln\left(\frac{\sqrt{x^2 + 9}}{3} - \frac{3}{x}\right) + C$$

10. $\int \sin 3x \cos 5x\, dx = I$

$\sin 3x \cos 5x = \frac{1}{2}[\sin 8x - \sin 2x]$

$I = \frac{1}{2} \int \sin 8x\, dx - \frac{1}{2} \int \sin 2x\, dx$

$= -\frac{1}{16} \cos 8x + \frac{1}{4} \cos 2x + C$

EXERCISE III

Find the indefinite integrals.

i. $\displaystyle\int \frac{dx}{x^4\sqrt{x^2 - 5}}$

11. $\int \sqrt{2x - x^2}\, dx = I$

$2x - x^2 = 1 - (x - 1)^2$

Let $u = x - 1$; $dx = du$

$I = \int \sqrt{1 - u^2}\, du$

$$= \frac{u}{2} \sqrt{1 - u^2} + \frac{1}{2} \text{Arcsin}\, u + C$$

$$= \frac{x - 1}{2} \sqrt{2x - x^2} + \frac{1}{2} \text{Arcsin}\,(x - 1) + C$$

12. $\int x^2 \sqrt{3 - x^2}\, dx = I$

Let $x = \sqrt{3} \sin z$, $dx = \sqrt{3} \cos z\, dz$

$x^2 = 3 \sin^2 z$, $3 - x^2 = 3 \cos^2 z$

$I = 9 \int \sin^2 z \cos^2 z\, dz$

$= \frac{9}{4} \int \sin^2 2z\, dz$

$= \frac{9}{8} \left[\frac{1}{2}(2z) - \frac{1}{4} \sin 4z\right] + C$

$$I = \frac{9}{8} \text{Arcsin}\, \frac{x}{\sqrt{3}} - \frac{1}{8}(3x - 2x^3) \sqrt{3 - x^2} + C$$

**ANSWERS**

| | |
|---|---|
| 1. $e^{ax}\left(\frac{x^2}{a} - \frac{2x}{a^2} + \frac{2}{a^3}\right) + C$ | 1 |
| 2. $\frac{1}{2} x^2 \ln x - \frac{1}{4} x^2 + C$ | 2 |
| 3. $\frac{1}{2} e^x (\sin x - \cos x) + C$ | 3 |
| 4. $\frac{1}{2} \tan x \sec x + \frac{1}{2} \ln(\sec x + \tan x) + C$ | 4 |
| 5. $\frac{3}{8} x - \frac{1}{8} \sin 4x + \frac{1}{64} \sin 8x + C$ | 5 |
| 6. $\frac{1}{5} \sec^5 x - \frac{1}{3} \sec^3 x + C$ | 6 |
| 7. $-\frac{1}{6} \cot^6 x - \frac{1}{8} \cot^8 x + C$ | 7 |
| 8. $\ln(x + \sqrt{x^2 - 4}) - \frac{\sqrt{x^2 - 4}}{x} + C$ | 8 |
| 9. $\frac{1}{3} \ln\left(\frac{\sqrt{x^2 + 9}}{3} - \frac{3}{x}\right) + C$ | 9 |
| 10. $-\frac{1}{16} \cos 8x + \frac{1}{4} \cos 2x + C$ | 10 |
| 11. $\frac{x - 1}{2} \sqrt{2x - x^2} + \frac{1}{2} \text{Arcsin}\,(x - 1) + C$ | 11 |
| 12. $\frac{9}{8} \text{Arcsin}\, \frac{x}{\sqrt{3}} - \frac{1}{8}(3x - 2x^3) \sqrt{3 - x^2} + C$ | 12 |

Solutions to the exercises are on page 102.

## BASIC FACTS

The process of integration is one of associating a given integral with one in a standard form whose anti-derivative is known. We have already exhibited some elementary formulas in Chapters 7 and 13. In this Chapter we shall present some formal techniques.

### Integration by Parts

$$\int u\,dv = uv - \int v\,du$$

We add the following hints:

1. The indicated differential ($dx$) is a part of $dv$
2. It must be possible to integrate the differential $dv$
3. If possible let $u$ contain expressions of higher power which will be reduced by differentiation
4. Repeat the process if necessary.

### Trigonometric Differentials

A. $\int \sin^m u \cos^n u\,du$

I. Either $m$ or $n$ odd and positive. Use $\sin^2 u + \cos^2 u = 1$ and change to $\int$(sum of cosine terms) $\sin u\,du$ or $\int$(sum of sine terms) $\cos u\,du$.

II. Both $m$ and $n$ even and $\geq 0$. Use $\sin u \cos u = \frac{1}{2}\sin 2u$

$$\sin^2 u = \frac{1}{2} - \frac{1}{2}\cos 2u$$

$$\cos^2 u = \frac{1}{2} + \frac{1}{2}\cos 2u$$

$$\int \sin^2 u\,du = \frac{1}{2}u - \frac{1}{4}\sin 2u + C$$

$$\int \cos^2 u\,du = \frac{1}{2}u + \frac{1}{4}\sin 2u + C.$$

B. $\int \tan^m u \sec^n u\,du$

I. If $n$ is even and positive, factor out $\sec^2 u$ and change remaining secants by identity $\sec^2 u = 1 + \tan^2 u$.

II. If $m$ is odd and positive, factor out $\sec u \tan u$ and change remaining tangents by $\tan^2 u = \sec^2 u - 1$.

## ADDITIONAL INFORMATION

The techniques of formal integration presented in this chapter rely upon a thorough knowledge of the trigonometric identies and addition formulas. See p 155. It is recommended that these be committed to memory.

In the application of the technique of **integration by parts**, the student is cautioned not to "go in a circle." This can happen when integration by parts is used more than once in the same problem. Consider, for example, the integral $\int e^x \sin x\,dx$. If we let $u = e^x$ and $dv = \sin x\,dx$, then $du = e^x dx$ and $v = -\cos x\,dx$ and the given integral becomes $-e^x\cos x + \int e^x \cos x\,dx$. If we now let $u = \cos x$ and $dv = e^x dx$ in the second integral, we have $du = -\sin x\,dx$ and $v = e^x$. When these are substituted into the complete integral, we have

$$\int e^x \sin x\,dx = -e^x\cos x + e^x \cos x + \int e^x \sin x\,dx = \int e^x \sin x\,dx,$$

and we are right back where we started. The second substitution simply cancelled the first integration by parts. The correct procedure is to let $u = e^x$ and $dv = \cos x\,dx$ in the second integral. This will yield $du = e^x dx$, $v = \sin x$ and

$$\int e^x \cos x\,dx = e^x \sin x - \int e^x \sin x\,dx.$$

Although we have returned to the original integral, the terms do not cancel, and we can solve the equation for the given integral by transposing, thus $2\int e^x \sin x\,dx = e^x(\sin x - \cos x)$ and then divide by 2.

The **trigonometric substitutions** are used to derive the integrals involving radicals of the form $\sqrt{a^2 - u^2}$ given in Chapter 13. We shall illustrate by deriving the formula for $\int \sqrt{u^2 \pm a^2}\,du$. This formula was not given in Chapter 13 and the student might like to add it to the others that should have been committed to memory. Let us consider $\sqrt{u^2 + a^2}$ and substitute $u = a\tan z$. We then have $du = a\sec^2 z\,dz$ and $u^2 + a^2 = a^2\tan^2 z + a^2 = a^2(\tan^2 z + 1) = a^2\sec^2 z$. Thus we have

$$\int \sqrt{u^2 + a^2}\,du = \int s \sec z(a\sec^2 z\,dz) = a^2 \int \sec^3 z\,dz.$$

We now apply integration by parts to the last integral. Let $u = \sec z$ and $dv = \sec^2 z\,dz$, then $du = \sec z \tan z\,dz$ and $v = \tan z$. The second integral in the formula for integration by parts now becomes

$$\int \tan^2 z \sec z\,dz = \int \sec^3 z\,dz - \int \sec z\,dz$$
$$= \int \sec^3 z\,dz - \ln(\sec z + \tan z)$$

Substitute into the formula for integration by parts to obtain

$$\int \sec^3 z\,dz = \sec z \tan z - \int \sec^3 z\,dz + \ln(\sec z + \tan z)$$

and we solve this equation for the integral of $\sec^3 z\,dz$. There remains to transform this expression from $z$ back into terms involving $u$. Consider the triangle of Figure 17-1. It is formed from the definition $u = a\tan z$ and obtaining the hypotenuse by use

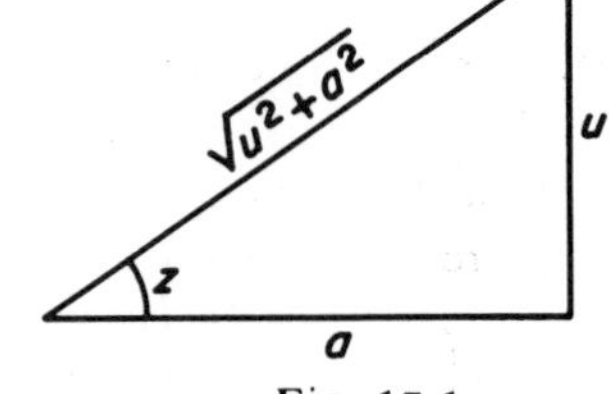

Fig. 17-1

**C. $\int \cot^m u \csc^n u\, du$**

Apply the same procedure used for $\tan^m u \sec^n u$ using the identity

$$1 + \cot^2 u = \csc^2 u.$$

**D. $\int \tan^n u\, du$ or $\int \cot^n u\, du$**

Change the integrand by

$$\tan^n u = \tan^{n-2} u(\sec^2 u - 1)$$

$$\cot^n u = \cot^{n-2} u(\csc^2 u - 1).$$

**E. $\int \sec^n u\, du$ or $\int \csc^n u\, du$**

If $n$ is a positive even integer change the integrand by

$$\sec^n u = (\tan^2 u + 1)^{n-4} \sec^2 u$$

$$\csc^n u = (\cot^2 u + 1)^{n-4} \csc^2 u.$$

**F. $\int \sin ku \cos ru\, du,\ k \neq r$**

Use the identity

$$\sin A \cos B = \tfrac{1}{2}[\sin(A+B) + \sin(A-B)].$$

**G. $\int \sin ku \sin ru\, du,\ k \neq r$**

Use the identity

$$\sin A \sin B = \tfrac{1}{2}[\cos(A-B) - \cos(A+B)].$$

**H. $\int \cos ku \cos ru\, du,\ k \neq r$**

Use the identity

$$\cos A \cos B = \tfrac{1}{2}[\cos(A+B) + \cos(A-B)].$$

**Trigonometric Substitutions**

For $\sqrt{a^2 - u^2}$ let $u = a \sin z$.

For $\sqrt{a^2 + u^2}$ let $u = a \tan z$.

For $\sqrt{u^2 - a^2}$ let $u = a \sec z$.

Use a right triangle with $z$ as an acute angle to obtain the needed expressions for the other trigonometric functions.

**Quadratic Functions**

The form $ax^2 + bx + c$ is changed to the form $y^2 + k^2$ by completing the square.

$$a\left(x + \frac{b}{2a}\right)^2 + c - \frac{b^2}{4a} = y^2 + k^2 .$$

of the Pythagorean Theorem. Then $\sec z = \dfrac{\sqrt{u^2 + a^2}}{a}$ and

$$\sec z \tan z + \ln(\sec z + \tan z) = \frac{u}{a}\frac{\sqrt{u^2 + a^2}}{a} + \ln\left(\frac{\sqrt{u^2 + a^2}}{a} + \frac{u}{a}\right).$$

Substituting these expressions into the original integral,

$$\int \sqrt{u^2 + a^2}\, du = \frac{a^2}{2}\left[\frac{u\sqrt{u^2 + a^2}}{a^2} + \ln\left(\sqrt{u^2 + a^2} + u\right) - \ln a\right] + C$$

$$= \frac{u}{2}\sqrt{u^2 + a^2} + \frac{a^2}{2}\ln(u + \sqrt{u^2 + a^2}) + C$$

in which we have combined $\dfrac{-a^2}{2}\ln a$ with $C$, the constant of integration, to form a new $C$. For the integral involving $\sqrt{u^2 - a^2}$ we substitute $u = a \sec z$ and proceed in the same manner.

$$\int \sqrt{u^2 \pm a^2}\, du = \frac{u}{2}\sqrt{u^2 \pm a^2} \pm \frac{a^2}{2}\ln(u + \sqrt{u^2 \pm a^2}) + C$$

A formula for $\int \ln x\, dx$ can be obtained by integration by parts letting $u = \ln x$ and $dv = dx$ to obtain

$$\int \ln x\, dx = x\ln x - \int \frac{x}{x}\, dx = x \ln x - x + C.$$

We recommend that this formula also be included among the elementary forms.

Although the techniques derived in this chapter are powerful tools for evaluating integrals, the student should not forget the simple substitutions learned in the earlier chapters. As an example it is not necessary to use the formula for integration by parts to evaluate $\int \frac{\ln x}{x}\, dx$ since by letting $u = \ln x$ then $du = \dfrac{dx}{x}$ and we have the simple form $\int u\, du$ which may be used to yield $\frac{1}{2}\ln^2 x + C$.

## SOLUTIONS TO THE EXERCISES

I

i. $\int e^{2x}\sin 3x\,dx = I$

Let $u = e^{2x}$; $dv = \sin 3x\,dx$

$du = 2e^{2x}dx$; $v = -\frac{1}{3}\cos 3x$

$I = -\frac{1}{3}e^{2x}\cos 3x + \frac{2}{3}\int e^{2x}\cos 3x\,dx$

$u = e^{2x}$ and $dv = \cos 3x\,dx$

$du = 2e^{2x}dx$; $v = \frac{1}{3}\sin 3x$

$I = -\frac{1}{3}e^{2x}\cos 3x + \frac{2}{9}e^{2x}\sin 3x - \frac{4}{9}I$

$13I = e^{2x}(-3\cos 3x + 2\sin 3x) + C$

$I = \frac{e^{2x}}{13}(2\sin 3x - 3\cos 3x) + C$

ii. $\int \arctan x\,dx = I$

$u = \arctan x$ and $dv = dx$

$du = \frac{dx}{1+x^2}$ and $v = x$

$I = x\arctan x - \int \frac{x\,dx}{1+x^2}$

$= x\arctan x - \frac{1}{2}\ln(1+x^2) + C$

iii. $\int \sin x\cos 3x\,dx = I_1$

$u = \sin x$; $dv = \cos 3x\,dx$

$du = \cos x\,dx$; $v = \frac{1}{3}\sin 3x$

$I_1 = \frac{1}{3}\sin x\sin 3x$

$-\frac{1}{3}\int \cos x\sin 3x\,dx$

$u = \cos x$ and $dv = \sin 3x\,dx$

$du = -\sin x$ and $v = -\frac{1}{3}\cos 3x$

$I_2 = -\frac{1}{3}\left(-\frac{1}{3}\cos x\cos 3x - \frac{1}{3}I_1\right)$

$I_1 = \frac{1}{3}\sin x\sin 3x$

$+\frac{1}{9}\cos x\cos 3x + \frac{1}{9}I_1$

$8I_1 = 3\sin x\sin 3x + \cos x\cos 3x$

$I_1 = \frac{1}{8}(3\sin x\sin 3x$

$+\cos x\cos 3x) + C$

II

i. $\int \sin^2 x\cos^4 x\,dx = I$

$\sin^2 x\cos^4 x = \cos^4 x - \cos^6 x$

$= \frac{1}{4}(1+\cos 2x)^2 - \frac{1}{8}(1+\cos 2x)^3$

$= \frac{1}{4}(1 + 2\cos 2x + \cos^2 2x)$

$-\frac{1}{8}(1 + 3\cos 2x + 3\cos^2 2x$

$+\cos^3 2x)$

$= \frac{1}{8} + \frac{1}{8}\cos 2x - \frac{1}{8}\cos^2 2x - \frac{1}{8}\cos^3 2x$

$= \frac{1}{8}\left[1 + \cos 2x - \frac{1}{2}(1+\cos 4x)\right.$

$\left. - \cos 2x(1 - \sin^2 2x)\right]$

$= \frac{1}{16}[1 - \cos 4x + 2\sin^2 2x\cos 2x]$

$I = \frac{1}{16}x - \frac{1}{64}\sin 4x + \frac{1}{16}\frac{\sin^3 2x}{3} + C$

$= \frac{1}{16}x - \frac{1}{64}\sin 4x + \frac{1}{48}\sin^3 2x + C$

ii. $\int \cot^2 x\,dx = \int \csc^2 x\,dx - \int dx$

$= -\cot x - x + C$

III

i. $\int \frac{dx}{x^4\sqrt{x^2-5}} = I$

Let $x = \sqrt{5}\sec z$, $x^4 = 25\sec^4 z$

$\sqrt{x^2-5} = \sqrt{5}\tan z$

$dx = \sqrt{5}\sec z\tan z\,dz$

$\text{Integrand} = \frac{\sqrt{5}\sec z\tan z\,dz}{25\sec^4 z\sqrt{5}\tan z}$

$= \frac{\cos^3 z}{25}dz$

$I = \frac{1}{25}\int \cos^3 z\,dz$

$= \frac{1}{25}\left[\sin z - \frac{\sin^3 z}{3}\right] + C$

$= \frac{1}{25}\left(\frac{\sqrt{x^2-5}}{x} - \frac{(x^2-5)^{\frac{3}{2}}}{3x^3}\right) + C$

$= \frac{\sqrt{x^2-5}}{x^3}\left(\frac{2x^2+5}{75}\right) + C$

# 18 METHODS OF INTEGRATION II

## SELF-TEST

DIRECTIONS: Write your answers in the numbered regions to the right. To check your answers, turn the page. Study the solutions to the problems you missed, and do the exercises following any problem or group of problems in which you had errors.

Evaluate the following integrals.

1. $\displaystyle\int \frac{x^3 + 2}{x^3 - x}\,dx$

2. $\displaystyle\int \frac{(3x^2 + 4x + 2)\,dx}{x^3 + 3x^2 + 4x + 2}$

3. $\displaystyle\int \frac{(3x^2 - 10x + 5)\,dx}{6x^3 + 11x^2 - 80x + 75}$

4. $\displaystyle\int \frac{x^3 + 3x^2 + 3}{(x^2 + 3x - 2)^2}\,dx$

5. $\displaystyle\int \frac{x^{\frac{3}{4}}\,dx}{1 + \sqrt{x}}$

6. $\displaystyle\int \frac{dx}{x^3\sqrt{1 + x^2}}$

7. $\displaystyle\int_0^{\frac{\pi}{2}} \frac{dx}{2 + \sin x}$

1

2

3

4

5

6

7

## SOLUTIONS

> See Additional Information for methods used to find the values of $A$, $B$, $C$, etc.

1. $\int \frac{x^3+2}{x^3-x}\,dx = \int dx + \int \frac{x+2}{x^3-x}\,dx = I$

By factoring the denominator

$$\frac{x+2}{x^3-x} = \frac{A}{x} + \frac{B}{x-1} + \frac{C}{x+1}.$$

$A = -2,\ B = \frac{3}{2},\ C = \frac{1}{2}$

$$I = x - 2\int \frac{dx}{x} + \frac{3}{2}\int \frac{dx}{x-1} + \frac{1}{2}\int \frac{dx}{x+1}$$

$$= x - 2\ln x + \frac{3}{2}\ln(x-1) + \frac{1}{2}\ln(x+1) + C$$

2. $\int \frac{(3x^2+4x+2)\,dx}{x^3+3x^2+4x+2} = \int \frac{P(x)}{Q(x)}\,dx = I$

$Q(x) = (x+1)(x^2+2x+2)$

$$\frac{P(x)}{Q(x)} = \frac{A}{x+1} + \frac{Bx+C}{x^2+2x+2}$$

$A = 1,\ B = 2,\ C = 0$

$$\int \frac{dx}{x+1} = \ln(x+1) + c_1 = I_1$$

$$\int \frac{2x\,dx}{x^2+2x+2} = I_2 + I_3$$

$$I_2 = \int \frac{(2x+2)\,dx}{x^2+2x+1}$$

$$= \ln(x^2+2x+2) + c_2$$

$$I_3 = -\int \frac{2\,dx}{(x+1)^2+1}$$

$$= -2\text{ Arctan }(x+1) + c_3$$

$$I = I_1 + I_2 + I_3 + C$$

3. $\int \frac{P(x)}{Q(x)}\,dx = I$ where

$P(x) = 3x^2 - 10x + 5$ and

$Q(x) = 6x^3 + 11x^2 - 80x + 75$

$= (x+5)(2x-3)(3x-5)$

$$\frac{P(x)}{Q(x)} = \frac{A}{x+5} + \frac{B}{2x-3} + \frac{C}{3x-5}$$

$A = \frac{1}{2},\ B = 1,\ C = -\frac{3}{2}$

$$I = \frac{1}{2}\ln(x+5) + \frac{1}{2}\ln(2x-3) - \frac{1}{2}\ln(3x-5) + C$$

$$= \frac{1}{2}\ln\frac{(x+5)(2x-3)}{3x-5} + C$$

4. $\int \frac{(x^3+3x^2+3)\,dx}{(x^2+3x-2)^2} = I$

$$\frac{P(x)}{Q(x)} = \frac{Ax+B}{x^2+3x-2} + \frac{Cx+D}{(x^2+3x-2)^2}$$

$A = 1,\ B = 0,\ C = 2,\ D = 3$

$$\int \frac{x\,dx}{x^2+3x-2} = \frac{1}{2}\int \frac{(2x+3)\,dx}{x^2+3x-2} - \frac{3}{2}\int \frac{dx}{\left(x+\frac{3}{2}\right)^2 - \frac{17}{4}}$$

$$= \frac{1}{2}\ln(x^2+3x-2) - \frac{3}{2}\left(\frac{1}{\sqrt{17}}\right)\ln\frac{2x+3-\sqrt{17}}{2x+3+\sqrt{17}}$$

$$\int \frac{(2x+3)\,dx}{(x^2+3x-2)^2} = -(x^2+3x-2)^{-1}$$

$$I = \frac{1}{2}\ln(x^2+3x-2) - (x^2+3x-2)^{-1} - \frac{3}{2\sqrt{17}}\ln\frac{2x+3-\sqrt{17}}{2x+3+\sqrt{17}} + C$$

> **EXERCISES I**
>
> Evaluate the integrals.
>
> i. $\int \frac{(x^2+x+4)\,dx}{x^3-x^2-8x+12}$
>
> ii. $\int \frac{(2x-5)\,dx}{x^2-5x-14}$
>
> iii. $\int \frac{(10x^2-10x+4)\,dx}{(x+1)(x^3+1)}$

5. $\int \frac{x^{\frac{3}{4}}\,dx}{1+\sqrt{x}} = I$

Let $u = x^{\frac{1}{4}},\ u^2 = \sqrt{x},\ dx = 4u^3\,du$

$$I = \int \frac{u^3}{1+u^2}(4u^3\,du)$$

$$= 4\int \left(u^4 - u^2 + 1 - \frac{1}{1+u^2}\right)du$$

$$= 4\left(\frac{u^5}{5} - \frac{u^3}{3} + u - \text{Arctan }u\right) + C$$

$$= \frac{4}{5}x^{\frac{5}{4}} - \frac{4}{3}x^{\frac{3}{4}} + 4x^{\frac{1}{4}} - 4\text{ Arctan }x^{\frac{1}{4}} + C$$

6. $\int \frac{dx}{x^3\sqrt{1+x^2}} = I$

Let $u^2 = 1 + x^2$, $u\,du = x\,dx$,
$x^2 = u^2 - 1$, $x^4 = (u^2 - 1)^2$

$$I = \int \frac{x\,dx}{x^4\sqrt{1 + x^2}} = \int \frac{u\,du}{u(u^2 - 1)^2}$$

Use partial fractions and integrate.

$$I = \tfrac{1}{4}\ln(u + 1) - \tfrac{1}{4}\ln(u - 1) - \tfrac{1}{4}\frac{1}{u + 1} - \tfrac{1}{4}\frac{1}{u - 1} + C$$

$$= \tfrac{1}{4}\ln\frac{u + 1}{u - 1} - \tfrac{1}{4}\frac{2u}{u^2 - 1} + C$$

$$= \tfrac{1}{2}\ln\frac{\sqrt{1 + x^2} + 1}{x} - \tfrac{1}{2}\frac{\sqrt{1 + x^2}}{x^2} + C$$

**EXERCISES II**

i. $\displaystyle\int \frac{(3x - 1)\,dx}{(x + 2)^{\frac{2}{3}}}$

ii. $\displaystyle\int_0^1 \frac{dx}{\sqrt{x}\,(9 + \sqrt[3]{x})}$

7. $\displaystyle\int_0^{\frac{\pi}{2}} \frac{dx}{2 + \sin x} = I$

$$2 + \sin x = 2 + \frac{2z}{1 + z^2}$$

$$dI = \frac{1 + z^2}{2(z^2 + z + 1)}\;\frac{2dz}{1 + z^2}$$

$$= \frac{dz}{\left(z + \frac{1}{2}\right)^2 + \frac{3}{4}}$$

$$I = \frac{2}{\sqrt{3}}\operatorname{Arctan}\frac{\left(z + \frac{1}{2}\right)2}{\sqrt{3}}\Bigg|_a^b$$

Since $z = \tan\frac{x}{2}$, we have $z = 0$ at $x = 0$ and $z = 1$ at $x = \frac{\pi}{2}$ so that $a = 0$ and $b = 1$.

$$I = \frac{2}{\sqrt{3}}\left(\operatorname{Arctan}\sqrt{3} - \operatorname{Arctan}\frac{1}{\sqrt{3}}\right)$$

$$= \frac{2}{\sqrt{3}}\left(\frac{\pi}{3} - \frac{\pi}{6}\right) = \frac{\pi}{3\sqrt{3}}$$

Solutions to the exercises are on page 108.

**ANSWERS**

1. $x - 2\ln x + \frac{3}{2}\ln(x - 1) + \frac{1}{2}\ln(x + 1) + C$ — 1

2. $\ln(x + 1) + \ln(x^2 + 2x + 2) - 2\operatorname{Arctan}(x + 1) + C$ — 2

3. $\frac{1}{2}\ln\frac{(x + 5)(2x - 3)}{3x - 5} + C$ — 3

4. $\frac{1}{2}\ln(x^2 + 3x - 2) - (x^2 + 3x - 2)^{-1} - \frac{3}{2\sqrt{17}}\ln\frac{2x + 3 - \sqrt{17}}{2x + 3 + \sqrt{17}} + C$ — 4

5. $\frac{4}{5}x^{\frac{5}{4}} - \frac{4}{3}x^{\frac{3}{4}} + 4x^{\frac{1}{4}} - 4\operatorname{Arctan}x^{\frac{1}{4}} + C$ — 5

6. $\frac{1}{2}\ln\frac{\sqrt{1 + x^2} + 1}{x} - \frac{1}{2}\frac{\sqrt{1 + x^2}}{x^2} + C$ — 6

7. $\frac{\pi}{3\sqrt{3}}$ — 7

**EXERCISES III**

i. $\displaystyle\int_0^{\frac{\pi}{2}} \frac{dx}{3 + 5\sin x}$

ii. $\displaystyle\int_a^b \frac{\sin x\,dx}{1 + \sin x}$

$a = \frac{\pi}{3}$ and $b = \frac{\pi}{2}$

## BASIC FACTS

**Rational Functions.** If $P(x)$ and $Q(x)$ are polynomials in the variable $x$, then $\frac{P(x)}{Q(x)}$ is called a rational function. The integration of a function of this form depends upon the ability to factor $Q(x)$ into the product of linear and quadratic factors each of which has real coefficients. A theorem to this effect is proved in advanced mathematics. A rational function can be decomposed into the sum of partial fractions whose denominators are factors of $Q(x)$. These fall into four cases.

CASE I. **Distinct linear factors**

Fractions of form $\frac{A_i}{x - a_i}$

CASE II. **Repeated linear factors**

Fractions of form $\frac{A_i}{(x - a_j)^i}$

CASE III. **Distinct quadratic factors**

Fractions of form $\frac{A_ix + B_i}{x^2 + b_ix + c_i}$

CASE IV. **Repeated quadratic factors**

Fractions of form $\frac{A_ix + B_i}{(x^2 + bx + c)^i}$

The rational function $\frac{P(x)}{Q(x)}$ can be decomposed into any one or combinations of these four cases. The values for $A_i$ and $B_i$ are found by equating the coefficients of like powers of $x$ contained in $P(x)$ to those that are in the numerator of the partial fractions when they are combined over the lowest common denominator (which is $Q(x)$). This is based on the theorem that two polynomials in $x$ are equal for all $x$ if and only if the coefficients of like powers are equal.

The **integral of the rational function** is expressed as the sum of the integrals of its partial fractions. The procedure is:

1. Reduce the degree of $P(x)$ to be

## ADDITIONAL INFORMATION

Theoretically any rational function of the form $\frac{P(x)}{Q(x)}$ where $P(x)$ and $Q(x)$ are polynomials can be integrated. The difficulty arises in finding the factors of $Q(x)$ and the numerators of the corresponding **partial fractions**. The linear factors of $Q(x)$ may be found by the Factor Theorem, "If $r$ is a zero of the polynomial $Q(x)$, then $(x - r)$ is a factor of $Q(x)$," and synthetic division to find the zeros. Consider, for example, $Q(x) = 2x^4 + 3x^3 - 3x^2 - 5x - 6$. The rational zeros of $Q(x)$ are contained in the ratios formed by the factors of 6 divided by the factors of 2, thus $\pm 1, \pm 2, \pm 3, \pm 6, \pm\frac{1}{2}, \pm\frac{3}{2}$. As we try these by synthetic division we find

$$\begin{array}{rrrrr|l} 2 & 3 & -3 & -5 & -6 & \underline{-2} \\ & -4 & 2 & 2 & 6 & \\ \hline 2 & -1 & -1 & -3 & 0 & \\ & 3 & 3 & 3 & & \underline{\tfrac{3}{2}} \\ \hline 2 & 2 & 2 & 0 & & \end{array}$$

Therefore $x + 2$ is a factor.

Therefore $x - \frac{3}{2}$ is a factor.

The remaining factor is $2x^2 + 2x + 2$ which has no linear factors. Thus $Q(x) = 2\left(x - \frac{3}{2}\right)(x + 2)(x^2 + x + 1)$, and we move the coefficient 2 into the first factor and write $(2x - 3)$.

There are two methods that may be used to obtain the constants in the numerators of the partial fractions. We shall illustrate each method by considering the following example.

$$\frac{P(x)}{Q(x)} = \frac{9x^3 + 12x^2 + 6x}{(x + 2)(x - 1)(x^2 + x + 1)} = \frac{A}{x + 2} + \frac{B}{x - 1} + \frac{Cx + D}{x^2 + x + 1}$$

Combine the last expression over its lowest common denominator and set the two numerators equal to each other.

$$\begin{aligned} &9x^3 + 12x^2 + 6x \\ &= A(x - 1)(x^2 + x + 1) + B(x + 2)(x^2 + x + 1) + (Cx + D)(x + 2)(x - 1) \\ &= (A + B + C)x^3 + (3B + C + D)x^2 + (3B - 2C + D)x - A + 2B - 2D \end{aligned}$$

**Method I.** Equate the coefficients of like powers of $x$.

$$\begin{array}{rl|rl} x^3: & (1)\ A + B + C = 9 & (2) - (3): & 3C = 6,\ C = 2 \\ x^2: & (2)\ 3B + C + D = 12 & 2(3) + (4): & -A + 8B = 12 + 4C = 20 \\ x: & (3)\ 3B - 2C + D = 6 & (1): & A + B = 9 - C = 7 \\ \hline x_0: & (4)\ -A + 2B - 2D = 0 & \text{Add:} & 9B = 27,\ B = 3 \end{array}$$

Then $A = 9 - B - C = 9 - 3 - 2 = 4$; $D = 12 - 3B - C = 1$.

**Method II.** Substitute values of $x$ into both sides of the equation. First use values which will make a linear factor equal to zero; then use other small integral values including $x = 0$. For the right member use the first expression.

$$x = 1:\ 9 + 12 + 6 = 27 = 0 + B(3)(3) + 0 = 9B;\ B = 3$$

$$x = -2:\ -72 + 48 - 12 = -36 = A(-3)(3) = -9A;\ A = 4$$

$$x = 0:\ 0 = -A + 2B - 2D;\ 2D = 2B - A = 6 - 4 = 2;\ D = 1$$

less than the degree of $Q(x)$ by long division.

2. Find the partial fractions.
3. Integrate each partial fraction.

If the fraction contains **repeated linear factors** in the denominator, then the decomposition has the form

$$\frac{B}{(x-a)^i} = \frac{A_1}{x-a} + \frac{A_2}{(x-a)^2} + \cdots + \frac{A_i}{(x-a)^i}.$$

Repeated quadratic factors are treated in a similar manner.

**Rationalizing.** If integrand contains

1) fractional powers of $x$ only, let $x = z^n$ where $n$ is the LCD of the fractional exponents of $x$.
2) fractional powers of $(a + bx)$ only, let $a + bx = z^n$ where $n$ is the LCD of the fractional exponents of $(a + bx)$.
3) expression of the form $x^m(a + bx^n)^{\frac{r}{s}}$ and
   a) $\frac{m+1}{n}$ is an integer or zero let $a + bx^n = z^s$.
   b) $\frac{m+1}{n} + \frac{r}{s}$ is an integer or zero let $a + bx^n = z^s x^n$.

**Trigonometric Differentials**

A trigonometric differential can be transformed into another form by

$$\sin u = \frac{2z}{1+z^2}, \quad \cos u = \frac{1-z^2}{1+z^2}$$

$$du = \frac{2dz}{1+z^2}, \text{ and } z = \tan\frac{u}{2}.$$

The last formula is used in changing the results from expression in $z$ back to the original variable $u$. It may also be used for changing the limits of integration in a definite integral. If $a$ and $b$ are the limits for the variable $u$, then $c = \tan\frac{a}{2}$ and $d = \tan\frac{b}{2}$ are the limits for the variable $z$.

$$x = -1: -9 + 12 - 6 = -3 = -2A + B + 2C - 2D;$$
$$2C = 2D + 2A - B - 3 = 2 + 8 - 3 - 3 = 4; \; C = 2.$$

It is also possible to use a combination of the two methods.

**Rationalizing the Integrand**

The most general binomial differential is of the form $dI = x^m(a + bx^n)^{\frac{r}{s}}\,dx$ where $m$, $n$, $r$, and $s$ are integers and $n > 0$. There are two substitutions which may be applied to rationalize this differential.

CASE I. Let $a + bx^n = z^s$, then $(a + bx^n)^{\frac{r}{s}} = z^r$ and

$$x = \left(\frac{z^s - a}{b}\right)^{\frac{1}{n}}, \quad x^m = \left(\frac{z^s - a}{b}\right)^{\frac{m}{n}}, \quad dx = \frac{s}{bn} z^{s-1} \left(\frac{z^s - a}{b}\right)^{\frac{1}{n}-1} dz$$

$$dI = \frac{s}{bn} z^{r+s-1} \left(\frac{z^s - a}{b}\right)^{\frac{m+1}{n}-1} dz$$ which is a rational expression if $\frac{m+1}{n}$ is an integer or zero.

CASE II. Let $a + bx^n = x^m z^s$, then $x^m = \frac{a}{z^s - b}$ and

$$a + bx^n = \frac{az^s}{z^s - b}, \quad (a + bx^n)^{\frac{r}{s}} = a^{\frac{r}{s}} (z^s - b)^{-\frac{r}{s}} z^r$$

$$x = \left(\frac{a}{z^s - b}\right)^{\frac{1}{n}}, \quad dx = -\frac{s}{n} a^{\frac{1}{n}} z^{s-1} (z^s - b)^{-\frac{1}{n}-1} dz$$

When these are substituted into the expression for $dI$, we obtain a factor $(z^s - b)^{-\left(\frac{m+1}{n} + \frac{r}{s} + 1\right)}$ which is a rational expression if $\frac{m+1}{n} + \frac{r}{s}$ is an integer or zero.

In applying these substitutions to a given integral, we check the condition first and then apply the appropriate substitution. We summarize by considering some examples.

| integrand | $m$ | $n$ | $\frac{r}{s}$ | condition | substitution |
|---|---|---|---|---|---|
| $\frac{x^3}{(1+x^3)^{\frac{4}{3}}}$ | 3 | 3 | $-\frac{4}{3}$ | $\frac{m+1}{n} + \frac{r}{s} = \frac{4}{3} - \frac{4}{3} = 0$ | $1 + x^3 = x^3z^3$ |
| $x^5(7 + x^3)^{\frac{3}{2}}$ | 5 | 3 | $\frac{3}{2}$ | $\frac{m+1}{n} = \frac{5+1}{3} = 2$ | $7 + x^3 = z^2$ |
| $\frac{2\sqrt{1+x^4}}{x^3}$ | $-3$ | 4 | $\frac{1}{2}$ | $\frac{m+1}{n} + \frac{r}{s} = \frac{-2}{4} + \frac{1}{2} = 0$ | $1 + x^4 = x^4z^2$ |

The methods of integration discussed in the last two chapters will not of course cover all possible integrands. Tables of integrals have been published and are used for extensive integration. An abbreviated table is given in the Appendix on page 157.

## SOLUTIONS TO EXERCISES

I

i. $$\int \frac{(x^2 + x + 4)\,dx}{x^3 - x^2 - 8x + 12} = I$$

$$\frac{P(x)}{Q(x)} = \frac{A}{x-2} + \frac{B}{(x-2)^2} + \frac{C}{x+3}$$

$A = \frac{3}{5}$, $B = 2$, $C = \frac{2}{5}$

$$\frac{3}{5}\int \frac{dx}{x-2} = \frac{3}{5}\ln(x-2) + c_1$$

$$2\int \frac{dx}{(x-2)^2} = -2(x-2)^{-1} + c_2$$

$$\frac{2}{5}\int \frac{dx}{x+3} = \frac{2}{5}\ln(x+3) + c_3$$

$$I = \frac{1}{5}\ln(x-2)^3(x+3)^2 - \frac{2}{x-2} + C$$

ii. $$\int \frac{(2x-5)\,dx}{x^2 - 5x - 14} = I$$

$$\frac{P(x)}{Q(x)} = \frac{1}{x-7} + \frac{1}{x+2}$$

$$I = \int \frac{dx}{x-7} + \int \frac{dx}{x+2}$$

$$= \ln(x-7)(x+2) + C$$

iii. $$\int \frac{(10x^2 - 10x + 4)\,dx}{(x+1)(x^3+1)}$$

$$\frac{P(x)}{Q(x)} = -\frac{2}{x+1} + \frac{8}{(x+1)^2} + \frac{2x-2}{x^2-x+1}$$

$$\int \frac{2\,dx}{x+1} = 2\ln(x+1) = I_1$$

$$\int \frac{8\,dx}{(x+1)^2} = -8(x+1)^{-1} = I_2$$

$$\int \frac{(2x-1)\,dx}{x^2-x+1} = \ln(x^2-x+1) = I_3$$

$$\int \frac{dx}{\left(x-\frac{1}{2}\right)^2 + \frac{3}{4}} = \frac{1}{\sqrt{3}}\ln\frac{2x-1-\sqrt{3}}{2x-1+\sqrt{3}} = I_4$$

$$I = -I_1 + I_2 + I_3 - I_4 + C$$

II

i. $$\int \frac{(3x-1)\,dx}{(x+2)^{\frac{2}{3}}} = I$$

Let $u = (x+2)^{\frac{1}{3}}$, $x = u^3 - 2$,
$dx = 3u^2\,du$, $3x - 1 = 3u^3 - 7$

$$I = \int \frac{(3u^3-7)\,3u^2\,du}{u^2} = 3\int (3u^3 - 7)\,du$$

$$= \frac{9}{4}u^4 - 21u + C$$

$$= \frac{9}{4}(x+2)^{\frac{4}{3}} - 21(x+2)^{\frac{1}{3}} + C$$

ii. $$\int_0^1 \frac{dx}{\sqrt{x}\,(9 + \sqrt[3]{x})} = I$$

Let $u = x^{\frac{1}{6}}$, $dx = 6u^5\,du$

$$I = 6\int_0^1 \frac{u^2\,du}{9+u^2}$$

$$= 6u - 54\left(\frac{1}{3}\text{Arctan}\,\frac{u}{3}\right)\Big|_0^1$$

$$= 6 - 18\,\text{Arctan}\,\frac{1}{3}$$

III

i. $$\int_0^{\frac{\pi}{2}} \frac{dx}{3 + 5\sin x} = I$$

$$dI = \frac{(1+z^2)(2\,dz)}{(3z^2+10z+3)(1+z^2)}$$

$$= \frac{3}{4}\frac{dz}{3z+1} - \frac{1}{4}\frac{dz}{z+3}$$

$$I = \frac{1}{4}\ln(3z+1) - \frac{1}{4}\ln(z+3)\Big|_0^1$$

$$= \frac{1}{4}(\ln 4 - \ln 1 - \ln 4 + \ln 3)$$

$$= \frac{1}{4}\ln 3$$

ii. $$\int_a^b \frac{\sin x\,dx}{1 + \sin x} = I,\ a = \frac{\pi}{3},\ b = \frac{\pi}{2}$$

$$dI = \frac{4z\,dz}{(1+z^2)(z+1)^2} = \frac{P(z)}{Q(z)}\,dz$$

$$\frac{P(z)}{Q(z)} = \frac{A}{z+1} + \frac{B}{(z+1)^2} + \frac{Cz+D}{z^2+1}$$

$A = C = 0$, $B = -2$, $D = 2$

$$\int_c^d dI = \int_c^d \frac{-2\,dz}{(z+1)^2} + \int_c^d \frac{2\,dz}{z^2+1}$$

$$= \frac{2}{z+1} + 2\,\text{Arctan}\,z\,\Big|_c^d$$

$$= \sqrt{3} - 2 + \frac{\pi}{6}$$

as $c = \tan\frac{\pi}{6} = \frac{\sqrt{3}}{3}$, $d = \tan\frac{\pi}{4} = 1$

# 19 APPLICATIONS OF INTEGRATION

## SELF-TEST

DIRECTIONS: Write your answers in the numbered regions to the right. To check your answers, turn the page. Study the solutions to the problems you missed, and do the exercises following any problem or group of problems in which you had errors.

It is recommended that the student sketch the curves and the solids involved in the problems.

1. Find the volume of the solid generated by revolving the region bounded by $y = 2x - x^2$, $y = 0$ about $X$-axis.

2. Find the volume of the solid generated by revolving the region bounded by $y = 4\sqrt{2x}$ and $y = x^3$ about the $Y$-axis.

3. Find the area of the surface formed by revolving the upper half of the hypocycloid $x^{\frac{2}{3}} + y^{\frac{2}{3}} = a^{\frac{2}{3}}$ about the $X$-axis.

4. Find the mean value of $f(x) = \ln x$ in $[1, 5]$.

5. Find the centroid of the area bounded by $y = \sin x$ and the lines $x = 0$, $x = \frac{\pi}{2}$, and $y = 0$.

6. Find the centroid of the solid formed by revolving about the $X$-axis the area bounded by the lines $x = 0$, $x = a$, $y = 0$ and the hyperbola

$$b^2x^2 - a^2y^2 + a^2b^2 = 0.$$

7. Find the $y$ value of the centroid of the arc of the cycloid $x = a(\theta + \sin\theta)$, $y = a(1 - \cos\theta)$ for $0 \leq \theta \leq \pi$.

8. Use Simpson's Rule with $n = 10$ to approximate the value of $\int_0^{10} x^3\,dx$.

1

2

3

4

5

6

7

8

## SOLUTIONS

1. The curve $y = 2x - x^2$ intersects $y = 0$ at $x = 0$ and $x = 2$.

$$V = \pi \int_0^2 y^2\,dx = \pi \int_0^2 (2x - x^2)^2\,dx$$

$$= \pi \left(4\,\frac{x^3}{3} - 4\,\frac{x^4}{4} + \frac{x^5}{5}\right)\Bigg|_0^2$$

$$= 16\pi \left(\tfrac{2}{3} - 1 + \tfrac{2}{5}\right) = \frac{16\pi}{15}$$

2. The curves $y = 4\sqrt{2x}$ and $y = x^3$ intersect at $(0,0)$ and $(2,8)$. We use the shell method.

$$V = 2\pi \int_0^2 x\,(4\sqrt{2x} - x^3)\,dx$$

$$= 2\pi \left[4\sqrt{2}\,x^{\frac{5}{2}}\left(\tfrac{2}{5}\right) - \frac{x^5}{5}\right]\Bigg|_0^2$$

$$= 2\pi \left[4\sqrt{2}\,(4\sqrt{2})\left(\tfrac{2}{5}\right) - \tfrac{32}{5}\right] = \frac{64\pi}{5}$$

> **EXERCISES I**
>
> Find the volume of the solids generated by revolving the areas:
>
> i. Bounded by $y = \cos 2x$, $x = 0$, $x = \frac{\pi}{4}$, $y = 0$ about the $X$-axis.
>
> ii. In the first quadrant and bounded by $x^2 + y^2 = 25$, $x + y = 5$ about the $Y$-axis.

3. The hypocycloid $x^{\frac{2}{3}} + y^{\frac{2}{3}} = a^{\frac{2}{3}}$ is defined in the interval $[-a, a]$ and is symmetric about the $Y$-axis. By implicit differentiation:

$$y' = -\frac{y^{\frac{1}{3}}}{x^{\frac{1}{3}}}.$$

$$ds = \sqrt{1 + (y')^2}\,dx = x^{-\frac{1}{3}}\,a^{\frac{1}{3}}\,dx$$

$$A(S) = 2\,(2\pi) \int_0^a y\,ds$$

$$= 4\pi \int_0^a (a^{\frac{2}{3}} - x^{\frac{2}{3}})^{\frac{3}{2}}\,ds$$

$$= -6\pi\,a^{\frac{1}{3}} \left[\tfrac{2}{5}\,(a^{\frac{2}{3}} - x^{\frac{2}{3}})^{\frac{5}{2}}\right]_0^a$$

$$= \frac{12}{5}\,\pi\,a^2$$

> **EXERCISES II**
>
> Find the area of the surfaces generated by:
>
> i. Revolving the arc $y^2 = 4x$ from $(3, 2\sqrt{3})$ to $(8, 4\sqrt{2})$ about $X$-axis.
>
> ii. Revolving about the $X$-axis the arc of $x = e^t \cos t$, $y = e^t \sin t$ from $t = 0$ to $t = \frac{\pi}{2}$.

4. $f(x) = \ln x$ in the interval $[1, 5]$.

$$f_m = \frac{1}{5 - 1} \int_1^5 \ln x\,dx$$

$$= \tfrac{1}{4}\,[x \ln x - x]_1^5 = \tfrac{5}{4} \ln 5 - 1$$

> **EXERCISES III**
>
> Find the mean value of the functions.
>
> i. $f(x) = \sqrt{r^2 - x^2}$ in $[0, r]$
>
> ii. $f(x) = \tan x$ in $\left[0, \frac{\pi}{4}\right]$

5. $y = \sin x$, $x = 0$, $x = \frac{\pi}{2}$, and $y = 0$.

$$M_y = \int_0^{\frac{\pi}{2}} xy\,dx = \int_0^{\frac{\pi}{2}} x \sin x\,dx$$

$$= \sin x - x \cos x \,\Big|_0^{\frac{\pi}{2}} = 1$$

$$M_x = \tfrac{1}{2} \int_0^{\frac{\pi}{2}} y^2\,dx = \tfrac{1}{2} \int_0^{\frac{\pi}{2}} \sin^2 x\,dx$$

$$= \tfrac{1}{2}\left(\tfrac{x}{2} - \tfrac{1}{4} \sin 2x\right)\Bigg|_0^{\frac{\pi}{2}} = \frac{\pi}{8}$$

$$A = \int_0^{\frac{\pi}{2}} y\,dx = \int_0^{\frac{\pi}{2}} \sin x\,dx = 1$$

$$\overline{x} = \frac{M_y}{A} = \tfrac{1}{1} = 1; \qquad \overline{y} = \frac{M_x}{A} = \frac{\pi}{8}$$

6. $b^2x^2 - a^2y^2 + a^2b^2 = 0$, $x = 0$, $x = a$, and $y = 0$. Since we are revolving about the $X$-axis, $\overline{y} = 0$.

$$M_y = \int_0^a \pi\,xy^2\,dx = \pi \int_0^a \frac{b^2}{a^2}\,(x^3 + a^2x)\,dx$$

$$= \frac{b^2}{a^2}\,\pi \left[\frac{x^4}{4} + \frac{a^2x^2}{2}\right]_0^a = a^2b^2\pi\,\left(\tfrac{3}{4}\right)$$

$$V = \int_0^a \frac{\pi}{a^2}[b^2x^2 + a^2b^2]\,dx$$

$$= \frac{b^2}{a^2}\pi\left[\frac{x^3}{3} + a^2x\right]_0^a = ab^2\pi\left(\frac{4}{3}\right)$$

$$\overline{x} = \frac{M_y}{V} = \frac{a^2b^2\pi\left(\frac{3}{4}\right)}{ab^2\pi\left(\frac{4}{3}\right)} = \frac{9}{16}a$$

7. $y = a(1 - \cos\theta)$, $x = a(\theta + \sin\theta)$ for the interval $0 \leq \theta \leq \pi$.
$y' = a\sin\theta$ and $x' = a(1 + \cos\theta)$
$ds = \sqrt{(y')^2 + (x')^2}\,d\theta$
$ds = a\sqrt{2}\sqrt{1 + \cos\theta}\,d\theta = 2a\cos\frac{\theta}{2}\,d\theta$

$$s = 2a\int_0^\pi \cos\frac{\theta}{2}\,d\theta = 4a\left[\sin\frac{\theta}{2}\right]_0^\pi$$

$$= 4a$$

$$\int_0^\pi y\,ds = 2a^2\int_0^\pi (1 - \cos\theta)\cos\frac{\theta}{2}\,d\theta$$

$$= 4a^2\int_0^\pi \sin^2\frac{\theta}{2}\cos\frac{\theta}{2}\,d\theta$$

$$= \left|\frac{8a^2}{3}\sin^3\frac{\theta}{2}\right|_0^\pi = \frac{8a^2}{3}$$

$$\overline{y} = \frac{8a^2}{3}\cdot\frac{1}{4a} = \frac{2}{3}a$$

8. $\int_0^{10} x^3\,dx$, $n = 10$

| $x$ | $f(x)$ | $S$ | Product |
|---|---|---|---|
| 0 | 0 | 1 | 0 |
| 1 | 1 | 4 | 4 |
| 2 | 8 | 2 | 16 |
| 3 | 27 | 4 | 108 |
| 4 | 64 | 2 | 128 |
| 5 | 125 | 4 | 500 |
| 6 | 216 | 2 | 432 |
| 7 | 343 | 4 | 1372 |
| 8 | 512 | 2 | 1024 |
| 9 | 729 | 4 | 2916 |
| 10 | 1000 | 1 | 1000 |

Sum = 7500

Since $\Delta x = x_2 - x_1 = 1$

$$\int_0^{10} x^3dx = \frac{1}{3}(\text{sum}) = \frac{7500}{3} = 2500$$

**ANSWERS**

1. $\frac{16\pi}{15}$
2. $\frac{64\pi}{5}$
3. $\frac{12}{5}\pi a^2$
4. $\frac{5}{4}\ln 5 - 1$
5. $\overline{x} = 1$, $\overline{y} = \frac{\pi}{8}$
6. $\overline{x} = \frac{9}{16}a$, $\overline{y} = 0$.
7. $\overline{y} = \frac{2}{3}a$
8. 2500

1 2 3 4 5 6 7 8

EXERCISE IV

Find the indicated coordinates of the centroid.

i. $\overline{x}$ of the arc of $x = \frac{1}{2}t^2$ and $y = \frac{1}{3}(2t + 1)^{\frac{3}{2}}$, $0 \leq t \leq 4$.

Solutions to the exercises are on page 114.

## BASIC FACTS

**Volume—Disc Method.** Let $f(x) \geq 0$ for $a \leq x \leq b$ and let a region (area) $R$ be defined by $a \leq x \leq b$ and $0 \leq y \leq f(x)$. Then the volume of the solid $S$ obtained by revolving $R$ about the $X$-axis is given by $V = \pi \int_a^b [f(x)]^2 dx$. Similarly, for a rotation of $R'$: $c \leq y \leq d$, $0 \leq x \leq g(y)$ about the $Y$-axis, we have $V = \pi \int_c^d [g(y)]^2 dy$.

**Volume—Shell Method.** If a region $R$ is revolved about the $Y$-axis, then

$$V = 2\pi \int_a^b x f(x)\, dx.$$

If $R'$ is revolved about the $X$-axis, then $V = 2\pi \int_c^d y\, g(y)\, dy$.

**Areas of Surfaces of Revolution**

If an arc of a continuous curve $C$ is revolved about a line, it generates a surface, and the area is given by

$$A(S) = 2\pi \int_C r ds$$

where $r$ is the distance from the line to a point of the curve $C$ and $ds$ is the differential of arc length. If $C$ is defined by $y = f(x)$, $a \leq x \leq b$ is revolved about the $X$-axis, then

$$A(S) = 2\pi \int_a^b f(x)\sqrt{1 + [f'(x)]^2}dx.$$

If $C$ is revolved about the $Y$-axis,

$$A(S) = 2\pi \int_a^b x\sqrt{1 + (y')^2}\, dx.$$

If $C$ is defined by parametric equations in $t$, then $ds = \sqrt{(D_t x)^2 + (D_t y)^2}dt$.

The **mean value** of a function $f$ over the interval $[a, b]$ is given by

$$f_m = \frac{1}{b-a}\int_a^b f(x)\, dx.$$

The **moments of an area** $A$ bounded by the curve $y = f(x)$, $x = a$, $x = b$

## ADDITIONAL INFORMATION

If a force $F$ is acting at a point $P$, then the moment of the force about a line is equal to the product of $F$ by the perpendicular distance from $P$ to the line. By making proper assumptions of the forces we can reduce the problems to finding moments of regions, arc lengths, and volumes and their associated centroids. It is important to understand the basic principle which permits the use of integration in solving these mechanics problems.

Consider the problem of finding the **centroid** of the area shown in Figure 19-1. The area is bounded by $y = f(x)$, $x = a$, and the $X$-axis. Construct the rectangle of sides $y$ and $dx$ as shown in the figure. The center of mass of this rectangle is located at its geometric center $P\left(x, \frac{y}{2}\right)$. The element of area is $y\, dx$, the area of the rectangle, and we let this be the element of mass. The moments of this element of mass about the coordinate axes are then given by $dM_y = (ydx)\, x = xdA$ (since $x$ is the distance from the $Y$-axis) and $dM_x = (ydx)\frac{y}{2} = \frac{y}{2}\, dA$ (since $\frac{y}{2}$ is the distance of the center of the rectangle from the $X$-axis.) We obtain the moments by summing these elements and using the fundamental theorem of integral calculus, we can express them in terms of integrals $M_y = \int_0^a xydx$ and $M_x = \frac{1}{2}\int_0^a y^2\, dx$ as we let $dx \longrightarrow 0$ and move the elements over the interval $[0, a]$. We recall that $A = \int_0^a y\, dx$. The integrals are all evaluated by making the substitution $f(x)$ for $y$. The coordinates of the centroid are then found by the formulas given in the Basic Facts. The student should also keep in mind the fact that *the centroid of a plane region lies on any axis of symmetry of that region.* The use of the element of area is especially essential when the region is bounded by two curves. Again we emphasize the advantage of sketching the configuration for any problem.

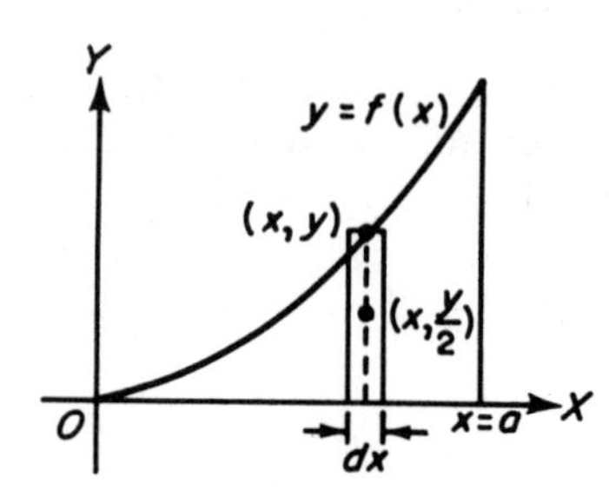


Fig. 19-1

The concept of considering elements is also used in finding the areas of surfaces of revolution and the **volumes** of solids of revolution. Let us consider a general formula for the volume of a solid of revolution. The procedure is:

> To find the volume of the solid generated by revolving a plane area about a line which lies in the plane of the area but does not intersect it, write an expression for the volume of an elementary cylindrical shell $dV = 2\pi rh\, dr$, where $r$ is the radius of revolution and $h$ is the height of the elementary shell. Then sum these volumes by integration $V = 2\pi \int_{r_1}^{r_2} rh\, dr$.

about the $X$-axis and $Y$-axis are

$$M_x = \frac{1}{2}\int_a^b y^2 dx \text{ and } M_y = \int_a^b xydx.$$

The **centroid** (center of mass of a plane region) of the area $A$ has the coordinates given by

$$\bar{x} = \frac{M_y}{A} \text{ and } \bar{y} = \frac{M_x}{A}$$

where $A = \int_a^b y\,dx$.

**Centroid of a Solid of Revolution**

Let the solid be generated by revolving the curve $C$ defined by $y = f(x)$, $a \leq x \leq b$, about the $X$-axis, then $\bar{y} = 0$ and for $\bar{x}$ we have

$$\bar{x} = \frac{M_y}{V} = \frac{\pi}{V}\int_a^b xy^2\,dx$$

and $V = \pi\int_a^b y^2\,dx$.

**Centroid of an Arc**

$$\bar{x} = \frac{\int_a^b xds}{\int_a^b ds} \text{ and } \bar{y} = \frac{\int_a^b yds}{\int_a^b ds}$$

where $ds$ is the differential of arc length between the limits $a$ and $b$.

**Approximate Integration.** Consider the integral $\int_a^b f(x)\,dx$. Divide the interval $[a,b]$ into $n$ equal subintervals with the abscissas of the end points being $x_0 = a, x_1, x_2, \ldots, x_n = b$. The interval length is $\Delta x = \frac{b-a}{n} = x_i - x_{i-1}$.

Calculate the corresponding ordinates using $y = f(x)$ to obtain $y_0 = f(x_0)$, $y_1 = f(x_1)$, $y_2 = f(x_2)$, $\ldots$, $y_n = f(x_n)$.

**Trapezoidal Rule**

$$\int_a^b f(x)\,dx \doteq \left(\tfrac{1}{2}y_0 + y_1 + y_2 + \ldots + y_{n-1} + \tfrac{1}{2}y_n\right)\Delta x$$

**Simpson's Rule.** See Additional Information.

It should be clear that $h$ and $r$ are both variables defined by the given region and the location of the axis of revolution. Furthermore, either $h$ is expressible as a function of $r$ or both $h$ and $r$ are functions of the same parameter $t$ which then becomes the variable of integration. The formulas in the Basic Facts are given for the coordinate axes as the lines about which the area is revolved.

The centroid of the area may also be used to find the surface areas and volumes. The procedure is based on two theorems of Pappus.

"If a plane area is revolved about a line lying in its plane but not intersecting it, then the volume of the solid generated is equal to the plane area multiplied by the length of the path traversed by the centroid of that area."

"If a plane arc revolves about a line in its plane but not intersecting it, then the area of the surface generated is equal to the length of the arc multiplied by the length of the path traversed by the centroid of the arc."

In both cases the length of the path is the circumference of a circle whose radius is the distance from the centroid to the line. If the axis of revolution is the $X$-axis, then the length of the path is $2\pi\bar{y}$ and for the $Y$-axis it is $2\pi\bar{x}$.

**Simpson's Rule.** $\int_a^b f(x)\,dx = \frac{h}{3}(y_0 + 4y_1 + 2y_2 \ldots + 4y_{n-1} + y_n)$

Let us divide the interval $[a, b]$ into $n$ (where $n$ is an even number) equal subintervals and write the integral as

$$\int_a^b f(x)\,dx = \int_{x_0}^{x_2} f(x)\,dx + \int_{x_2}^{x_4} f(x)\,dx + \ldots + \int_{x_{n-2}}^{x_n} f(x)\,dx.$$

The interval of the first integral is $[x_0, x_2]$, and we have the three points $(x_0, y_0)$, $(x_1, y_1)$, and $(x_2, y_2)$ on the curve $y = f(x)$. Let us approximate the curve $y = f(x)$ in this interval by a parabola $y = ax^2 + bx + c$ and let the three points be $(-h, y_0)$, $(0, y_1)$, and $(h, y_2)$ since $x_2 - x_1 = x_1 - x_0 = h$. Then we have $y_0 = ah^2 - bh + c$, $y_1 = c$, and $y_2 = ah^2 + bh + c$

and $$\int_{-h}^{h}(ax^2 + bx + c)\,dx = \frac{h}{3}(2ah^2 + 6c) = \frac{h}{3}(y_0 + 4y_1 + y_2)$$

since $y_0 + y_2 = 2ah^2 + 2c$ and $4c = 4y_1$. Thus the integral

$$\int_{x_0}^{x_2} f(x)\,dx = \int_{-h}^{h}(ax^2 + bx + c)\,dx = \frac{h}{3}(y_0 + 4y_1 + y_2)$$

to the degree that the parabola approximates the original curve in the interval $[x_0, x_2]$. We continue in the same manner for the rest of the integrals, and adding the results we obtain Simpson's Rule

The integrals of many functions which occur in scientific problems cannot be expressed in terms of elementary functions.*

*See Kaj L. Nielsen, *Methods in Numerical Analysis* (2nd ed.; New York: Macmillan, 1964), pp. 119–139.

## SOLUTIONS TO THE EXERCISES

I

i. $y = \cos 2x,\ x = 0,\ x = \frac{\pi}{4},\ y = 0$

$$V = \pi \int_0^{\frac{\pi}{4}} \cos^2 2x\, dx$$

$$= \frac{\pi}{2}\left[\frac{2x}{2} + \frac{1}{4}\sin 4x\right]_0^{\frac{\pi}{4}} = \frac{\pi^2}{8}$$

ii. $x^2 + y^2 = 25$ and $x + y = 5$ intersect at $(0, 5)$ and $(5, 0)$.

$$V = 2\pi \int_0^5 x[\sqrt{25 - x^2} - (5 - x)]\,dx$$

$$= 2\pi\left[-\frac{1}{2}(25 - x^2)^{\frac{3}{2}}\left(\frac{2}{3}\right) - \frac{5}{2}x^2 + \frac{x^3}{3}\right]_0^5$$

$$= 250\pi\left[\frac{2}{3} - \frac{1}{2}\right] = \frac{125}{3}\pi$$

II

i. $y^2 = 4x,\ (3, 2\sqrt{3}),\ (8, 4\sqrt{2})$

$a = 3,\ b = 8,\ y = 2\sqrt{x},\ y' = x^{-\frac{1}{2}}$

$$ds = \sqrt{1 + y'^2}\,dx = \frac{\sqrt{1 + x}}{\sqrt{x}}\,dx$$

$$A(S) = 2\pi \int_a^b y\,ds = 2\pi \int_3^8 2\sqrt{x + 1}\,dx$$

$$= 4\pi\left(\frac{2}{3}\right)(x + 1)^{\frac{3}{2}}\Big|_3^8 = \frac{152}{3}\pi$$

ii. $x = e^t \cos t,\ y = e^t \sin t,\ 0 \leq t \leq \frac{\pi}{2}$

$x' = e^t(\cos t - \sin t)$ and

$y' = e^t(\sin t + \cos t)$

$ds = e^t\sqrt{2}\,dt = \sqrt{x'^2 + y'^2}\,dt$

$$A(S) = 2\pi \int_0^{\frac{\pi}{2}} e^t \sin t\, e^t\sqrt{2}\,dt$$

$$= 2\sqrt{2}\,\pi \int_0^{\frac{\pi}{2}} e^{2t}\sin t\,dt$$

$$= 2\sqrt{2}\,\pi\left[\frac{1}{5}e^{2t}(2\sin t - \cos t)\right]_0^{\frac{\pi}{2}}$$

$$= \frac{2\sqrt{2}}{5}\pi(2e^{\pi} + 1)$$

III

i. $f(x) = \sqrt{r^2 - x^2}$ in $[0, r]$

$$f_m = \frac{1}{r}\int_0^r \sqrt{r^2 - x^2}\,dx$$

$$= \frac{1}{r}\left[\frac{x}{2}\sqrt{r^2 - x^2} + \frac{r^2}{2}\text{Arcsin}\,\frac{x}{r}\right]_0^r$$

$$= \frac{1}{r}\left[0 + \frac{r^2}{2}\,\frac{\pi}{2} - 0\right] = \frac{\pi r}{4}$$

ii. $f(x) = \tan x$ in $\left[0, \frac{\pi}{4}\right]$

$$f_m = \frac{4}{\pi}\int_0^{\frac{\pi}{4}} \tan x\,dx$$

$$= \frac{4}{\pi}\ln \sec x\Big|_0^{\frac{\pi}{4}}$$

$$= \frac{4}{\pi}[\ln\sqrt{2} - \ln 1] = \frac{2}{\pi}\ln 2$$

IV

i. $x = \frac{1}{2}t^2,\ y = \frac{1}{3}(2t + 1)^{\frac{3}{2}},\ [0, 4]$

$x' = t$ and $y' = (2t + 1)^{\frac{1}{2}}$

$ds = \sqrt{x'^2 + y'^2}\,dt = (t + 1)\,dt$

$$s = \int_0^4 (t + 1)\,dt = 12$$

$$\int_0^4 x\,ds = \frac{1}{2}\int_0^4 (t^3 + t^2)\,dt$$

$$= \frac{1}{2}\left[\frac{t^4}{4} + \frac{t^3}{3}\right]_0^4 = 32\left(\frac{4}{3}\right)$$

$$\overline{x} = \frac{32\left(\frac{4}{3}\right)}{12} = \frac{32}{9}$$

# 20 INDETERMINATE FORMS IMPROPER INTEGRALS

## SELF-TEST

DIRECTIONS: Write your answers in the numbered regions to the right. To check your answers, turn the page. Study the solutions to the problems you missed, and do the exercises following any problem or group of problems in which you had errors.

In **1–10** define a value for the indicated limits.

**1.** $\lim_{x \to 0} \left( \dfrac{\sin x^2}{\sin^2 x} \right)$

**2.** $\lim_{x \to \infty} \left( \dfrac{x^2}{e^x} \right)$

**3.** $\lim_{x \to 3} \left( \dfrac{x^2 - x - 6}{x^3 - x^2 - 7x + 3} \right)$

**4.** $\lim_{x \to \infty} xe^{-x}$

**5.** $\lim_{x \to 0} \left( \csc x - \dfrac{1}{1 - \cos x}, \right)$

**6.** $\lim_{x \to 0} x^x$

**7.** $\lim_{x \to 0} (1 + x)^{\frac{1}{x}}$

**8.** $\lim_{x \to 1} \left( \dfrac{1}{1 - x} \right)^{1-x}$

**9.** $\lim_{x \to 2} \left( 1 - \dfrac{x}{2} \right) \sin \pi x$

**10.** $\lim_{x \to 0} \left( \dfrac{e^x - e^{-x} - 2 \sin x}{x - \sin x} \right)$

**11.** Evaluate $\displaystyle\int_1^{\infty} \frac{dx}{x^2}$.

**12.** Evaluate $\displaystyle\int_{-\infty}^{1} e^x dx$.

| |
|---|
| 1 |
| 2 |
| 3 |
| 4 |
| 5 |
| 6 |
| 7 |
| 8 |
| 9 |
| 10 |
| 11 |
| 12 |

## SOLUTIONS

| Functions and Derivatives | $\lim_{x\to a}$ |
|---|---|
| 1. $\dfrac{\sin x^2}{\sin^2 x}$, $(a = 0)$ | $\dfrac{0}{0}$ |
| $\dfrac{2x \cos x^2}{\sin 2x}$ | $\dfrac{0}{0}$ |
| $\dfrac{-4x^2 \sin x^2 + 2 \cos x^2}{2 \cos 2x}$ | $\dfrac{2}{2} = 1$ |

| Functions and Derivatives | $\lim_{x\to a}$ |
|---|---|
| 2. $\dfrac{x^2}{e^x}$, $(a = \infty)$ | $\dfrac{\infty}{\infty}$ |
| $\dfrac{2x}{e^x}$ | $\dfrac{\infty}{\infty}$ |
| $\dfrac{2}{e^x}$ | $\dfrac{2}{\infty} = 0$ |

| Functions and Derivatives | $\lim_{x\to a}$ |
|---|---|
| 3. $\dfrac{x^2 - x - 6}{x^3 - x^2 - 7x + 3}$, $(a = 3)$ | $\dfrac{0}{0}$ |
| $\dfrac{2x - 1}{3x^2 - 2x - 7}$ | $\dfrac{5}{14}$ |

### EXERCISE I

Evaluate.

i. $\lim_{x\to 0} \left[\dfrac{(2 - x^2)\sin x - 2x \cos x}{x^3}\right]$

ii. $\lim_{x\to 4} \left(\dfrac{2 - \sqrt{x}}{x^2 - 16}\right)$

iii. $\lim_{x\to 1} \left[\dfrac{\ln x - \sin (x - 1)}{(x - 1)^3}\right]$

iv. $\lim_{x\to 2} \left[\dfrac{\sqrt{2x} - \sqrt{6 - x}}{3x - \sqrt{3x + 30}}\right]$

v. $\lim_{x\to 0} \left[\dfrac{\ln \cot x}{\ln \cot 2x}\right]$

| Functions and Derivatives | $\lim_{x\to a}$ |
|---|---|
| 4. $x e^{-x}$, $(a = \infty)$ | $\infty\ 0$ |
| $\dfrac{x}{e^x}$ | $\dfrac{\infty}{\infty}$ |
| $\dfrac{1}{e^x}$ | $0$ |

| Functions and Derivatives | $\lim_{x\to a}$ |
|---|---|
| 5. $\csc x - \dfrac{1}{1 - \cos x}$, $(a = 0)$ | $\infty - \infty$ |
| $\dfrac{1}{\sin x} - \dfrac{1}{1 - \cos x}$ | $\infty - \infty$ |
| $\dfrac{1 - \cos x - \sin x}{\sin x - \sin 2x}$ | $\dfrac{0}{0}$ |
| $\dfrac{\sin x - \cos x}{\cos x - \cos 2x}$ | $\dfrac{-1}{0} = \infty$ |

| Functions and Derivatives | $\lim_{x\to a}$ |
|---|---|
| 6. $y = x^x$, $(a = 0)$ | $0^0$ |
| $\ln y = x \ln x$ | $0\ (-\infty)$ |
| $= \dfrac{\ln x}{x^{-1}}$ | $\dfrac{-\infty}{\infty}$ |
| $\dfrac{x^{-1}}{-x^{-2}} = -x$ | $0$ |

$\ln x^x = 0$; $x^x = e^0 = 1$ as $x\to 0$

| Functions and Derivatives | $\lim_{x\to a}$ |
|---|---|
| 7. $(1 + x)^{\frac{1}{x}}$, $(a = 0)$ | $1^\infty$ |
| $\ln y = \dfrac{\ln (1 + x)}{x}$ | $\dfrac{0}{0}$ |
| $\dfrac{1}{1 + x}$ | $1$ |

$\ln y = 1$; $(1 + x)^{\frac{1}{x}} = e$ as $x\to 0$

| Functions and Derivatives | $\lim_{x\to a}$ |
|---|---|
| 8. $\left(\dfrac{1}{1 - x}\right)^{1-x}$, $(a = 1)$ | $\infty^0$ |
| $\ln y = (1 - x) \ln (1 - x)^{-1}$ | $0 \cdot \infty$ |
| $\dfrac{\ln (1 - x)^{-1}}{(1 - x)^{-1}}$ | $\dfrac{\infty}{\infty}$ |

$$\frac{\frac{1}{(1-x)^{-1}}}{\frac{1}{(1-x)^2}} = (1-x)^3 \qquad 0$$

$\ln y = 0;\ y = e^0 = 1$ as $x \to 1$

**EXERCISE II**

Evaluate.

i. $\lim_{x\to 0} \left(\frac{1}{x}\right)^{\tan x}$ ii. $\lim_{x\to 0} (1-x^3)^{\frac{1}{x}}$

**9.** $\left(1-\frac{x}{2}\right)^{\sin \pi x}$, $(a = 2)$ $\qquad 0^0$

$$\ln y = \sin \pi x \ln \left(1 - \frac{x}{2}\right)$$

$$= \frac{\ln(2-x) - \ln 2}{\csc \pi x} \qquad \frac{-\infty}{\infty}$$

$$\frac{-(2-x)^{-1}}{-\pi \csc \pi x \cot \pi x}$$

$$= \frac{\sin^2 \pi x}{\pi(2-x)\cos \pi x} \qquad \frac{0}{0}$$

$$\frac{\pi \sin 2\pi x}{\pi[-\cos \pi x - \pi(2-x)\sin \pi x]} \qquad \frac{0}{-\pi} = 0$$

$\ln y = 0,\ y = 1$ as $x \to 2$

**10.** $\frac{e^x - e^{-x} - 2\sin x}{x - \sin x}$, $(a = 0)$ $\qquad \frac{0}{0}$

$$\frac{e^x + e^{-x} - 2\cos x}{1 - \cos x} \qquad \frac{0}{0}$$

$$\frac{e^x - e^{-x} + 2\sin x}{\sin x} \qquad \frac{0}{0}$$

$$\frac{e^x + e^{-x} + 2\cos x}{\cos x} \qquad 4$$

**11.** $\int_1^\infty \frac{dx}{x^2} = I$

$$\int_1^b x^{-2}\,dx = -\frac{1}{x}\Bigg|_1^b = -\frac{1}{b} + 1$$

$$\lim_{b\to\infty}\left(-\frac{1}{b} + 1\right) = 0 + 1 = 1$$

**ANSWERS**

| | |
|---|---|
| 1. 1 | 1 |
| 2. 0 | 2 |
| 3. $\frac{5}{14}$ | 3 |
| 4. 0 | 4 |
| 5. $\infty$ | 5 |
| 6. 1 | 6 |
| 7. $e$ | 7 |
| 8. 1 | 8 |
| 9. 1 | 9 |
| 10. 4 | 10 |
| 11. 1 | 11 |
| 12. $e$ | 12 |

**12.** $\int_{-\infty}^1 e^x dx = \lim_{a\to -\infty} \int_a^1 e^x dx = I$

$$I = \lim_{a\to-\infty} [e - e^a] = e - 0 = e$$

**EXERCISE III**

Evaluate.

i. $\int_1^\infty \frac{dx}{x}$ ii. $\int_0^8 \frac{dx}{\sqrt{8x - x^2}}$

Solutions to the exercises are on page 120.

## BASIC FACTS

A function of the form $\frac{P(x)}{Q(x)}$ is defined for all values of $x$ for which $P(x)$ and $Q(x)$ are real numbers except values for which $Q(x) = 0$. An expression of the form $\frac{0}{0}$ is in general meaningless. However, if for a value $x = a$ we obtain $\frac{P(a)}{Q(a)} = \frac{0}{0}$, it may be advantageous to define this fraction at $x = a$. This was the case for $\frac{\sin x}{x}$ at $x = 0$.

**Definition.** Expressions of the form

$$\frac{0}{0}, \frac{\infty}{\infty}, \infty - \infty, 0\,\infty, 0^0, 1^\infty, \infty^0$$

are called indeterminate forms.

This chapter deals with methods for evaluating these forms in order that a function which assumes one of them at $x = a$ may be defined there.

1. Forms $\frac{0}{0}$ and $\frac{\infty}{\infty}$. If $\frac{f(a)}{F(a)} = \frac{0}{0}$ or $\frac{\infty}{\infty}$

and if $\lim_{x \to a} \frac{f'(x)}{F'(x)} = L$, then

$$\lim_{x \to a} \frac{f(x)}{F(x)} = \lim_{x \to a} \frac{f'(x)}{F'(x)} = L.$$

*Rule.* Differentiate the numerator $f(x)$ for a new numerator, $f'(x)$ and the denominator $F(x)$ for a new denominator, $F'(x)$. The value of this new fraction at $x = a$ will be the assigned value of the original fraction as $x \to a$. If the new fraction is also an indeterminate form *of this type*, repeat the process on the new fraction.

2. **The form $0 \cdot \infty$.** If $f(x)\,\phi(x)$ takes the form $0 \cdot \infty$ for $x = a$, write the expression in the form

$$\frac{f(x)}{\frac{1}{\phi(x)}} \text{ or } \frac{\phi(x)}{\frac{1}{f(x)}}$$

## ADDITIONAL INFORMATION

The evaluation of indeterminate forms is based on l'Hopital's Rule and the **General Theorem of the Mean.**

If $f$ and $F$ are continuous functions of $x$ in the interval $a \leq x \leq b$ and each has a derivative at all interior points of $(a, b)$ with $F'(x) \neq 0$ for $a < x < b$, then there is at least one value $x_1$ of $x$ between $a$ and $b$ such that

$$\frac{f(b) - f(a)}{F(b) - F(a)} = \frac{f'(x_1)}{F'(x_1)}.$$

PROOF. The equation of the conclusion is defined for $x$ in $(a, b)$ since by the hypothesis $F'(x) \neq 0$ for $a < x < b$ so $F'(x_1) \neq 0$, and by the First Theorem of the Mean Value (see p. 28) we have $F(b) - F(a) = F'(x_0)(b - a)$ for $x_0$ in $(a, b)$. Now since $F'(x_0) \neq 0$ and $b - a \neq 0$ then $F(b) - F(a) \neq 0$. Form the function

$$\phi(x) = \frac{f(b) - f(a)}{F(b) - F(a)}[F(x) - F(a)] - [f(x) - f(a)].$$

$$\text{At } x = a\text{: } \phi(a) = \frac{f(b) - f(a)}{F(b) - F(a)}[F(a) - F(a)] - [f(a) - f(a)] = 0$$

$$\text{At } x = b\text{: } \phi(b) = \frac{f(b) - f(a)}{F(b) - F(a)}[F(b) - F(a)] - [f(b) - f(a)]$$

$$= f(b) - f(a) - [f(b) - f(a)] = 0$$

Thus the function $\phi$ satisfies the conditions of Rolle's Theorem (see p. 28), and its derivative must equal zero for some value $x_1$ in $(a, b)$. Differentiating and setting $x = x_1$ we obtain

$$\phi'(x_1) = \frac{f(b) - f(a)}{F(b) - F(a)}[F'(x_1)] - f'(x_1) = 0.$$

If we now divide by $F'(x_1) \neq 0$ and transpose the last term, we obtain the conclusion of the theorem.

**Theorem (l'Hopital's Rule).** If $f$ and $F$ satisfy the hypothesis of the General Theorem of the Mean in an interval about $a$ and if

$$\lim_{x \to a} f(x) = 0, \lim_{x \to a} F(x) = 0 \text{ and } \lim_{x \to a} \frac{f'(x)}{F'(x)} = L, \text{ then}$$

$$\lim_{x \to a} \frac{f(x)}{F(x)} = \lim_{x \to a} \frac{f'(x)}{F'(x)} = L.$$

PROOF. Let $x$ be in the interval $a < x < a + h$. Then

$$\frac{f(a+h) - f(a)}{F(a+h) - F(a)} = \frac{f(a+h)}{F(a+h)} = \frac{f'(x_1)}{F'(x_1)}, \quad a < x_1 < a + h$$

since $f(a) = F(a) = 0$. Now as $x \longrightarrow a$, $h \longrightarrow 0$, and $x_1 \longrightarrow a$ and

$$\lim_{x \to a} \frac{f(x)}{F(x)} = \lim_{h \to 0} \frac{f(a+h)}{F(a+h)} = \lim_{x_1 \to a} \frac{f'(x_1)}{F'(x_1)} = L.$$

If we take $x$ in the interval $(a - h) < x < a$, we can repeat the above steps with the appropriate changes in sign and thus complete the proof.

which then assumes the form

$$\frac{0}{0} \text{ or } \frac{\infty}{\infty} \text{ at } x = a.$$

**3. The form $\infty - \infty$.** Expressions which result in forms $\infty - \infty$ can usually be transformed by trigonometric identities or algebraic operations to the form $\frac{0}{0}$ or $\frac{\infty}{\infty}$.

**4. Forms $0^0$, $1^\infty$, $\infty^0$.** The expression $f(x)^{\phi(x)}$ may result in one of these forms for $x = a$. Let $y = f(x)^{\phi(x)}$, then $\ln y = \phi(x) \ln f(x)$ which will be of the form $0 \infty$ at $x = a$. Evaluate this expression (Method 2 above) and obtain $\ln y = k$, then $y = e^k$.

**Improper Integrals.** If the limits on the right exist, then we have

$$\int_a^\infty f(x)\,dx = \lim_{b\to\infty} \int_a^b f(x)\,dx$$

$$\int_{-\infty}^b f(x)\,dx = \lim_{a\to-\infty} \int_a^b f(x)\,dx$$

$$\int_{-\infty}^\infty f(x)dx = \int_{-\infty}^0 f(x)dx + \int_0^\infty f(x)dx.$$

If $f(x)$ is continuous in $[a, b]$ except at $a$, then we have

$$\int_a^b f(x)dx = \lim_{\varepsilon\to 0} \int_{a+\varepsilon}^b f(x)dx$$

if this limit exists. Similarly,

$$\int_a^b f(x)dx = \lim_{\varepsilon\to 0} \int_a^{b-\varepsilon} f(x)dx.$$

If $f(x)$ is continuous in $[a, b]$ except at $c$ for $a < c < b$, then

$$\int_a^b f(x)dx = \lim_{\varepsilon\to 0} \int_a^{c-\varepsilon} f(x)dx$$

$$+ \lim_{\varepsilon\to 0} \int_{c+\varepsilon}^b f(x)dx.$$

The proof for l'Hopital's Rule for the indeterminate form $\frac{\infty}{\infty}$ is more difficult and requires knowledge which is beyond the usual college course in calculus.

If we use the properties of logarithms on the expression $y = f(x)^{\phi(x)}$, we may write $\ln y = \phi(x) \ln f(x)$. The limit of $y$ as $x \longrightarrow a$ may take any of the indeterminate forms $0^0$, $1^\infty$, $\infty^0$ depending upon $\lim_{x\to a} f(x)$ and $\lim_{x\to a} \phi(x)$. We evaluate these indeterminate forms by considering the limit of $\ln y$ and then solving for the corresponding value of $y$. The procedure is based on the theorem "The limit of the logarithm of $f(x)$ is the logarithm of the limit of $f(x)$ providing that this limit is positive." By considering the logarithmic function (see p. 70), it should be clear that as $x \longrightarrow 0$ from the positive side (remember that the function is not defined for $x < 0$), the function $\longrightarrow -\infty$ and as $x \longrightarrow \infty$, the function $\longrightarrow \infty$. The student should also review the exponential function (see p. 70) and see that as $x \longrightarrow \infty$, $e^x \longrightarrow \infty$ and as $x \longrightarrow -\infty$, $e^x \longrightarrow 0$.

The evaluation of expressions involving the trigonometric functions depend upon frequent substitutions using the identities, and the student is again referred to p. 155. In addition, the following formulas are convenient.

$$\frac{d}{dx}(\sin^2 x) = 2 \sin x \cos x = \sin 2x$$

$$\frac{d}{dx}(\cos^2 x) = -2 \sin x \cos x = -\sin 2x$$

$$\frac{d^2}{dx^2}(\sin^2 x) = 2 \cos 2x, \text{ and } \frac{d^2}{x^2}(\cos^2 x) = -2 \cos 2x$$

$$\frac{d}{dx}(\sec^2 x) = 2 \sec^2 x \tan x = 2(\tan x + \tan^3 x)$$

$$\frac{d}{dx}(\csc^2 x) = -2 \csc^2 c \cot x = -2(\cot x + \cot^3 x)$$

Improper integrals that occur because the integrand is discontinuous in the interval of integration are often missed because the continuity of the integrand is not checked. If the integrand contains trigonometric functions, it should be checked for inadmissible values of the argument. If the integrand is a fraction, set the denominators equal to zero and solve for the values of the variable. If these values are in the interval of integration, the integral is improper.

| Review of Elementary Limits | | |
|---|---|---|
| $\lim_{x\to a} (u \pm v) = \lim_{x\to a} u \pm \lim_{x\to a} v$ | $\lim_{x\to 0} \frac{c}{x} = \infty$ | $\lim_{x\to\infty} \frac{c}{x} = 0$ |
| $\lim_{x\to a} (u\,v) = \left(\lim_{x\to a} u\right)\left(\lim_{x\to a} v\right)$ | $\lim_{x\to\infty} \frac{x}{c} = \infty$ | $\lim_{x\to\infty} e^x = \infty$ |
| $\lim_{x\to a} \frac{u}{v} = \frac{\lim_{x\to a} u}{\lim_{x\to a} v}$ | $\lim_{x\to\infty} (cx) = \infty$<br>$\lim_{x\to 0} \ln x = -\infty$ | $\lim_{x\to 0} e^x = 1$<br>$\lim_{x\to\infty} \ln x = \infty$ |

## SOLUTIONS TO THE EXERCISES

I

| Functions and derivatives | $\lim\limits_{x\to a}$ |
|---|---|
| i. $\dfrac{(2-x^2)\sin x - 2x\cos x}{x^3}$, $(a=0)$ | $\dfrac{0}{0}$ |
| $\dfrac{-x^2\cos x}{3x^2} = -\dfrac{\cos x}{3}$ | $-\dfrac{1}{3}$ |
| ii. $\dfrac{2-\sqrt{x}}{x^2-16}$, $(a=4)$ | $\dfrac{0}{0}$ |
| $\dfrac{-\frac{1}{2}x^{-\frac{1}{2}}}{2x} = -\dfrac{1}{4x\sqrt{x}}$ | $-\dfrac{1}{32}$ |
| iii. $\dfrac{\ln x - \sin(x-1)}{(x-1)^3}$, $(a=1)$ | $\dfrac{0}{0}$ |
| $\dfrac{\frac{1}{x} - \cos(x-1)}{3(x-1)^2}$ | $\dfrac{0}{0}$ |
| $\dfrac{-x^{-2} + \sin(x-1)}{6(x-1)}$ | $\infty$ |
| iv. $\dfrac{\sqrt{2x}-\sqrt{6-x}}{3x-\sqrt{3x+30}}$, $(a=2)$ | $\dfrac{0}{0}$ |
| $\dfrac{(2x)^{-\frac{1}{2}} + \frac{1}{2}(6-x)^{-\frac{1}{2}}}{3-\frac{1}{2}(3)(3x+30)^{-\frac{1}{2}}}$ | $\dfrac{3}{11}$ |
| v. $\dfrac{\ln\cot x}{\ln\cot 2x}$, $(a=0)$ | $\dfrac{\infty}{\infty}$ |
| $\dfrac{\tan x(-\csc^2 x)}{\tan 2x(2)(-\csc^2 2x)} = \dfrac{\sin 4x}{2\sin 2x}$ | $\dfrac{0}{0}$ |
| $\dfrac{4\cos 4x}{4\cos 2x}$ | $1$ |

II

| Functions and derivatives | $\lim\limits_{x\to a}$ |
|---|---|
| i. $\left(\dfrac{1}{x}\right)^{\tan x}$, $(a=0)$ | $\infty^0$ |
| $\ln y = \tan x \ln\dfrac{1}{x}$ | $0\cdot\infty$ |
| $= -\dfrac{\ln x}{\cot x}$ | $\dfrac{\infty}{\infty}$ |
| $-\dfrac{1}{x}\left(\dfrac{1}{-\csc^2 x}\right) = \dfrac{\sin^2 x}{x}$ | $\dfrac{0}{0}$ |
| $\dfrac{2\sin 2x}{1}$ | $0$ |

$\ln y = 0$ and $y = 1$

| Functions and derivatives | $\lim\limits_{x\to a}$ |
|---|---|
| ii. $(1-x^3)^{\frac{1}{x}}$, $(a=0)$ | $1^\infty$ |
| $\dfrac{\ln(1-x^3)}{x}$ | $\dfrac{0}{0}$ |
| $\left(\dfrac{-3x^2}{1-x^3}\right)\left(\dfrac{1}{1}\right)$ | $0$ |

$\ln y = 0$ and $y = 1$

III

i. $\displaystyle\int_1^\infty \frac{dx}{x} = I$

$$\int_1^b \frac{dx}{x} = \ln x\Big|_1^b = \ln b - 0$$

$\lim\limits_{b\to\infty}(\ln b) \longrightarrow \infty$. $I$ is divergent.

ii. $\displaystyle\int_0^8 \frac{dx}{\sqrt{8x-x^2}} = I$

Discontinuous at $x = 8$ and $x = 0$

$$\int_0^8 dI = \lim_{\varepsilon\to 0}\int_{0+\varepsilon}^{8-\varepsilon} dI$$

$$\int dI = \int\frac{dx}{\sqrt{16-(x-4)^2}} = \text{Arcsin}\,\frac{x-4}{4}$$

$$x = 8-\varepsilon:\ \lim_{\varepsilon\to 0}\text{Arcsin}\,\frac{8-\varepsilon-4}{4} = \frac{\pi}{2}$$

$$x = 0+\varepsilon:\ \lim_{\varepsilon\to 0}\text{Arcsin}\,\frac{0+\varepsilon-4}{4} = -\frac{\pi}{2}$$

$$I = \frac{\pi}{2} - \left(-\frac{\pi}{2}\right) = \pi$$

# 21 SERIES

## SELF-TEST

DIRECTIONS: Write your answers in the numbered regions to the right. To check your answers, turn the page. Study the solutions to the problems you missed, and do the exercises following any problem or group of problems in which you had errors.

In **1-5** test the series for convergence.

**1.** $1 + \frac{1}{\sqrt{2}} + \frac{1}{\sqrt{3}} + \ldots + \frac{1}{\sqrt{n}} + \ldots$

**2.** $\frac{2}{3} + \left(\frac{2}{3}\right)^2 + \left(\frac{2}{3}\right)^3 + \ldots + \left(\frac{2}{3}\right)^n + \ldots$

**3.** $1 - \frac{1}{2!} + \frac{1}{3!} - \frac{1}{4!} + \ldots + (-1)^{n+1}\left(\frac{1}{n!}\right) + \ldots$

**4.** $\sum_{n=1}^{\infty} \frac{5^n}{n(n+1)(n+2)}$

**5.** $\sum_{n=1}^{\infty} \frac{n}{n^4 + 1}$

In **6-8** find the interval of convergence.

**6.** $1 + \frac{x^2}{2!} + \frac{x^4}{4!} + \frac{x^6}{6!} + \ldots + \frac{x^{2n}}{(2n)!} + \ldots$

**7.** $x - \frac{x^3}{3} + \frac{x^5}{5} - \frac{x^7}{7} + \ldots$

**8.** $(x - 1) + (x - 1)^2 + (x - 1)^3 + \ldots$

In **9-10** find the Maclaurin series.

**9.** $f(x) = \sin x$

**10.** $f(x) = e^{2x}$

**11.** Find the Taylor series for $\ln x$ about the value $x = a = 1$.

**12.** Is it possible to expand $\ln x$ by a Maclaurin series?

| |
|---|
| 1 |
| 2 |
| 3 |
| 4 |
| 5 |
| 6 |
| 7 |
| 8 |
| 9 |
| 10 |
| 11 |
| 12 |

## SOLUTIONS

1. $1 + \frac{1}{\sqrt{2}} + \frac{1}{\sqrt{3}} + \ldots + \frac{1}{\sqrt{n}} + \ldots$

$$\int_0^\infty x^{-\frac{1}{2}}\,dx = \lim_{b\to\infty} 2b^{\frac{1}{2}} \longrightarrow \infty$$

The series diverges.

2. $\frac{2}{3} + \left(\frac{2}{3}\right)^2 + \ldots + \left(\frac{2}{3}\right)^n + \ldots$

$u_{n+1} = \left(\frac{2}{3}\right)^{n+1}$ and $u_n = \left(\frac{2}{3}\right)^n$

$$\lim_{n\to\infty}\left(\frac{u_{n+1}}{u_n}\right) = \lim_{n\to\infty}\left(\frac{2}{3}\right) = \frac{2}{3}$$

Since $L < 1$, series converges.

3. $\sum_{n=1}^{\infty} (-1)^{n+1}\frac{1}{n!}$

This is an alternating series and $\lim_{n\to\infty}\frac{1}{n!} = 0$. Therefore the series converges.

4. $\sum_{n=1}^{\infty} \frac{5^n}{n(n+1)(n+2)}$

$$\frac{u_{n+1}}{u_n} = \frac{5^{n+1}(n)(n+1)(n+2)}{5^n(n+1)(n+2)(n+3)}$$

$$= \frac{5n}{n+3} = \frac{5}{1+\frac{3}{n}}$$

$$\lim_{n\to\infty}\frac{u_{n+1}}{u_n} = 5 > 1 \text{ Diverges.}$$

---

**EXERCISES I**

Show that the following series are divergent.

i. $\sum_{n=1}^{\infty} \frac{3n}{2n-7}$ ii. $\sum_{n=1}^{\infty} \frac{(n+2)(n-1)}{3n^2-4}$

---

5. $\sum_{n=1}^{\infty} \frac{n}{n^4+1}$ The ratio test fails.

$$\int \frac{x\,dx}{x^4+1} = \frac{1}{2}\text{ Arctan } x^2 = I$$

$$I\Big|_1^\infty = \lim_{b\to\infty}\frac{1}{2}\left[\text{Arctan } b^2 - \frac{\pi}{4}\right]$$

$$= \frac{1}{2}\left[\frac{\pi}{2} - \frac{\pi}{4}\right] = \frac{\pi}{8} \text{ Converges.}$$

6. $\sum_{n=1}^{\infty} \frac{x^{2n}}{(2n)!}$

$$\frac{u_{n+1}}{u_n} = \frac{x^{2n+2}}{(2n+2)!}\,\frac{(2n)!}{x^{2n}}$$

$$= x^2\left[\frac{1}{(2n+1)(2n+2)}\right]$$

$$\lim_{n\to\infty}\left|\frac{u_{n+1}}{u_n}\right| = x^2(0) = 0$$

Series converges for all $x$.

7. $\sum_{n=1}^{\infty} (-1)^{n+1}\frac{x^{2n-1}}{2n-1}$

$$\left|\frac{u_{n+1}}{u_n}\right| = \left|\frac{x^{2n+1}}{2n+1}\,\frac{2n-1}{x^{2n-1}}\right| = x^2\left|\frac{2n-1}{2n+1}\right|$$

$$\lim_{n\to\infty}\left|\frac{2n-1}{2n+1}\right|x^2 = x^2 < 1 \text{ for } -1 < x < 1$$

At $x = -1$: $\sum_{n=1}^{\infty} (-1)^{n+1}\left(\frac{-1}{2n-1}\right)$

an alternating series with $u_n \longrightarrow 0$.

At $x = 1$: $\sum_{n=1}^{\infty} (-1)^{n+1}\left(\frac{1}{2n-1}\right)$

an alternating series with $u_n \longrightarrow 0$. Interval of convergence: $-1 \le x \le 1$.

8. $\sum_{n=1}^{\infty} (x-1)^n$, $\left|\frac{(x-1)^{n+1}}{(x-1)^n}\right| = |x-1|$

$\lim_{n\to\infty} |x-1| = |x-1| < 1$ for $0 < x < 2$. At $x = 0$ we have a series

with $|u_n| = 1$, and at $x = 2$ we have $u_n = 1$. Both series diverge.
Convergence interval: $0 < x < 2$.

EXERCISES II

Find the interval of convergence.

i. $\sum_{n=1}^{\infty} \frac{n\, x^n}{2^{n-1} n(n+1)}$

ii. $\sum_{n=1}^{\infty} \frac{(n+1)x^n}{2^n 3^{n+1}}$

9. $f(x) = \sin x \qquad f(0) = 0$
$f'(x) = \cos x \qquad f'(0) = 1$
$f''(x) = -\sin x \qquad f''(0) = 0$
$f'''(x) = -\cos x \qquad f'''(0) = -1$
$f^{iv}(x) = \sin x \qquad f^{iv}(0) = 0$

$$\sin x = x - \frac{x^3}{3!} + \frac{x^5}{5!} - \frac{x^7}{7!} + \ldots$$

10. $f(x) = e^{2x} \qquad f(0) = 1$
$f'(x) = 2e^{2x} \qquad f'(0) = 2$
$f''(x) = 4e^{2x} \qquad f''(0) = 4$
$f'''(x) = 8e^{2x} \qquad f'''(x) = 8$

$$e^{2x} = 1 + 2x + \frac{4x^2}{2!} + \frac{8x^3}{3!} + \ldots$$

$$= 1 + 2x + \frac{(2x)^2}{2!} + \frac{(2x)^3}{3!} + \ldots$$

EXERCISES III

i. Find the Taylor series for $f(x) = \cos x$ and $a = \frac{\pi}{3}$.

ii. Find the Taylor series for $f(x) = \sqrt{x}$ and $a = 9$.

iii. Compare the Maclaurin series for $f(x) = (1 + x)^{-1}$ with its binomial ex-expansion.

11. $f(x) = \ln x \qquad f(1) = 0$
$f'(x) = x^{-1} \qquad f'(1) = 1$

**ANSWERS**

| | |
|---|---|
| 1. Diverges | 1 |
| 2. Converges | 2 |
| 3. Converges | 3 |
| 4. Diverges | 4 |
| 5. Converges | 5 |
| 6. All $x$ | 6 |
| 7. $-1 \leq x \leq 1$ | 7 |
| 8. $0 < x < 2$ | 8 |
| 9. $\sum_{n=1}^{\infty} (-1)^{n+1} \frac{x^{2n-1}}{(2n-1)!}$ | 9 |
| 10. $\sum_{n=0}^{\infty} \frac{(2x)^n}{n!}$ | 10 |
| 11. $\sum_{n=1}^{\infty} (-1)^{n+1} \frac{(x-1)^n}{n}$ | 11 |
| 12. No. | 12 |

$f''(x) = -x^{-2} \qquad f''(1) = -1$
$f'''(x) = 2x^{-3} \qquad f'''(1) = 2$

$$\ln x = x - 1 - \frac{1}{2}(x-1)^2 + \frac{1}{3}(x-1)^3 + \ldots$$

12. No, since $f'(x) = \frac{1}{x}$ and $f'(0)$ does not exist.

## BASIC FACTS

An ordered set of objects is called a **sequence**; $a_1, a_2, \ldots, a_n, \ldots$ The individual objects are called **terms** or **elements** of the sequence. If a sequence has a first and last term it is a **finite sequence**; otherwise it is an **infinite sequence**. An infinite sequence $\{a_i\}$ is said to approach the constant $c$ as a limit if for each $\varepsilon > 0$ there exists a positive integer $N$ such that $|a_n - c| < \varepsilon$ for $n > N$. If a sequence approaches a limit, it is said to be **convergent**; if it does not approach a limit it is said to be **divergent**.

The sum of the terms of a sequence is called a **series**.

$$\sum_{i=1}^{\infty} a_i = a_1 + a_2 + \cdots + a_n + \cdots$$

The quantity $S_k = \Sigma_{i=1}^{k} a_i$ is called the $k$th **partial sum** of the series. The sum of an infinite series is defined by $S = \lim_{n\to\infty} S_n$. If this limit exists, the series is said to converge to the sum $S$. If the limit does not exist, the series is said to diverge.

**Theorem.** If the series $\Sigma_{i=1}^{\infty} a_i$ converges, then $\lim_{n\to\infty} a_n = 0$.

**Corollary.** If $a_n$ does not tend to zero as $n \longrightarrow \infty$, then the series is divergent.

**Theorem.** If $S_1$ and $S_2$ are two converging series and $k$ is any constant, then $kS_1$, $kS_2$, and $S_1 \pm S_2$ all converge.

### Tests for Convergence

**Integral Test.** Let $f(x)$ be a continuous, nonnegative, and nonincreasing function for which $f(n) = u_n$ in the series $\Sigma u_n$ for each $n \geq 1$, then the series converges if $\int_1^\infty f(x)\,dx$ exists and diverges if the integral does not exist.

**Comparison Test.** Given two series

(1) $a_1 + a_2 + a_3 + \ldots + a_n + \ldots$

(2) $b_1 + b_2 + b_3 + \ldots + b_n + \ldots$

## ADDITIONAL INFORMATION

The sum of the infinite geometric progression $\Sigma_{i=0}^{\infty} ar^i$ forms the geometric series. The ratio test gives $ar^{n+1} \div ar^n$ which is equal to $r$ and the series converges for $|r| < 1$ and diverges for all other values of $r$. We recall from algebra that when $|r| < 1$, the sum is given by $S = a(1 - r)^{-1}$.

THEOREM. The $p$-series. The series defined by

$$\sum_{n=1}^{\infty} \frac{1}{n^p} = 1 + \frac{1}{2^p} + \frac{1}{3^p} + \frac{1}{4^p} + \ldots + \frac{1}{n^p} + \ldots$$

is convergent if $p > 1$ and divergent if $p \leq 1$.

PROOF. Let $f(x) = \dfrac{1}{x^p}$ for $p > 0$, then we have the integral

$$\int_1^\infty \frac{dx}{x^p} = \lim_{b\to\infty} \int_1^b \frac{dx}{x^p} = \lim_{b\to\infty} \left(\frac{x^{1-p}}{1-p}\right)\Bigg|_1^b = \lim_{b\to\infty} \left(\frac{b^{1-p}}{1-p} - \frac{1}{1-p}\right)$$

providing $p \neq 1$. If $p > 1$, the first term approaches zero as $b \longrightarrow \infty$ and the limit exists and is the second term. If $p < 1$, the limit does not exist as the first term tends to infinity. If $p = 1$, the integral yields $\ln b - \ln 1 = \ln b$ which tends to infinity as $b \longrightarrow \infty$.

The $p$-series with $p = 1$ is also known as the **harmonic series**, $1 + \frac{1}{2} + \frac{1}{3} + \frac{1}{4} + \cdots + \frac{1}{n} + \cdots$. We showed above that this series diverges even though $\lim_{n\to\infty} u_n = \lim_{n\to\infty} \left(\frac{1}{n}\right) = 0$. A second proof is accomplished by grouping the terms as follows:

$$1 + \tfrac{1}{2} + \left(\tfrac{1}{3} + \tfrac{1}{4}\right) + \left(\tfrac{1}{5} + \tfrac{1}{6} + \tfrac{1}{7} + \tfrac{1}{8}\right) + \text{(next 8 terms)} + \cdots.$$

Now consider $1 + \frac{1}{2} = \frac{3}{2}$ and $\frac{1}{3} + \frac{1}{4} = \frac{7}{12} > \frac{1}{4} + \frac{1}{4} = \frac{1}{2}$

$$\frac{1}{5} + \frac{1}{6} + \frac{1}{7} + \frac{1}{8} > \frac{1}{8} + \frac{1}{8} + \frac{1}{8} + \frac{1}{8} = \frac{4}{8} = \frac{1}{2}.$$

We can continue this grouping, taking twice the number of terms each time, and show that each group is greater than $\frac{1}{2}$. Thus the sum continuously grows and the series does not converge to a finite value.

General Directions For Testing a Series, $\Sigma_{n=1}^{\infty} u_n$.

1. Test $\lim_{n\to\infty} (u_n) = A$. If $A \neq 0$, the series diverges.
2. If the series is an alternating series and $A = 0$, then the series converges.
3. Test the ratio $\lim_{n\to\infty} \left|\dfrac{u_{n+1}}{u_n}\right| = L$ and if $L < 1$, the series converges; if $L > 1$, the series diverges; if $L = 1$, the test fails.
4. Use the integral test $\int_1^\infty f(x)\,dx$. If the integral exists, the series converges; if it does not exist, the series diverges.

If $b_n \geq 0$ and $b_n \leq a_n$ for all $n$ and (1) converges, then the series (2) converges. If $a_n \geq 0$ and series (1) diverges and $b_n \geq a_n$ for all $n$, then the series (2) diverges.

**Ratio Test.** For the series

$a_1 + a_2 + \ldots + a_n + a_{n+1} + \ldots$

with $a_n > 0$, we have

$$\text{convergence if } \lim_{n\to\infty}\left(\frac{a_{n+1}}{a_n}\right) = L < 1$$

$$\text{divergence if } \lim_{n\to\infty}\left(\frac{a_{n+1}}{a_n}\right) = L > 1$$

divergence if $L = +\infty$

no information if $L = 1$.

**Theorem.** If the series $\sum_{i=1}^{\infty} |a_i|$ converges, then $\sum_{i=1}^{\infty} a_i$ converges.

**Alternating Series.** If $a_n$ are alternately positive and negative, $|a_{n+1}| < |a_n|$ for every $n$, and $\lim_{n\to\infty} (a_n) = 0$, then the series is convergent.

A series is said to be **absolutely convergent** if the series formed from it by replacing all its terms by their absolute values is convergent.

**Power Series.** A series of the form

$a_0 + a_1 x + a_2 x^2 + \ldots + a_n x^n + \ldots$

where $x$ is a variable and $a_i$ are constants is called a power series.

**Maclaurin series** is a power series where $a_i = \dfrac{f^{(i)}(0)}{i!}$ and represents $f(x)$ in its interval of convergence.

**Taylor Series.** Let $a$ be a constant,

$$f(x) = \sum_{i=0}^{\infty} \frac{f^{(i)}(a)}{i!}(x - a)^i.$$

5. Compare the series with a known series such as
   a) Convergent series
      Geometric: $a + ar + ar^2 + \cdots ; r < 1$
      $p$-series: $1 + 2^{-p} + 3^{-p} + \cdots ; p > 1$.
   b) Divergent series
      Harmonic: $1 + \frac{1}{2} + \frac{1}{3} + \frac{1}{4} + \cdots$
      $p$-series: $1 + 2^{-p} + 3^{-p} + \cdots ; p < 1$.

Convergent power series may be differentiated and integrated. These properties are based on two theorems.

THEOREM. If the series $\sum_{n=0}^{\infty} a_n x^n$ converges for $|x| < k$ and $k > 0$, then the series obtained by a term-by-term differentiation of the given series converges for $|x| < k$.

THEOREM. If a function is represented by $f(x) = \sum_{n=0}^{\infty} a_n x^n$ and the series converges for $|x| < k$ and $k > 0$, then

$$F(x) = \int_0^x f(t)\,dt = \sum_{n=0}^{\infty} a_n \left(\frac{x^{n+1}}{n+1}\right)$$

holds for $|x| < k$.

**Power series** may also be added, subtracted, multiplied and divided. The addition and subtraction simply collects like powered terms into a new series. To illustrate consider

$$f = \sum_{n=0}^{\infty} a_n x^n = a_0 + a_1 x + a_2 x^2 + a_3 x^3 + \cdots + a_n x^n + \cdots$$

$$g = \sum_{n=0}^{\infty} b_n x^n = b_0 + b_1 x + b_2 x^2 + b_3 x^3 + \cdots + b_n x^n + \cdots$$

$$f + g = (a_0 + b_0) + (a_1 + b_1)x + \cdots + (a_n + b_n)x^n + \cdots$$

The product of two power series requires the multiplication of each term of one by each term of the other. This can be arranged in the order of increasing powers of the variable. Consider the two series $f$ and $g$ given above.

$$f \cdot g = c_0 + c_1 x + c_2 x^2 + c_3 x^3 + \cdots c_n x^n + \cdots = \sum_{n=0}^{\infty} c_n x^n$$

where

$$c_n = a_n b_0 + a_{n-1} b_1 + \cdots + a_0 b_n = \sum_{k=0}^{n} a_{n-k} b_k$$

This product is called the Cauchy Product.

The division of one series by another is set up in the same manner as long division of two polynomials.

Some well known series are listed in the appendix.

## SOLUTIONS TO THE EXERCISES

I

i. Given $\sum_{n=1}^{\infty} u_n$ with $u_n = \dfrac{3n}{2n-7}$

$$\lim_{n\to\infty} u_n = \lim_{n\to\infty}\left(\frac{3}{2-\frac{7}{n}}\right) = \frac{3}{2}$$

Since $u_n$ does not approach zero as $n \longrightarrow \infty$ the series diverges.

ii. $\sum_{n=1}^{\infty} \dfrac{(n+2)(n-1)}{3n^2-4}$

$$u_n = \frac{(n+2)(n-1)}{3n^2-4} = \frac{\left(1+\frac{2}{n}\right)\left(1-\frac{1}{n}\right)}{3-4n^{-2}}$$

$$\lim_{n\to\infty} u_n = \frac{(1)(1)}{3} = \frac{1}{3}$$

Series diverges.

II

i. $\sum_{n=1}^{\infty} \dfrac{nx^n}{2^{n-1}n(n+1)}$

$$\frac{u_{n+1}}{u_n} = \frac{n+1}{2(n+2)}\,x$$

$$\lim_{n\to\infty} |x| \left|\frac{n+1}{2(n+2)}\right| = |x|\,\frac{1}{2} = L$$

$L < 1$ for $-2 < x < 2$

At $x = -2$: $\sum_{n=1}^{\infty} (-1)^n \dfrac{2}{n+1}$

an alternating series with $u_n \longrightarrow 0$.

At $x = 2$: $\sum_{n=1}^{\infty} \dfrac{2}{n+1}$

The integral test shows divergence.
Convergence interval: $-2 \leq x < 2$.

ii. $\sum_{n=1}^{\infty} \dfrac{(n+1)x^n}{2^n\,3^{n+1}}$; $\dfrac{u_{n+1}}{u_n} = \dfrac{x(n+2)}{6(n+1)}$

$$\lim_{n\to\infty} \left|\frac{u_{n+1}}{u_n}\right| < 1 \text{ for } -6 < x < 6$$

At $x = -6$, we have an alternating series which diverges. At $x = 6$, the series also diverges.
Convergence interval: $-6 < x < 6$.

III

i.

$f(x) = \cos x \qquad f(a) = \frac{1}{2}$

$f'(x) = -\sin x \qquad f'(a) = -\frac{\sqrt{3}}{2}$

$f''(x) = -\cos x \qquad f''(a) = -\frac{1}{2}$

$f'''(x) = \sin x \qquad f'''(a) = \frac{\sqrt{3}}{2}$

$f^{iv}(x) = \cos x \qquad f^{iv}(a) = \frac{1}{2}$

$$\cos x = \frac{1}{2} - \frac{\sqrt{3}}{2}\left(x-\frac{\pi}{3}\right) - \frac{1}{4}\left(x-\frac{\pi}{3}\right)^2 + \frac{\sqrt{3}}{12}\left(x-\frac{\pi}{3}\right)^3 + \cdots$$

ii.

$f(x) = \sqrt{x} \qquad f(9) = 3$

$f'(x) = \frac{1}{2}x^{-\frac{1}{2}} \qquad f'(9) = \frac{1}{6}$

$f''(x) = -\frac{1}{4}x^{-\frac{3}{2}} \qquad f''(9) = -\frac{1}{108}$

$f'''(x) = \frac{3}{8}x^{-\frac{5}{2}} \qquad f'''(9) = \frac{1}{648}$

$$\sqrt{x} = 3 + \frac{1}{6}(x-9) - \frac{1}{216}(x-9)^2 + \frac{1}{3888}(x-9)^3 + \cdots$$

iii.

$f(x) = (1+x)^{-1} \qquad f(0) = 1$
$f'(x) = -(1+x)^{-2} \qquad f'(0) = -1$
$f''(x) = 2(1+x)^{-3} \qquad f''(0) = 2$
$f'''(x) = -6(1+x)^{-4} \qquad f'''(0) = -6$

$$\frac{1}{1+x} = 1 - x + x^2 - x^3 + x^4 + \cdots$$

The binomial series is the same.

# 22 SOLID ANALYTIC GEOMETRY

## SELF-TEST

DIRECTIONS: Write your answers in the numbered regions to the right. To check your answers, turn the page. Study the solutions to the problems you missed, and do the exercises following any group of problems in which you had incorrect answers.

1. Find the distance between $P_1\left(\frac{3}{2}, \frac{1}{3}, -\frac{2}{3}\right)$ and $P_2\left(\frac{1}{3}, -\frac{1}{2}, \frac{3}{2}\right)$

2. Do the lines joining $P_1(1,1,-5)$, $P_2(6,-9,10)$, and $P_3(6,11,0)$ form a right triangle?

3. Find the equation of the locus of a point which moves so that it is always equidistant from the points (4,1,3) and (2,0,1).

4. Find the direction cosines and the length of the radius vector of the point $P_1(-2,3,5)$.

5. Find the direction cosines of the lines joining $P_1$ and $P_2$, (5,−3,2) and (3,−2,−4).

Find the equation of the planes satisfying the conditions given in **6–8**.

6. The plane through the points (1,−3,2), (3,1,4), and (−1,−1,−2)

7. The plane through (4,2,−3) and perpendicular to the radius vector of that point

8. The plane whose $XY$-trace is $2x + 3y = 6$ and $YZ$-trace is $3y - 4z = 6$

9. Match the given equations with the listed names.

(A) $\frac{x^2}{4} + \frac{y^2}{9} + \frac{z^2}{16} = 1$
(B) $4x^2 + 9y^2 - z^2 = 36$
(C) $2x^2 + 3y^2 = 6z$
(D) $x^2 + y^2 + z^2 = 16$
(E) $3x^2 - 2y^2 = z$
(F) $x^2 + y^2 = 16$
(G) $x^2 - y^2 - z^2 = 1$

(a) hyperboloid of one sheet
(b) hyperboloid of two sheets
(c) sphere
(d) elliptic paraboloid
(e) ellipsoid
(f) hyperbolic paraboloid
(g) cylinder
(h) none of these

10. Find the distance from $P(3,-4,3\sqrt{5})$ to the plane

$$4x + 8y + 2\sqrt{5}z - 5 = 0.$$

1 ______
2 ______
3 ______
4 ______
5 ______
6 ______
7 ______
8 ______
9 (A) ______ (B) ______ (C) ______ (D) ______ (E) ______ (F) ______ (G) ______
10 ______

## SOLUTIONS

1. Given: $P_1\left(\frac{3}{2}, \frac{1}{3}, -\frac{2}{3}\right)$ and $P_2\left(\frac{1}{3}, -\frac{1}{2}, \frac{3}{2}\right)$

$$d = \sqrt{\left(\frac{3}{2}-\frac{1}{3}\right)^2 + \left(\frac{1}{3}+\frac{1}{2}\right)^2 + \left(-\frac{2}{3}-\frac{3}{2}\right)^2}$$

$$= \sqrt{\frac{49}{36}+\frac{25}{36}+\frac{169}{36}} = \frac{1}{6}\sqrt{243} = \frac{3}{2}\sqrt{3}$$

2. Given: $P_1(1,1,-5), P_2(6,-9,10), P_3(6,11,0)$
$\overline{P_1P_2}^2 = 25 + 100 + 225 = 350$
$\overline{P_1P_3}^2 = 25 + 100 + 25 = 150$
$\overline{P_1P_2}^2 + \overline{P_1P_3}^2 = 350 + 150 = 500 = \overline{P_2P_3}^2$
$\therefore \Delta P_1P_2P_3$ is a right triangle.

> **EXERCISES I**
>
> In i–ii, find the distance between the pairs of points.
>
> i. $P_1(5,-3,2)$ and $P_2(3,-2,-4)$
>
> ii. $P_1(1,2,8)$ and $P_2(-1,-1,2)$

3. Let $P(x, y, z)$ be the point moving at equal distances from $(4,1,3)$ and $(2,0,1)$.
$\sqrt{(x-4)^2 + (y-1)^2 + (z-3)^2}$
$= \sqrt{(x-2)^2 + (y-0)^2 + (z-1)^2}$
$x^2 - 8x + 16 + y^2 - 2y + 1 + z^2 - 6z + 1$
$= x^2 - 4x + 4 + y^2 + z^2 - 2z + 1$
$-4x + 12 - 2y + 1 - 4z = 0$
$4x + 2y + 4z = 13$

4. $p = \sqrt{x^2 + y^2 + z^2} = \sqrt{4 + 9 + 25} = \sqrt{38}$
$\cos\alpha = \frac{x}{p} = -\frac{2}{38}\sqrt{38} = -\frac{1}{19}\sqrt{38}$
$\cos\beta = \frac{y}{p} = \frac{3}{38}\sqrt{38}$, $\cos\gamma = \frac{z}{p} = \frac{5}{38}\sqrt{38}$

5. $d = \sqrt{41}$.
$\cos\alpha = \frac{x_2 - x_1}{d} = \frac{3-5}{41}\sqrt{41} = -\frac{2}{41}\sqrt{41}$
$\cos\beta = \frac{y_2 - y_1}{d} = \frac{-2+3}{41}\sqrt{41} = \frac{\sqrt{41}}{41}$
$\cos\gamma = \frac{z_2 - z_1}{d} = \frac{-4-2}{41}\sqrt{41} = -\frac{6}{41}\sqrt{41}$

> **EXERCISES II**
>
> In i–ii, find the direction cosines of the lines joining the pairs of points.
>
> i. $P_2(1,2,8)$ and $P_2(-1,-1,2)$
>
> ii. $P_1(8,3,6)$ and $P_2(5,3,2)$
>
> iii. A point is at a distance of 4 from the origin and its radius vector has $\cos\alpha = \frac{1}{2}$ and $\cos\beta = \frac{3}{4}$. Find its coordinates.
>
> iv. Find the angle between the line joining the points $P_1(3,1,2)$ and $P_2(1,0,4)$ and the line joining $P_3(5,-1,7)$ and $P_4(4,-2,7)$.
>
> v. Are the lines $\overline{P_1P_2}$ and $\overline{P_3P_4}$ perpendicular if $P_1(3,-1,2)$, $P_2(2,-1,1)$, $P_3(4,-2,3)$, and $P_4(2,-1,3)$?

6. Given: $P_1(1,-3,2), P_2(3,1,4), P_3(-1,-1,-2)$
Substituting into $Ax + By + Cz + D = 0$,

$$A - 3B + 2C + D = 0$$
$$3A + B + 4C + D = 0$$
$$-A - B - 2C + D = 0.$$

Let $D = 1$ and solve for $A = -\frac{5}{2}$, $B = \frac{1}{2}$, and $C = \frac{3}{2}$. The equation simplifies into $-5x + y + 3z + 2 = 0$.

7. The equation of the plane through $(4,2,-3)$ and $\perp \overline{OP}$ is
$a(x - x_1) + b(y - y_1) + c(z - z_1) = 0$
$4(x-4) + 2(y-2) - 3(z+3) = 0$
$4x + 2y - 3z - 11 = 0.$

8. Equation of a plane: $Ax + By + Cz + D = 0$
$XY$-trace: $Ax + By + D = 0$
$2x + 3y - 6 = 0$; $A = 2$, $B = 3$, $D = -6$
$YZ$-trace: $By + Cz + D = 0$
$3y - 4z - 6 = 0$; $B = 3$, $C = -4$, $D = -6$
Plane is $2x + 3y - 4z - 6 = 0$

### EXERCISES III

Find the equation of the planes satisfying the conditions of i–ii.

i. The plane parallel to the $XY$-plane and at a distance of 7 above it

ii. The plane through the line $x + y = 4$, $y - 3z = 2$ and perpendicular to the plane $3x - 3y + z = 5$

9. (A) $\frac{x^2}{4} + \frac{y^2}{9} + \frac{z^2}{16} = 1$: ellipsoid
   (B) $4x^2 + 9y^2 - z^2 = 36$: hyperboloid of one sheet
   (C) $2x^2 + 3y^2 = 6z$: elliptic paraboloid
   (D) $x^2 + y^2 + z^2 = 16$: sphere
   (E) $3x^2 - 2y^2 = z$: hyperbolic paraboloid
   (F) $x^2 + y^2 = 16$: cylinder
   (G) $x^2 - y^2 - z^2 = 1$ hyperboloid of two sheets

10. $4x + 8y + 2\sqrt{5}z = 5$, $P(3, -4, 3\sqrt{5})$

$$d = \frac{|ax_1 + by_1 + cz_1 + d|}{\sqrt{a^2 + b^2 + c^2}} = \frac{12 - 32 + 30 - 5}{\sqrt{16 + 64 + 20}} = \frac{5}{10} = \frac{1}{2}$$

### EXERCISES IV

i. Find the $XY$-trace of the surface
$$4x^2 - 8x + 9y^2 + 18y = z^2 - 2z$$
and name the curve.

ii. Find the three traces of the right circular cylinder whose elements are parallel to the $Z$-axis and base is a circle of radius 4 and center at the origin.

iii. Find the distance from the point $P(e, \pi, \sqrt{2})$ to the plane defined by
$$ex + \pi y + \sqrt{2}z = 0.$$

Solutions to the exercises appear on page 132.

## ANSWERS

1. $\frac{3}{2}\sqrt{3}$ — 1
2. yes — 2
3. $4x + 2y + 4z = 13$ — 3
4. $-\frac{\sqrt{38}}{19}$, $\frac{3\sqrt{38}}{38}$, $\frac{5\sqrt{38}}{38}$ — 4
5. $-\frac{2\sqrt{41}}{41}$, $\frac{\sqrt{41}}{41}$, $-\frac{6\sqrt{41}}{41}$ — 5
6. $5x - y - 3z = 2$ — 6
7. $4x + 2y - 3z = 11$ — 7
8. $2x + 3y - 4z = 6$ — 8
9. — 9
   (A)—(e) — (A) ______
   (B)—(a) — (B) ______
   (C)—(d) — (C) ______
   (D)—(c) — (D) ______
   (E)—(f) — (E) ______
   (F)—(g) — (F) ______
   (G)—(b) — (G) ______
10. $\frac{1}{2}$ — 10

## BASIC FACTS

A point $P(x, y, z)$, may be located in three dimensions by three real numbers called its coordinates.

The **three-dimensional coordinate system** has three axes, the $X$-, $Y$-, and $Z$-axes; and three coordinate planes, the $XY$-plane which contains the $X$- and $Y$-axes, the $YZ$-plane which contains the $Y$- and $Z$-axes, and the $XZ$-plane which contains the $X$- and $Z$-axis.

The coordinate planes divide the three-dimensional space into eight parts called **octants.** The intersection of the three axes is called the **origin,** $O(0, 0, 0)$.

The **distance,** $d$, between two points, $P_1(x_1, y_1, z_1)$ and $P_2(x_2, y_2, z_2)$, is given by

$$\sqrt{(x_2 - x_1)^2 + (y_2 - y_1)^2 + (z_2 - z_1)^2}.$$

The distance from the origin to any point $P(x, y, z)$ is given by

$$p = \sqrt{x^2 + y^2 + z^2}.$$

The **direction angles,** $\alpha$, $\beta$, and $\gamma$, of the line $\overline{OP}$ are the angles formed by $\overline{OP}$ and the $X$-, $Y$-, and $Z$-axes, respectively.

The **direction cosines** are

$$\cos\alpha = \frac{x}{p}, \cos\beta = \frac{y}{p}, \cos\gamma = \frac{z}{p}.$$

The direction cosines of a line through two points $P_1(x_1, y_1, z_1)$ and $P_2(x_2, y_2, z_2)$ are $\cos\alpha = \frac{x_2 - x_1}{d}$, $\cos\beta = \frac{y_2 - y_1}{d}$, and $\cos\gamma = \frac{z_2 - z_1}{d}$.

The sum of the squares of the direction cosines of any line equals one.

The **cosine of the angle,** $\theta$, **between two lines** is equal to the sum of the products of their corresponding direction cosines.

Two lines in space are perpendicular if $\cos\theta = 0$.

**A plane** is a surface such that a straight line joining any two points

## ADDITIONAL INFORMATION

One of the easiest ways to visualize a rectangular coordinate system in space is to look into the corner of a room. Here we have three mutually perpendicular planes, the two walls and the floor. We can locate a point as a certain distance from each of the walls and from the floor.

The positive and negative directions of the coordinate axes are shown in Figure 22-1.

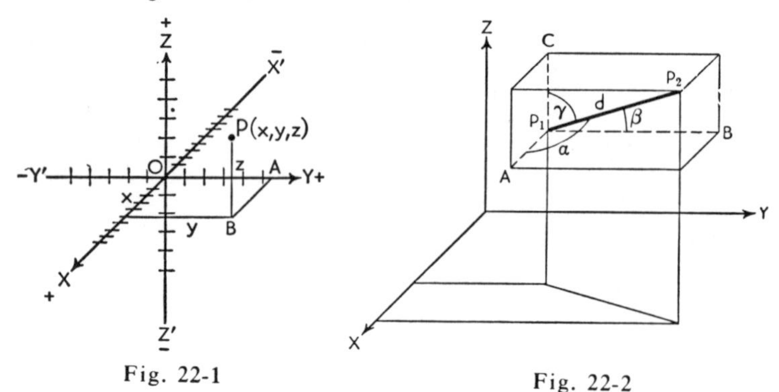


Fig. 22-1

Fig. 22-2

Consider the line between the two points $P_1(x_1, y_1, z_1)$ and $P_2(x_2, y_2, z_2)$. The **direction angles** of this line are shown in Figure 22-2. Let $a = x_2 - x_1$, $b = y_2 - y_1$, and $c = z_2 - z_1$. Then, since

$$\cos\alpha = \frac{x_2 - x_1}{d} = \frac{a}{d}, \quad \cos\beta = \frac{y_2 - y_1}{d} = \frac{b}{d}, \quad \cos\gamma = \frac{z_2 - z_1}{d} = \frac{c}{d},$$

we see that the numbers $a$, $b$, and $c$ are proportional to the **direction cosines.** They are called the **direction numbers** of the line, and we note that $d = \sqrt{a^2 + b^2 + c^2}$.

The formula for the cosine of **the angle between two lines** in space can be derived by considering the origin to be at the projected point of intersection of the two lines. Choose a point, $P_1$, on line $L_1$ and a point, $P_2$, on line $L_2$. Then the cosine law for triangle $OP_1P_2$ states that

$$\cos\theta = \frac{p_1^2 + p_2^2 - d^2}{2p_1p_2} \quad \text{where } \overline{OP_1} = p_1, \overline{OP_2} = p_2, \text{ and } d = \overline{P_1P_2}.$$

Since $p_1^2 = x_1^2 + y_1^2 + z_1^2$, $p_2^2 = x_2^2 + y_2^2 + z_2^2$, and $d$ is the distance formula, we have $p_1^2 + p_2^2 - d^2 = 2(x_1x_2 + y_1y_2 + z_1z_2)$. If the values $x_1 = p_1\cos\alpha_1$, $y_1 = p_1\cos\beta_1$, $z_1 = p_1\cos\gamma_1$, *etc.* are substituted into this formula and the result is substituted into the expression for the $\cos\theta$, we obtain

$$\cos\theta = \cos\alpha_1\cos\alpha_2 + \cos\beta_1\cos\beta_2 + \cos\gamma_1\cos\gamma_2.$$

If $\overline{OP_1} \perp \overline{OP_2}$ then the right side of this equation is zero and

$$a_1a_2 + b_1b_2 + c_1c_2 = 0.$$

The locus of a linear equation in three variables is a **plane.** If the equation is given we can find the **intercepts** by setting two of the

of the surface lies entirely in the surface. The locus of the equation

$$Ax + By + Cz + D = 0 \qquad (1)$$

is a plane.

An equation of the first degree in one variable, when considered in three dimensions, represents a plane parallel to the plane of the other two variables.

The **normal form** of the equation of a plane is

$$x \cos \alpha + y \cos \beta + z \cos \gamma = p$$

where $\alpha$, $\beta$, and $\gamma$ are the direction angles of the line from the origin perpendicular to the plane.

The distance from the point $P_1(x_1, y_1, z_1)$ to the plane given by $Ax + By + Cz + D = 0$ is

$$d = \frac{|Ax_1 + By_1 + Cz_1 + D|}{\sqrt{A^2 + B^2 + C^2}}$$

To reduce the equation of any plane given by (1) to the normal form, divide by $\sqrt{A^2 + B^2 + C^2}$ and choose signs so that the constant is positive in the right member.

Two planes are **perpendicular** if $A_1A_2 + B_1B_2 + C_1C_2 = 0$.

The intersections of a surface with the coordinate planes are called the **traces** of the surface. The traces of a plane are straight lines.

The locus of two simultaneous equations of the first degree is a **straight line.** The **general form** of the equations of a line is

$$A_1 x + B_1 y + C_1 z + D_1 = 0$$
$$A_2 x + B_2 y + C_2 z + D_2 = 0.$$

The **symmetric form** of the equations of a line through two points, $P_1$ and $P_2$, is given by

$$\frac{x - x_1}{x_2 - x_1} = \frac{y - y_1}{y_2 - y_1} = \frac{z - z_1}{z_2 - z_1}.$$

Let $u = 0$ and $v = 0$ be the equations of any line. Then $u + kv = 0$, where $k$ is a constant, represents a plane through the given line.

variables equal to zero and solving for the remaining one. The **traces** can be found by setting one of the variables equal to zero. The **normal form** of the equation of a plane can be found by dividing the equation by the square root of the sum of the squares of the coefficients of the variables and choosing the sign of the radical so that the constant is positive when placed in the right member of the equation.

The **equation of a plane through a point** $P_1(x_1, y_1, z_1)$ can be written in the form $A(x - x_1) + B(y - y_1) + C(z - z_1) = 0$. To find the equation of a plane through three given points, substitute the coordinates of the points into the general equation of a plane. There results three linear equations in the four unknowns $A$, $B$, $C$, $D$. Let $D = 1$ and solve for $A$, $B$, and $C$. Place these values into the general form and simplify the equation.

It takes more than one equation to define a **straight line in space.** Since two planes intersect in a straight line, the locus of two simultaneous equations of the first degree is a straight line. The **general form** of a system of two linear equations in three variables is therefore taken as the general form of the equations of a straight line. The **symmetric form** can be obtained by considering a line through a point $(x_1, y_1, z_1)$ with given direction numbers $a$, $b$, and $c$. Let $(x, y, z)$ be any point on this line; then the direction numbers are proportional to $x - x_1$, $y - y_1$, and $z - z_1$ and we write $\dfrac{x - x_1}{a} = \dfrac{y - y_1}{b} = \dfrac{z - z_1}{c}$.

If two points are given then $a = x_2 - x_1$, $b = y_2 - y_1$, and $c = z_2 - z_1$. Substitute these to symmetric form to obtain the equations of a line.

To reduce the equations of a line to a symmetric form eliminate any of two variables, obtaining two equations which have the third variable in common. Solve for this variable and equate the values. Finally, divide through by a number which will make the coefficient of each variable positive and equal to unity.

A surface which can be obtained by rotating a curve about a straight line is called a **surface of revolution.** The following table gives the equations and corresponding names of some surfaces.

| Equation | Surface |
|---|---|
| $A(x^2 + y^2 + z^2) + Gx + Hy + Iz + K = 0$ | sphere |
| $\frac{x^2}{a^2} + \frac{y^2}{b^2} + \frac{z^2}{c^2} = 1$ | ellipsoid |
| $\frac{x^2}{a^2} + \frac{y^2}{b^2} - \frac{z^2}{c^2} = 1$ | hyperboloid of one sheet |
| $\frac{x^2}{a^2} - \frac{y^2}{b^2} - \frac{z^2}{c^2} = 1$ | hyperboloid of two sheets |
| $\frac{x^2}{a^2} + \frac{y^2}{b^2} = \frac{z}{c}$ | elliptic paraboloid |
| $\frac{x^2}{a^2} - \frac{y^2}{b^2} = \frac{z}{c}$ | hyperbolic paraboloid |
| $\frac{x^2}{a^2} + \frac{y^2}{b^2} - \frac{z^2}{c^2} = 0$ | elliptic cone |

## SOLUTIONS TO EXERCISES

I

i. Given: $P_1(5,-3,2)$ and $P_2(3,-2,-4)$

$$d = \sqrt{(x_1 - x_2)^2 + (y_1 - y_2)^2 + (z_1 - z_2)^2}$$
$$= \sqrt{(5-3)^2 + (-3+2)^2 + (2+4)^2}$$
$$= \sqrt{4+1+36} = \sqrt{41}$$

ii. Given: $P_1(1,2,8)$ and $P_2(-1,-1,2)$

$$d = \sqrt{2^2 + 3^2 + 6^2} = \sqrt{49} = 7$$

II

i. $d = 7$

$$\cos\alpha = -\frac{2}{7},\ \cos\beta = -\frac{3}{7},\ \cos\gamma = -\frac{6}{7}$$

ii. Given: $P_1(8,3,6)$ and $P_2(5,3,2)$

$$d = 5,\ \cos\alpha = -\frac{3}{5},\ \cos\beta = 0,\ \cos\gamma = -\frac{4}{5}$$

iii. Given: $p = 4$, $\cos\alpha = \frac{1}{2}$, $\cos\beta = \frac{3}{4}$

$$x = p\cos\alpha = 4\left(\frac{1}{2}\right) = 2$$
$$y = p\cos\beta = 4\left(\frac{3}{4}\right) = 3$$
$$\cos^2\gamma = 1 - \cos^2\alpha - \cos^2\beta = 1 - \frac{1}{4} - \frac{9}{16} = \frac{3}{16}$$
$$\cos\gamma = \frac{\sqrt{3}}{4},\ z = p\cos\gamma = \sqrt{3}$$

iv. The line $\overline{P_1P_2}$: $a_1 = -2, b_1 = -1, c_1 = 2$
The line $\overline{P_3P_4}$: $a_2 = -1, b_2 = -1, c_2 = 0$
Let $N = a_1a_2 + b_1b_2 + c_1c_2 = 2 + 1 + 0 = 3$

$$d_1 = \sqrt{a_1^2 + b_1^2 + c_1^2} = \sqrt{4+1+4} = 3,\ d_2 = \sqrt{2}$$
$$\cos\phi = \frac{N}{d_1d_2} = \frac{3}{3\sqrt{2}} = \frac{\sqrt{2}}{2},\ \phi = 45°$$

v. $a_1a_2 + b_1b_2 + c_1c_2 = 2 + 0 + 0 = 2$
The lines are not perpendicular.

III

i. The plane parallel to the $XY$-plane and at a distance of 7 above it is $z = 7$.

ii. The family of planes on the given line is $x + (1+k)y - 3kz - 4 - 2k = 0$. The plane on the given line and $\perp$ to $3x - 3y + z - 5 = 0$ is $x - 4y + 15z + 6 = 0$ since $3(1) - 3(1+k) + (1)(-3k) - 5(-4-2k) = 0$, $k = -5$.

IV

i. $4x^2 - 8x + 9y^2 + 18y = z^2 - 2z$
For the $XY$-trace we have $z = 0$:
$4x^2 - 8x + 9y^2 + 18y = 0$
which is an ellipse with the center at $(1,-1)$.

ii. The equation of the right circular cylinder is $x^2 + y^2 = 16$ which is also the $XY$-trace. The $XZ$-trace is $x^2 = 16$ and the $YZ$-trace is $y^2 = 16$.

iii. The distance formula is

$$d = \frac{|Ax_1 + By_1 + Cz_1 + d|}{\sqrt{A^2 + B^2 + C^2}}$$
$$= \frac{e^2 + \pi^2 + 2}{\sqrt{e^2 + \pi^2 + 2}}$$
$$= \sqrt{e^2 + \pi^2 + 2}.$$

# 23 THREE DIMENSIONAL VECTORS

## SELF-TEST

DIRECTIONS: Write your answers in the numbered regions to the right. To check your answers, turn the page. Study the solutions to the problems you missed, and do the exercises following any problem or group of problems in which you had errors.

In **1–5** find the indicated quantity given the three vectors, $\mathbf{v} = 2\mathbf{i} - 3\mathbf{j} + \mathbf{k}$, $\mathbf{w} = \mathbf{i} + 2\mathbf{j} - 3\mathbf{k}$, $\mathbf{u} = 3\mathbf{i} - \mathbf{j} - 2\mathbf{k}$.

**1.** Find $\mathbf{r} = \mathbf{u} + \mathbf{v} - \mathbf{w}$.

**2.** Find $\mathbf{v} \cdot \mathbf{w} + \mathbf{u} \cdot \mathbf{v}$.

**3.** Find the angle between **u** and **v**.

**4.** Find $\mathbf{v} \times (\mathbf{u} + \mathbf{w})$.

**5.** Find the projection of **v** on **w**.

**6.** Find the equation of a plane through $A(1,-1,3)$, $B(2,4,-1)$, and $C(3,-2,-4)$.

**7.** Find $\mathbf{f}''(t)$ if $\mathbf{f}(t) = t^2\mathbf{i} - t^3\mathbf{j} + t^{-1}\mathbf{k}$.

**8.** Find $\mathbf{g}'(t)$ if $\mathbf{g}(t) = \mathbf{u}(t) \times \mathbf{v}(t)$ and $\mathbf{u}(t) = \sin t\,\mathbf{i} + \cos t\,\mathbf{j} + e^t\mathbf{k}$
$\mathbf{v}(t) = \cos t\,\mathbf{i} + \sin t\,\mathbf{j} - e^t\mathbf{k}$.

**9.** Find $h'(t)$ if $h(t) = \mathbf{f}(t) \cdot [\mathbf{u}(t) \times \mathbf{v}(t)]$ where $\mathbf{f}(t) = t\,\mathbf{i} - 2\mathbf{j} + t^2\mathbf{k}$, $\mathbf{u}(t) = 2\mathbf{i} - t\,\mathbf{j} + 2t\,\mathbf{k}$, and $\mathbf{v}(t) = t^2\mathbf{i} - 2t\,\mathbf{j} - \mathbf{k}$.

**10.** Find the length of the curve defined by $x = 2t$, $y = t^2$, $z = \frac{1}{3}t^3 + 1$ and $0 \leq t \leq 3$.

**11.** Find the unit tangent vector $\mathbf{T}(t)$ for
$\mathbf{r}(t) = e^t \sin t\,\mathbf{i} + e^{-t}\mathbf{j} + e^t \cos t\,\mathbf{k}$.

**12.** Find the equation of the osculating plane to the curve $x = \cos t$, $y = \sin t$, $z = 3t$ and $t_0 = \frac{\pi}{2}$.

1 ______
2 ______
3 ______
4 ______
5 ______
6 ______
7 ______
8 ______
9 ______
10 ______
11 ______
12 ______

## SOLUTIONS

1. $\mathbf{r} = \mathbf{u} + \mathbf{v} - \mathbf{w}$
$= (3 + 2 - 1)\mathbf{i} + (-1 - 3 - 2)\mathbf{j} + (-2 + 1 + 3)\mathbf{k}$
$= 4\mathbf{i} - 6\mathbf{j} + 2\mathbf{k}$

2. $\mathbf{v} \cdot \mathbf{w} = (2)(1) + (-3)(2) + (1)(-3) = -7$
$\mathbf{u} \cdot \mathbf{w} = (2)(3) + (-3)(-1) + (1)(-2) = 7$
$\mathbf{v} \cdot \mathbf{w} + \mathbf{u} \cdot \mathbf{v} = -7 + 7 = 0$

3. $\cos\theta = \dfrac{\mathbf{u} \cdot \mathbf{v}}{|\mathbf{u}||\mathbf{v}|}$
$|\mathbf{u}| = \sqrt{9 + 1 + 4} = \sqrt{14}$
$|\mathbf{v}| = \sqrt{4 + 9 + 1} = \sqrt{14}$
$\cos\theta = \dfrac{7}{\sqrt{14}\sqrt{14}} = \dfrac{7}{14} = \dfrac{1}{2}, \ \theta = \dfrac{\pi}{3}$

4. $\mathbf{u} + \mathbf{w} = 4\mathbf{i} + \mathbf{j} - 5\mathbf{k}$
$\mathbf{v} \times (\mathbf{u} + \mathbf{w}) = (2\mathbf{i} - 3\mathbf{j} + \mathbf{k}) \times (4\mathbf{i} + \mathbf{j} - 5\mathbf{k})$
$$= \begin{vmatrix} \mathbf{i} & \mathbf{j} & \mathbf{k} \\ 2 & -3 & 1 \\ 4 & 1 & -5 \end{vmatrix}$$
$= (15 - 1)\mathbf{i} - (-10 - 4)\mathbf{j} + (2 + 12)\mathbf{k}$
$= 14\mathbf{i} + 14\mathbf{j} + 14\mathbf{k}$

5. $|\mathbf{v}|\cos\theta = \dfrac{\mathbf{v} \cdot \mathbf{w}}{|\mathbf{w}|} = \dfrac{-7}{\sqrt{14}} = -\sqrt{\dfrac{7}{2}}$

### EXERCISES I

$\mathbf{v} = 3\mathbf{i} + \mathbf{j} - 7\mathbf{k}$, $\mathbf{u} = -\mathbf{i} - 3\mathbf{j} + 5\mathbf{k}$, $\mathbf{w} = 2\mathbf{i} + 5\mathbf{j} + 3\mathbf{k}$

i. Find $\mathbf{v} - \mathbf{u}$.

ii. Find $\mathbf{u} \cdot \mathbf{w}$.

iii. Find $\mathbf{v} \times \mathbf{w}$.

iv. Find $5|\mathbf{w}|$.

v. Find $(2\mathbf{v} - 3\mathbf{w}) \times \mathbf{u}$.

vi. Find projection of $\mathbf{v}$ on $\mathbf{w}$.

6. $A(1,-1,3)$, $B(2,4,-1)$, $C(3,-2,-4)$
$\overrightarrow{AB} = (2 - 1)\mathbf{i} + (4 + 1)\mathbf{j} + (-1 - 3)\mathbf{k}$
$= \mathbf{i} + 5\mathbf{j} - 4\mathbf{k}$
$\overrightarrow{AC} = (3 - 1)\mathbf{i} + (-2 + 1)\mathbf{j} + (-4 - 3)\mathbf{k}$
$= 2\mathbf{i} - \mathbf{j} - 7\mathbf{k}$
$\overrightarrow{AB} \times \overrightarrow{AC} = (\mathbf{i} + 5\mathbf{j} - 4\mathbf{k}) \times (2\mathbf{i} - \mathbf{j} - 7\mathbf{k})$
$= -39\mathbf{i} - \mathbf{j} - 11\mathbf{k}$
$-39(x - 1) - (y + 1) - 11(z - 3) = 0$
$39x + y + 11z = 71$

### EXERCISES II

Given $A(2,-1,3), B(1,3,-2), C(2,2,2)$

i. Find $2\overrightarrow{AC} + 3\overrightarrow{AB}$.

ii. Find $|\overrightarrow{BC}|$.

iii. Find the equation of a plane through $A$, $B$, and $C$.

7. $\mathbf{f}(t) = t^2\mathbf{i} - t^3\mathbf{j} + t^{-1}\mathbf{k}$
$\mathbf{f}'(t) = 2t\,\mathbf{i} - 3t^2\mathbf{j} - t^{-2}\mathbf{k}$
$\mathbf{f}''(t) = 2\mathbf{i} - 6t\,\mathbf{j} + 2t^{-3}\mathbf{k}$

8. $\mathbf{g}(t) = \mathbf{u}(t) \times \mathbf{v}(t)$
$$= \begin{vmatrix} \mathbf{i} & \mathbf{j} & \mathbf{k} \\ \sin t & \cos t & e^t \\ \cos t & \sin t & -e^t \end{vmatrix}$$
$= \mathbf{i}(-e^t\cos t - e^t\sin t) - \mathbf{j}(-e^t\sin t - e^t\cos t) + \mathbf{k}(\sin^2 t - \cos^2 t)$
$= -e^t(\cos t + \sin t)\mathbf{i} + e^t(\sin t + \cos t)\mathbf{j} - \cos 2t\,\mathbf{k}$
$\mathbf{g}'(t) = -e^t(\cos t + \sin t - \sin t + \cos t)\mathbf{i} + e^t(\sin t + \cos t + \cos t - \sin t)\mathbf{j} + 2\sin 2t\,\mathbf{k}$
$= -2e^t\cos t\,\mathbf{i} + 2e^t\cos t\,\mathbf{j} + 2\sin 2t\,\mathbf{k}$

9. $\mathbf{f}(t) = t\,\mathbf{i} - 2\mathbf{j} + t^2\mathbf{k}$
$\mathbf{u}(t) = 2\mathbf{i} - t\,\mathbf{j} + 2t\,\mathbf{k}$
$\mathbf{v}(t) = t^2\mathbf{i} - 2t\,\mathbf{j} - \mathbf{k}$
$$\mathbf{u}(t) \times \mathbf{v}(t) = \begin{vmatrix} \mathbf{i} & \mathbf{j} & \mathbf{k} \\ 2 & -t & 2t \\ t^2 & -2t & -1 \end{vmatrix}$$
$= (t + 4t^2)\mathbf{i} + (2 + 2t^3)\mathbf{j} + (t^3 - 4t)\mathbf{k}$
$h(t) = \mathbf{f}(t) \cdot [\mathbf{u}(t) \times \mathbf{v}(t)]$
$= t^5 - 4t^3 + t^2 - 4$
$h'(t) = 5t^4 - 12t^2 + 2t$

10. $\mathbf{r}(t) = x(t)\mathbf{i} + y(t)\mathbf{j} + z(t)\mathbf{k}$
$x = 2t, \ y = t^2, \ z = \frac{1}{3}t^3 + 1$
$x' = 2, \ y' = 2t, \ z' = t^2$
$x'^2 + y'^2 + z'^2 = (t^2 + 2)^2$
$|\mathbf{r}'(t)| = t^2 + 2$
$l(C) = \displaystyle\int_0^3 |\mathbf{r}'(t)|\,dt = \int_0^3 (t^2 + 2)\,dt$
$= \frac{1}{3}t^3 + 2t\,\big|_0^3 = 9 + 6 = 15$

11. $\mathbf{r}(t) = e^t \sin t\, \mathbf{i} + e^{-t}\mathbf{j} + e^t \cos t\, \mathbf{k}$
$\mathbf{r}'(t) = e^t(\sin t + \cos t)\mathbf{i} - e^{-t}\mathbf{j}$
$\quad + e^t(\cos t - \sin t)\mathbf{k}$
$a^2 = e^{2t}(\sin^2 t + 2 \sin t \cos t + \cos^2 t)$
$b^2 = e^{-2t}$
$c^2 = e^{2t}(\cos^2 t - 2 \sin t \cos t + \sin^2 t)$
$a^2 + b^2 + c^2 = e^{2t}(2) + e^{-2t}$
$|\mathbf{r}'(t)| = \sqrt{2e^{2t} + e^{-2t}}$
$$\mathbf{T}(t) = \frac{\mathbf{r}'(t)}{\sqrt{2e^{2t} + e^{-2t}}}$$

**EXERCISES III**

$\mathbf{f}(t) = 2t^2\mathbf{i} - 3t\,\mathbf{j} + (2t - 1)\mathbf{k}$
$\mathbf{u}(t) = 2e^t\mathbf{i} + 3e^t\mathbf{j} - 6e^t\mathbf{k}$
$\mathbf{v}(t) = \ln \sec t\, \mathbf{i} + \mathbf{j} - t\,\mathbf{k}$

i. Find $\mathbf{f}'(t) \times \mathbf{u}'(t)$.

ii. Find $h'(t)$ if $h(t) = \mathbf{f}'(t) \cdot \mathbf{v}'(t)$.

iii. Find $\mathbf{T}(t)$ for $\mathbf{v}(t)$.

iv. Find $|\mathbf{K}(t)|$ for $\mathbf{u}(t)$.

v. Find the equation of the osculating plane to $\mathbf{v}(t)$ at $t_0 = 0$.

12. $x = \cos t,\ y = \sin t,\ z = \sqrt{3}t$
$t_0 = \frac{\pi}{2},\ x_0 = 0,\ y_0 = 1,\ z_0 = \frac{\sqrt{3}\pi}{2}$
$x' = -\sin t,\ y' = \cos t,\ z' = \sqrt{3}$
$\mathbf{r}'(t) = -\sin t\, \mathbf{i} + \cos t\, \mathbf{j} + \sqrt{3}\mathbf{k}$
$|\mathbf{r}'(t)| = 2 = s'(t)$
$\mathbf{T}(t) = \frac{1}{2}(-\sin t\, \mathbf{i} + \cos t\, \mathbf{j} + \sqrt{3}\mathbf{k})$
$\mathbf{T}'(t) = \frac{1}{2}(-\cos t\, \mathbf{i} - \sin t\, \mathbf{j})$
$\mathbf{K}(t) = \frac{1}{4}(-\cos t\, \mathbf{i} - \sin t\, \mathbf{j})$
$|\mathbf{K}(t)| = \frac{1}{4}\sqrt{\cos^2 t + \sin^2 t} = \frac{1}{4}$
$\mathbf{N}(t) = (-\cos t\, \mathbf{i} - \sin t\, \mathbf{j})$
$\mathbf{B}(t) = \mathbf{T}(t) \times \mathbf{N}(t)$
$\quad = \frac{1}{2}(\sqrt{3} \sin t\, \mathbf{i} - \sqrt{3} \cos t\, \mathbf{j} + \mathbf{k})$
$\mathbf{B}(t_0) = \frac{\sqrt{3}}{2}\mathbf{i} - 0\,\mathbf{j} + \frac{1}{2}\mathbf{k}$
$$\frac{\sqrt{3}}{2}(x - 0) - 0(y - 1) + \frac{1}{2}\left(z - \frac{\sqrt{3}\pi}{2}\right) = 0$$
$2\sqrt{3}x + 2z = \sqrt{3}\pi$

Solutions to the exercises are on page 138.

## ANSWERS

| | |
|---|---|
| 1. $4\mathbf{i} - 6\mathbf{j} + 2\mathbf{k}$ | 1 |
| 2. 0 | 2 |
| 3. $\theta = \frac{\pi}{3}$ | 3 |
| 4. $14\mathbf{i} + 14\mathbf{j} + 14\mathbf{k}$ | 4 |
| 5. $-\sqrt{\frac{7}{2}}$ | 5 |
| 6. $39x + y + 11z = 71$ | 6 |
| 7. $2\mathbf{i} - 6t\,\mathbf{j} + 2t^{-3}\mathbf{k}$ | 7 |
| 8. $-2e^t\cos t\, \mathbf{i} + 2e^t \cos t\, \mathbf{j} + 2 \sin 2t\, \mathbf{k}$ | 8 |
| 9. $5t^4 - 12t^2 + 2t$ | 9 |
| 10. 15 | 10 |
| 11. $(2e^{2t} + e^{-2t})^{-\frac{1}{2}}[e^t(\sin t + \cos t)\mathbf{i} - e^{-t}\mathbf{j} + e^t(\cos t - \sin t)\mathbf{k}]$ | 11 |
| 12. $2\sqrt{3}\,x + 2z = \sqrt{3}\,\pi$ | 12 |

## BASIC FACTS

Review Chapter 16 for notation. Let i, j, k be unit vectors along the $X$, $Y$, and $Z$ axes respectively, then a vector v may be written

$$\mathrm{v} = a\,\mathrm{i} + b\,\mathrm{j} + c\,\mathrm{k}$$

where $a, b, c$ are scalar numbers and $\mathrm{i} = \mathrm{i}(1, 0, 0)$, $\mathrm{j} = \mathrm{j}(0, 1, 0)$, $\mathrm{k} = \mathrm{k}(0, 0, 1)$. The length of v is $|\mathrm{v}| = \sqrt{a^2 + b^2 + c^2}$. The vector from $P_1(x_1, y_1, z_1)$ to $P_2(x_2, y_2, z_2)$:

$$\mathrm{v} = (x_2 - x_1)\,\mathrm{i} + (y_2 - y_1)\,\mathrm{j} + (z_2 - z_1)\,\mathrm{k}.$$

**Operations.** Let $\mathrm{v} = a_1\mathrm{i} + b_1\mathrm{j} + c_1\mathrm{k}$ and $\mathrm{w} = a_2\mathrm{i} + b_2\mathrm{j} + c_2\mathrm{k}$, then:

$$\mathrm{v} \pm \mathrm{w} = (a_1 \pm a_2)\,\mathrm{i} + (b_1 \pm b_2)\,\mathrm{j} + (c_1 \pm c_2)\,\mathrm{k}$$

$$h\,\mathrm{v} = ha_1\mathrm{i} + hb_1\mathrm{j} + hc_1\mathrm{k}$$

$$\mathrm{v} \cdot \mathrm{w} = a_1a_2 + b_1b_2 + c_1c_2$$

$$\mathrm{v} \times \mathrm{w} = \begin{vmatrix} \mathrm{i} & \mathrm{j} & \mathrm{k} \\ a_1 & b_1 & c_1 \\ a_2 & b_2 & c_2 \end{vmatrix}$$

$$\mathrm{i} \times \mathrm{i} = \mathrm{j} \times \mathrm{j} = \mathrm{k} \times \mathrm{k} = 0$$

$$\mathrm{i} \times \mathrm{j} = -\mathrm{j} \times \mathrm{i} = \mathrm{k},\ \mathrm{j} \times \mathrm{k} = -\mathrm{k} \times \mathrm{j} = \mathrm{i}$$

$$\mathrm{k} \times \mathrm{i} = -\mathrm{i} \times \mathrm{k} = \mathrm{j},\ \mathrm{u} \times \mathrm{v} = -\mathrm{v} \times \mathrm{u}$$

$$\mathrm{u} \times (\mathrm{v} \pm \mathrm{w}) = \mathrm{u} \times \mathrm{v} \pm \mathrm{u} \times \mathrm{w}.$$

The angle between two vectors v and w is given by $\cos\theta = \dfrac{a_1a_2 + b_1b_2 + c_1c_2}{|\mathrm{v}||\mathrm{w}|}$.

If $\cos\theta = 0$, then $\mathrm{v} \perp \mathrm{w}$. The projection of v along w is given by $|\mathrm{v}| \cos\theta$. Let $\mathrm{f}(t) = f_1(t)\,\mathrm{i} + f_2(t)\,\mathrm{j} + f_3(t)\,\mathrm{k}$ be a vector function in variable $t$.

$$\mathrm{f}'(t) = f_1'(t)\,\mathrm{i} + f_2'(t)\,\mathrm{j} + f_3'(t)\,\mathrm{k}$$

If $f(t) = \mathrm{u}(t) \cdot \mathrm{v}(t)$, then $f'(t) = \mathrm{u}(t) \cdot \mathrm{v}'(t) + \mathrm{u}'(t) \cdot \mathrm{v}(t)$. If $\mathrm{w}(t) = \mathrm{u}(t) \times \mathrm{v}(t)$, then $\mathrm{w}'(t) = \mathrm{u}(t) \times \mathrm{v}'(t) + \mathrm{u}'(t) \times \mathrm{v}(t)$. Let $\mathrm{r}(t) = x(t)\,\mathrm{i} + y(t)\,\mathrm{j} + z(t)\,\mathrm{k}$ represent an arc $C$ in space, then the **length of** the arc is given by $l(C) = \int_a^b |\mathrm{r}'(t)|\,dt$ where the length

$$|\mathrm{r}'(t)| = \sqrt{(x')^2 + (y')^2 + (z')^2} = s'(t).$$

The **unit tangent vector** to $C$ is

$$\mathrm{T}(t) = \frac{\mathrm{r}'(t)}{|\mathrm{r}'(t)|} = \frac{\mathrm{r}'(t)}{s'(t)} = \frac{d\mathrm{r}}{ds}.$$

## ADDITIONAL INFORMATION

Vectors are represented by directed line segments to indicate a quantity that has direction and magnitude—the length of the line segment. If the base (initial point) of the vector is placed at the origin (see Figure 23-1), then the vector can be described by its components along the three coordinate axes. This yields the form $\mathrm{v} = a\,\mathrm{i} + b\,\mathrm{j} + c\,\mathrm{k}$ where i, j, k are unit vectors along the axes.

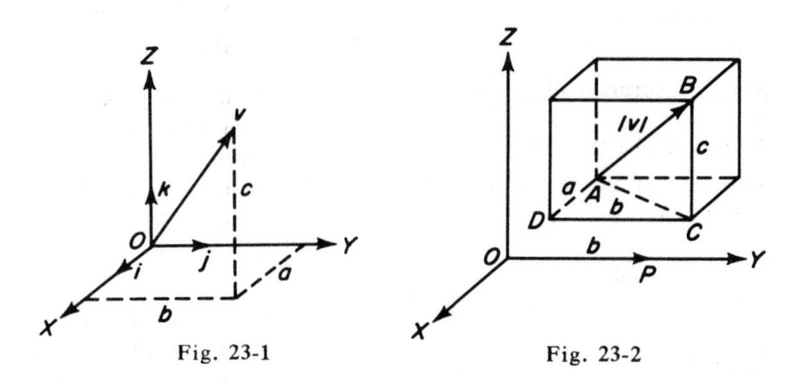


Fig. 23-1 Fig. 23-2

A vector need not be placed with one end at the origin. The vector from $A$ to $B$ shown in Figure 23-2 still has components along the axes as projections on lines parallel to the axes. Two vectors are said to be equal if they have the same length and same direction. It is not necessary that they be located in the same place in space. Thus in Figure 23-2, $\overrightarrow{DC}$ is equal to $\overrightarrow{OP}$. The formula for the length of a vector is derived in the same manner as the distance between two points. Refer again to Figure 23-2. Triangle $ADC$ is a right triangle with right angle at $D$ and by the Pythagorean Theorem $AC = \sqrt{a^2 + b^2}$. Now triangle $ACB$ is a right triangle with the right angle at $C$, and we have $\overline{AB}^2 = (\sqrt{a^2 + b^2})^2 + c^2$ or $\overline{AB} = \sqrt{a^2 + b^2 + c^2}$ which is the length of the vector v. If we let the coordinates of the points be $A(x_0, y_0, z_0)$ and $B(x_1, y_1, z_1)$, then we see that $a = x_1 - x_0$, $b = y_1 - y_0$, and $c = z_1 - z_0$. If we make the substitution into the formula for $|\mathrm{v}|$, we have the familiar formula for the distance between two points.

**The Vector Product,** $\mathrm{w} = \mathrm{u} \times \mathrm{v}$. Let **u** and **v** form a plane and let **n** be a unit vector perpendicular to this plane and pointing in the direction in which a right handed screw advances when its head is rotated from **u** to **v** through the angle $\theta$. See Figure 23-3. Then **u**, **v**, and **n** form a **right-handed triple.** The definition of the cross product of a vector **u** by a vector **v** is a vector **w** that is perpendicular to the plane determined by **u** and **v** and whose magnitude is given by $|\mathrm{w}| = |\mathrm{u}||\mathrm{v}| \sin\theta$ where $\theta$ is the angle between **u** and **v** and the direction is such that **u**, **v**, and **w** form a right-handed triple. We notice that the magnitude of **w** is the area of the parallelogram with adjacent sides **u** and **v** since the area of a parallelogram is the product of the lengths of the adjacent sides times the sine of the angle between them. Furthermore, we see that the area of the triangle $ABC$ is then equal to $\frac{1}{2}|\mathrm{u} \times \mathrm{v}|$. There are many formulas

The line through $P_0(x_0, y_0, z_0)$ and parallel to $\mathbf{T}(t_0)$ is the tangent line to the arc at $t_0$. Furthermore, if $\mathbf{T}(t_0) = a\,\mathbf{i} + b\,\mathbf{j} + c\,\mathbf{k}$, then the equations of the tangent line are

$$\frac{x - x_0}{a} = \frac{y - y_0}{b} = \frac{z - z_0}{c}.$$

Curvature vector: $\mathbf{K}(t) = \dfrac{d\mathbf{T}}{ds} = \dfrac{\mathbf{T}'(t)}{s'(t)}$

Normal vector: $\mathbf{N}(t) = \dfrac{\mathbf{K}(t)}{|\mathbf{K}(t)|}$

$\mathbf{T}'(t) = K(t)s'(t)\mathbf{N}(t)$

The center of curvature $C(t)$ is located at the end of the vector

$$\mathbf{w} = \overrightarrow{OC} = \mathbf{r}(t) + \frac{\mathbf{N}(t)}{|\mathbf{K}(t)|}.$$

Radius of curvature: $R(t) = \dfrac{1}{|\mathbf{K}(t)|}$

Binormal vector: $\mathbf{B}(t) = \mathbf{T}(t) \times \mathbf{N}(t)$

**B**, **T**, **N** form a mutually orthogonal triple of unit vectors and is called the **trihedral** at a point on $C$.

The **osculating plane** is the plane containing the tangent line and the center of curvature; its equation is

$$b_1(x-x_0) + b_2(y-y_0) + b_3(z-z_0) = 0$$

where $(x_0, y_0, z_0)$ is a point on the curve $C$ at $t = t_0$ and the binormal vector is $\mathbf{B}(t_0) = b_1\mathbf{i} + b_2\mathbf{j} + b_3\mathbf{k}$. The equation of the plane through three given points $A$, $B$, and $C$ may be found by using vectors. Using the coordinates of the points, calculate the vectors $\overrightarrow{AB}$, $\overrightarrow{AC}$ and the cross product $\overrightarrow{AB} \times \overrightarrow{AC} = a\,\mathbf{i} + b\,\mathbf{j} + c\,\mathbf{k}$. Then the equation of the plane is

$$a(x-x_0) + b(y-y_0) + c(z-z_0) = 0$$

where $(x_0, y_0, z_0)$ are coordinates of one of the given points.

If $t$ represents time, then for motion along the curve of $\mathbf{r}(t)$,

velocity vector: $\mathbf{r}'(t) = \mathbf{v}(t)$.

acceleration vector: $\mathbf{r}''(t) = \mathbf{v}'(t)$

$$= \mathbf{a}(t) = s''(t)\mathbf{T}(t) + \frac{|\mathbf{v}(t)|^2}{R(t)}\mathbf{N}(t).$$

in geometry and trigonometry that can be displayed in vector form. The formula for the vector product is derived from the above definition. We displayed it in determinant form in the Basic Facts since it is easier to remember in that form. If we expand the determinant, we obtain a form that appears in most textbooks,

$$\mathbf{u} \times \mathbf{v} = (b_1c_2 - b_2c_1)\mathbf{i} + (a_2c_1 - a_1c_2)\mathbf{j} + (a_1b_2 - a_2b_1)\mathbf{k}.$$

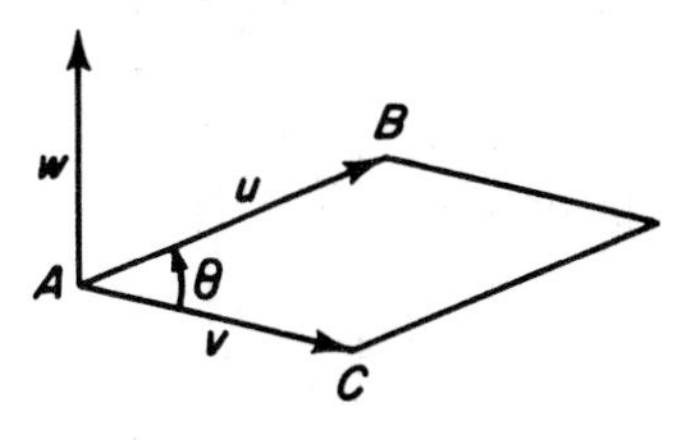

Fig. 23-3

[Test your algebra by verifying that the magnitude of this vector equals $|\mathbf{w}|$ using $\sin\theta = \sqrt{1 - \cos^2\theta}$ and the formula for $\cos\theta$ between **u** and **v**.]

**Planes and Lines.** The vector product may be used to find the equation of a plane under certain conditions. In Chapter 22 we saw that the equation of a plane through a point $P_1(x_1, y_1, z_1)$ can be written in the form $A(x - x_1) + B(y - y_1) + C(z - z_1) = 0$. This is also the condition when two lines, one with direction numbers $A$, $B$, and $C$ and the other with direction numbers $(x - x_1)$, $(y - y_1)$ and $(z - z_1)$, are perpendicular. If a plane contains two vectors **u** and **v**, then the normal to this plane will also be perpendicular to both vectors. The cross product $\mathbf{u} \times \mathbf{v}$ is such a perpendicular vector. Thus we can take $\mathbf{u} \times \mathbf{v}$ as the line which has direction numbers $A$, $B$, and $C$. Furthermore, if a vector is expressed in the form $a\,\mathbf{i} + b\,\mathbf{j} + c\,\mathbf{k}$, then the coefficients $a$, $b$, and $c$ are direction numbers of the line defined by the vector. Consequently, the coefficients of **i**, **j**, and **k** of $\mathbf{u} \times \mathbf{v}$ may be chosen as the $A$, $B$, and $C$ of the equation of the plane containing **u** and **v**.

Consider now two **skew lines** in space given by

$$L_1: \frac{x - x_0}{a_1} = \frac{y - y_0}{b_1} = \frac{z - z_0}{c_1} \quad \text{and} \quad L_2: \frac{x - x_1}{a_2} = \frac{y - y_1}{b_2} = \frac{z - z_1}{c_2}$$

The vector $\mathbf{v}_1 = a_1\mathbf{i} + b_1\mathbf{j} + c_1\mathbf{k}$ is along $L_1$ and the vector $\mathbf{v}_2 = a_2\mathbf{i} + b_2\mathbf{j} + c_2\mathbf{k}$ is along $L_2$ since $a_i$, $b_i$, and $c_i$ are the direction numbers of the lines. The cross product $\mathbf{v}_1 \times \mathbf{v}_2 = \mathbf{w}$ is a vector that is perpendicular to both $\mathbf{v}_1$ and $\mathbf{v}_2$. Let $P_0(x_0, y_0, z_0)$ be on $L_1$ and $P_1(x_1, y_1, z_1)$ be on $L_2$ and the vector $\overrightarrow{P_0P_1} = (x_1 - x_0)\mathbf{i} + (y_1 - y_0)\mathbf{j} + (z_1 - z_0)\mathbf{k}$. The projection of this vector is the perpendicular distance between the two skew lines; $d = \dfrac{\overrightarrow{P_0P_1} \cdot \mathbf{w}}{|\mathbf{w}|}$.

**Curvilinear motion** in three dimensions is very similar to curvilinear motion in two dimensions when the discussion is made in terms of vectors. We recommend rereading the information in Chapter 16.

## SOLUTIONS TO THE EXERCISES

I

i.
$$\begin{array}{r} \mathbf{v} = 3\mathbf{i} + \mathbf{j} - 7\mathbf{k} \\ \mathbf{u} = -\mathbf{i} - 3\mathbf{j} + 5\mathbf{k} \\ \hline \mathbf{v} - \mathbf{u} = 4\mathbf{i} + 4\mathbf{j} - 12\mathbf{k} \end{array}$$

ii. $\mathbf{w} = 2\mathbf{i} + 5\mathbf{j} + 3\mathbf{k}$

$\mathbf{u} \cdot \mathbf{w} = -2 - 15 + 15 = -2$

iii.
$$\mathbf{v} \times \mathbf{w} = \begin{vmatrix} \mathbf{i} & \mathbf{j} & \mathbf{k} \\ 3 & 1 & -7 \\ 2 & 5 & 3 \end{vmatrix}$$
$$= (3 + 35)\,\mathbf{i} - (9 + 14)\,\mathbf{j} + (15 - 2)\,\mathbf{k}$$
$$= 38\mathbf{i} - 23\mathbf{j} + 13\mathbf{k}$$

iv. $5\,|\mathbf{w}| = 5\sqrt{4 + 25 + 9} = 5\sqrt{38}$

v.
$$\begin{array}{r} 2\mathbf{v} = 6\mathbf{i} + 2\mathbf{j} - 14\mathbf{k} \\ 3\mathbf{w} = 6\mathbf{i} + 15\mathbf{j} + 9\mathbf{k} \\ \hline 2\mathbf{v} - 3\mathbf{w} = -13\mathbf{j} - 23\mathbf{k} \end{array}$$
$$(2\mathbf{v} - 3\mathbf{w}) \times \mathbf{u} = \begin{vmatrix} \mathbf{i} & \mathbf{j} & \mathbf{k} \\ 0 & -13 & -23 \\ -1 & -3 & 5 \end{vmatrix}$$
$$= -134\mathbf{i} + 23\mathbf{j} - 13\mathbf{k}$$

vi. $|\mathbf{v}| \cos\theta = \dfrac{\mathbf{v} \cdot \mathbf{w}}{|\mathbf{w}|} = \dfrac{6 + 5 - 21}{\sqrt{38}} = -\dfrac{10}{\sqrt{38}}$

II

$A(2, -1, 3)$, $B(1, 3, -2)$, $C(2, 2, 2)$

i. $\overrightarrow{AC} = (2 - 2)\,\mathbf{i} + (2 + 1)\,\mathbf{j} + (2 - 3)\mathbf{k}$
$= 3\mathbf{j} - \mathbf{k}$

$\overrightarrow{AB} = (1 - 2)\,\mathbf{i} + (3 + 1)\,\mathbf{j} + (-2 - 3)\mathbf{k}$
$= -\mathbf{i} + 4\mathbf{j} - 5\mathbf{k}$

$2\overrightarrow{AC} + 3\overrightarrow{AB} = -3\mathbf{i} + 18\mathbf{j} - 17\mathbf{k}$

ii. $\overrightarrow{BC} = \mathbf{i} - \mathbf{j} + 4\mathbf{k}$

$|\overrightarrow{BC}| = \sqrt{1 + 1 + 16} = 3\sqrt{2}$

iii.
$$\overrightarrow{AB} \times \overrightarrow{AC} = \begin{vmatrix} \mathbf{i} & \mathbf{j} & \mathbf{k} \\ -1 & 4 & -5 \\ 0 & 3 & -1 \end{vmatrix}$$
$$= 11\mathbf{i} - \mathbf{j} - 3\mathbf{k}$$

Equation of the plane is

$11(x - 2) - (y - 2) - 3(z - 2) = 0.$

$11x - y - 3z = 14$

III

$\mathbf{f}(t) = 2t^2\mathbf{i} - 3t\mathbf{j} + (2t - 1)\mathbf{k}$

$\mathbf{u}(t) = 2e^t\mathbf{i} + 3e^t\mathbf{j} - 6e^t\mathbf{k}$

$\mathbf{v}(t) = \ln\sec t\,\mathbf{i} + \mathbf{j} - t\mathbf{k}$

i. $\mathbf{f}'(t) = 4t\mathbf{i} - 3\mathbf{j} + 2\mathbf{k}$

$\mathbf{u}'(t) = 2e^t\mathbf{i} + 3e^t\mathbf{j} - 6e^t\mathbf{k}$

$$\mathbf{f}'(t) \times \mathbf{u}'(t) = \begin{vmatrix} \mathbf{i} & \mathbf{j} & \mathbf{k} \\ 4t & -3 & 2 \\ 2e^t & 3e^t & -6e^t \end{vmatrix}$$
$$= 12e^t\mathbf{i} + 28e^t\mathbf{j} + 18e^t\mathbf{k}$$

ii. $\mathbf{v}'(t) = \tan t\,\mathbf{i} - \mathbf{k}$

$\mathbf{f}'(t) \cdot \mathbf{v}'(t) = 4t\tan t - 2 = h(t)$

$h'(t) = 4\tan t + 4t\sec^2 t$

iii. $|\mathbf{v}'(t)| = \sqrt{\tan^2 t + 1} = \sec t$

$$\mathbf{T}(t) = \frac{\mathbf{v}'(t)}{|\mathbf{v}'(t)|} = \frac{\tan t}{\sec t}\,\mathbf{i} - \frac{1}{\sec t}\,\mathbf{k}$$
$$= \sin t\,\mathbf{i} - \cos t\,\mathbf{k}$$

iv. $|\mathbf{u}'(t)| = e^t\sqrt{4 + 9 + 36} = 7e^t = s'(t)$

$$\mathbf{T}(t) = \frac{\mathbf{u}'(t)}{|\mathbf{u}'(t)|} = \frac{1}{7}(2\mathbf{i} + 3\mathbf{j} - 6\mathbf{k})$$

$$\mathbf{K}(t) = \frac{\mathbf{T}'(t)}{s'(t)} = \frac{0}{7e^t} = 0$$

v. $\mathbf{B}(t) = \mathbf{T}(t) \times \mathbf{N}(t)$

$$= \begin{vmatrix} \mathbf{i} & \mathbf{j} & \mathbf{k} \\ \sin t & 0 & -\cos t \\ \cos t & 0 & \sin t \end{vmatrix}$$
$$= 0\,\mathbf{i} - (\sin^2 t + \cos^2 t)\,\mathbf{j} + 0\mathbf{k}$$
$$= -\mathbf{j}$$

$x_0 = \ln\sec 0 = 0$, $y_0 = 1$, $z_0 = 0$

Equation of the plane is

$0(x - 0) - (y - 1) + 0(z - 0) = 0.$

$y = 1$

# 24 PARTIAL DIFFERENTIATION

## SELF-TEST

DIRECTIONS: Write your answers in the numbered regions to the right. To check your answers, turn the page. Study the solutions to the problems you missed, and do the exercises following any problem or group of problems in which you had errors.

In 1-5 find the first partial derivatives.

1. $z = 4x^3 - 2xy + 5y^2$
2. $z = \sin^2(3t + 2r)$
3. $r = \alpha \cos \dfrac{\beta}{\alpha}$
4. $z = \ln \sin(x^2 + y^2)$
5. $u = \text{Arctan}\, \dfrac{xy^2}{z^3}$
6. Find the three second partial derivatives of problem 2.
7. Find $z_{xx}$ for $x^2 + y^2 + z^2 = a^2$
8. Express $\dfrac{du}{dt}$ as a function of $x$ and $y$ if $u = \sqrt{x^2 - y^2}$, $x = \sin t$ and $y = \cot t$.
9. The legs of a right triangle are measured to be 3.0 and 4.0 inches with a possible error of 0.1 inch in each measurement. Find the largest approximate error in computing the hypotenuse using the two sides.
10. Find the equation of the tangent plane to the surface $x^2 + 2y^2 + 3z^2 = 21$ at $P(1, 2, -2)$.
11. Find the critical values of $z = e^{-x} \sin^2 y$.

1

2

3

4

5

6

7

8

9

10

11

## SOLUTIONS

1. $z = 4x^3 - 2xy + 5y^2$

$z_x = 12x^2 - 2y \qquad z_y = -2x + 10y$

2. $z = \sin^2(3t + 2r)$

$z_t = 3\sin(6t + 4r)$

$z_r = 2\sin(6t + 4r)$

3. $r = \alpha \cos \dfrac{\beta}{\alpha}$

$$\frac{\partial r}{\partial \alpha} = \cos\frac{\beta}{\alpha} + \alpha\left(-\sin\frac{\beta}{\alpha}\right)\left(-\frac{\beta}{\alpha^2}\right)$$

$$= \cos\frac{\beta}{\alpha} + \frac{\beta}{\alpha}\sin\frac{\beta}{\alpha}$$

$$\frac{\partial r}{\partial \beta} = \alpha\left(-\sin\frac{\beta}{\alpha}\right)\left(\frac{1}{\alpha}\right) = -\sin\frac{\beta}{\alpha}$$

> **EXERCISES I**
>
> Verify that $f_{xy} = f_{yx}$ for the following functions.
>
> i. $f(x, y) = 3x^2y^3 - 2x^3y^2$
>
> ii. $f(x, y) = e^y\cos^2 x - e^x \sin^2 y$
>
> iii. $f(x, y) = \text{Arctan}(3x - 2y)$

4. $z = \ln \sin(x^2 + y^2)$

$$z_x = \frac{1}{\sin(x^2 + y^2)}[\cos(x^2 + y^2)](2x)$$

$$= 2x\cot(x^2 + y^2)$$

$z_y = 2y\cot(x^2 + y^2)$

5. $u = \text{Arctan}\left(\dfrac{xy^2}{z^3}\right)$

$$u_x = \frac{1}{1 + (xy^2z^{-3})^2}\left(\frac{y^2}{z^3}\right)$$

$$= \frac{z^6}{z^6 + x^2y^4}\left(\frac{y^2}{z^3}\right) = \frac{y^2z^3}{z^6 + x^2y^4}$$

$$u_y = \frac{z^6}{z^6 + x^2y^4}\left(\frac{2xy}{z^3}\right) = \frac{2xyz^3}{z^6 + x^2y^4}$$

$$u_z = \frac{z^6}{z^6 + x^2y^4}\left(-\frac{3xy^2}{z^4}\right) = -\frac{3xy^2z^2}{z^6 + x^2y^4}$$

6. $z = \sin^2(3t + 2r)$

$z_{tt} = 3\cos(6t + 4r)(6)$

$= 18\cos(6t + 4r)$

$z_{rt} = 3\cos(6t + 4r)(4)$

$= 12\cos(6t + 4r)$

$z_{rr} = 2\cos(6t + 4r)(4)$

$= 8\cos(6t + 4r)$

7. $x^2 + y^2 + z^2 = a^2$

$2x + 2zz_x = 0;\ z_x = -\dfrac{x}{z}$

$$z_{xx} = -\frac{z - xz_x}{z^2} = -\frac{z - x\left(-\frac{x}{z}\right)}{z^2}$$

$$= -\frac{z^2 + x^2}{z^3}$$

8. $u = \sqrt{x^2 - y^2};\ x = \sin t;\ y = \cot t$

$u_x = x(x^2 - y^2)^{-\frac{1}{2}};\ \dfrac{dx}{dt} = \cos t$

$u_y = -y(x^2 - y^2)^{-\frac{1}{2}};\ \dfrac{dy}{dt} = -\csc^2 t$

Since $\cot t = \dfrac{\cos t}{\sin t} = y$, $\cos t = xy$

$\csc^2 t = 1 + \cot^2 t = 1 + y^2$ and

$$\frac{du}{dt} = \frac{x}{\sqrt{x^2 - y^2}}\cos t + \frac{-y}{\sqrt{x^2 - y^2}}(-\csc^2 t)$$

$$= \frac{y(x^2 + y^2 + 1)}{\sqrt{x^2 - y^2}}$$

> **EXERCISES II**
>
> i. Find $\dfrac{du}{dt}$ if $u = \ln(x - y)$, $x = e^t$ and $y = e^{-t}$.
>
> ii. Find $F_y(x, y)$ if $F = \text{Arctan}\, uv$, $u = \sqrt{x} + \sqrt{y}$ and $v = \sqrt{x} - \sqrt{y}$.

9. Let $x$ and $y$ be the legs of the triangle, then $h = \sqrt{x^2 + y^2}$.

$h_x = x(x^2 + y^2)^{-\frac{1}{2}};\ h_x(3, 4) = \frac{3}{5}$

$h_y = y(x^2 + y^2)^{-\frac{1}{2}}$; $h_y(3, 4) = \frac{4}{5}$

$\Delta h = h_x \Delta x + h_y \Delta y$

$= \frac{3}{5}(0.1) + \frac{4}{5}(0.1) = \frac{7}{5}(0.1) = .14$

**10.** $x^2 + 2y^2 + 3z^2 = 21$; $(1, 2, -2)$

$2x + 6zz_x = 0$, $z_x = -\dfrac{x}{3z}$

$m_1 = z_x(1, 2, -2) = -\dfrac{1}{3(-2)} = \dfrac{1}{6}$

$4y + 6zz_y = 0$, $z_y = -\dfrac{2y}{3z}$

$m_2 = z_y(1, 2, -2) = -\dfrac{2(2)}{3(-2)} = \dfrac{4}{6}$

$z - z_0 = m_1(x - x_0) + m_2(y - y_0)$

$z + 2 = \frac{1}{6}(x - 1) + \frac{4}{6}(y - 2)$

$x + 4y - 6z = 21$

> EXERCISES III
>
> Find the equations of the tangent planes to the surfaces at $P_0$.
>
> i. $z = 2x^2 - 5y^2$, $P_0(3, -2, -2)$
>
> ii. $z = \text{Arctan}\,\dfrac{y}{x}$, $P_0\left(1, 1, \dfrac{\pi}{4}\right)$

**11.** $z = e^{-x}\sin^2 y$, $z_x = -e^{-x}\sin^2 y$, $z_y = e^{-x}\sin 2y$. Now $e^{-x}\sin^2 y = 0$ and $e^{-x}\sin 2y = 0$ simultaneously. Since $e^{-x} \neq 0$ for all $x$, we have $\sin^2 y = \sin 2y = 0$ which holds for $y = n\pi$ where $n$ is an integer. The critical points are $(x, n\pi)$; i.e., all the points on lines $y = n\pi$.

> EXERCISE IV
>
> i. Find the critical values of $z = 3x^2 + 6xy + 2y^2 - x + y$.

Solutions to the exercises are on page 144.

## ANSWERS

1. $z_x = 12x^2 - 2y$, $z_y = -2x + 10y$ — 1
2. $z_t = 3\sin(6t + 4r)$, $z_r = 2\sin(6t + 4r)$ — 2
3. $\dfrac{\partial r}{\partial \alpha} = \cos\dfrac{\beta}{\alpha} + \dfrac{\beta}{\alpha}\sin\dfrac{\beta}{\alpha}$; $\dfrac{\partial r}{\partial \beta} = -\sin\dfrac{\beta}{\alpha}$ — 3
4. $z_x = 2x\cot(x^2 + y^2)$, $z_y = 2y\cot(x^2 + y^2)$ — 4
5. $u_x = \dfrac{y^2 z^3}{z^6 + x^2 y^4}$, $u_y = \dfrac{2xyz^3}{z^6 + x^2 y^4}$, $u_z = -\dfrac{3xy^2 z^2}{z^6 + x^2 y^4}$ — 5
6. $z_{tt} = 18\cos(6t + 4r)$, $z_{rt} = 12\cos(6t + 4r)$, $z_{rr} = 8\cos(6t + 4r)$ — 6
7. $z_{xx} = -\dfrac{z^2 + x^2}{z^3}$ — 7
8. $\dfrac{du}{dt} = \dfrac{y(x^2 + y^2 + 1)}{\sqrt{x^2 - y^2}}$ — 8
9. 0.14 — 9
10. $x + 4y - 6z = 21$ — 10
11. $(x, n\pi)$ — 11

## BASIC FACTS

The collection of ordered pairs $(a, z)$ where the elements $a$ are themselves ordered pairs $(x, y)$ is a **function of two variables** if for every combination $(x, y)$ there exists a unique $z$. The set of ordered pairs $a$ is called the **domain** of the function and the totality of all possible values for $z$ is called the **range** of the function. Symbolic expressions for the function of two variables are $f(x,y)$, $F(x,y)$, $\phi(x,y)$, etc.

A function $f(x,y)$ approaches a limit $L$ as the pair $(x,y)$ tends to $(a,b)$ if and only if for each $\varepsilon > 0$ there is a $\delta > 0$ such that $|f(x,y) - L| < \varepsilon$ whenever $|x - a| < \delta$ and $|y - b| < \delta$. The function $f$ is **continuous** at $(a,b)$ if and only if $f(a,b)$ is defined and $f(x,y) \longrightarrow f(a,b)$ as $(x,y) \longrightarrow (a,b)$. If $f(x,y)$ is continuous at every point $(x,y)$ of a region $R$ of the $X,Y$-plane, then it is said to be **continuous in** $R$.

The **partial derivatives** of a function $f(x,y)$ are defined as

$$f_x(x,y) = \lim_{h \to 0} \frac{f(x+h,y) - f(x,y)}{h}$$

$$f_y(x,y) = \lim_{h \to 0} \frac{f(x,y+h) - f(x,y)}{h}$$

where $y$ is held fixed in the limit for $f_x$ and $x$ is held fixed in the limit for $f_y$. Notations for partial derivatives are

$$f_x, \frac{\partial f}{\partial x}, f_1(x,y), \frac{\partial z}{\partial x} [\text{if } z = f(x,y)].$$

To find $f_x$ if $z = f(x,y)$, hold $y$ constant and differentiate with respect to $x$ in the usual manner. **Implicit differentiation** is performed in the same manner as for a function of a single variable and using the rule of the previous sentence.

The **Partial derivatives of higher order** for the function $z = f(x,y)$ are

$$\frac{\partial}{\partial x}\left(\frac{\partial z}{\partial x}\right) = \frac{\partial^2 z}{\partial x^2} = z_{xx} = f_{xx}(x,y)$$

## ADDITIONAL INFORMATION

**Geometric Interpretation of the Partial Derivatives**

The equation $z = f(x,y)$ represents a surface in the three-dimensional space. A point $P_0(x_0, y_0, z_0)$ where $z_0 = f(x_0, y_0)$ is on the surface since the equation is satisfied. (See Figure 24-1.) If $y$ is held constant with $x$ and $z$ allowed to vary, then $y = y_0$ is a plane parallel to the $XZ$-plane and passing through the point $P_0$. This plane will intersect the surface $z = f(x,y)$ in a curve $C_1$ which is defined by the equation $z = f(x,y_0)$, a relation between $x$ and $z$ with $y_0$ as a constant parameter. The value of the partial derivative of $f$ with respect to $x$ at the point $P_0$ is $f_x(x_0, y_0)$ which, in the plane $y = y_0$, has the same definition as the slope of the tangent line $L_1$ to the curve $C_1$ at $P_0$. If we now hold $x$ constant and consider the plane $x = x_0$ which is parallel to the $YZ$-plane, then we have its intersection with the surface as the curve $C_2$ given by $z = f(x_0, y)$. The value of the partial derivative of $f$ with respect to $y$ at the point $P_0$ is given by $f_y(x_0,y_0)$ and is the slope of the tangent line $L_2$ to the curve $C_2$ at $P_0$. Since $L_1$ and $L_2$ pass through the same point $P_0$, they form a plane, which is defined as the tangent plane to the surface $z = f(x,y)$ at $P_0$. The equation of this **tangent plane** is given by $z - z_0 = f_x(x_0,y_0)(x - x_0) + f_y(x_0,y_0)(y - y_0)$. The line perpendicular to the tangent plane and passing through $P_0$ will have direction numbers $f_x(x_0,y_0)$, $f_y(x_0,y_0)$, and $-1$. This line is called the **normal line**, and its defining equations are $\dfrac{x - x_0}{f_x(x_0,y_0)} = \dfrac{y - y_0}{f_y(x_0,y_0)} = \dfrac{z - z_0}{-1}$.

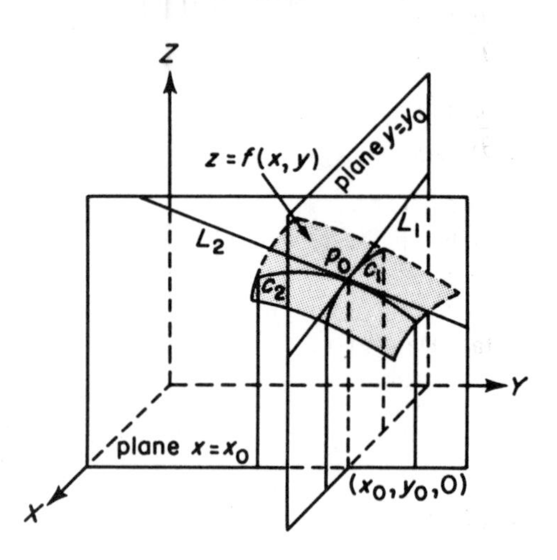


Fig. 24-1

If the surface is defined by the equation $F(x,y,z) = 0$, we consider a general curve $C$ lying on this surface and let the parametric equations of the curve be $x = \phi(t)$, $y = \psi(t)$, and $z = \theta(t)$. Since these values of the variables satisfy the equation $F(x,y,z) = 0$, the total differential is zero.

$$\frac{dF}{dt} = \frac{\partial F}{\partial x}\frac{dx}{dt} + \frac{\partial F}{\partial y}\frac{dy}{dt} + \frac{\partial F}{\partial z}\frac{dz}{dt} = 0$$

Interpreted geometrically this is the condition that a line with direction numbers $\dfrac{dx}{dt}$, $\dfrac{dy}{dt}$, and $\dfrac{dz}{dt}$ is perpendicular to a line with direction numbers $\dfrac{\partial F}{\partial x}$, $\dfrac{\partial F}{\partial y}$, and $\dfrac{\partial F}{\partial z}$. Since the derivatives with respect

$$\frac{\partial}{\partial x}\left(\frac{\partial z}{\partial y}\right) = \frac{\partial^2 z}{\partial x \partial y} = z_{yx} = f_{yx}(x,y)$$

$$\frac{\partial}{\partial y}\left(\frac{\partial z}{\partial x}\right) = \frac{\partial^2 z}{\partial y \partial x} = z_{xy} = f_{xy}(x,y)$$

$$\frac{\partial}{\partial y}\left(\frac{\partial z}{\partial y}\right) = \frac{\partial^2 z}{\partial y^2} = z_{yy} = f_{yy}(x,y).$$

If the partial derivatives are continuous, then $f_{yx} = f_{xy}$, and for third order partial derivatives $f_{xxy} = f_{xyx} = f_{yxx}$, etc.

**Total Derivative.** If $z = f(x,y)$ and $x = x(t)$ and $y = y(t)$, then

$$\frac{dz}{dt} = z_x \frac{dx}{dt} + z_y \frac{dy}{dt}.$$

**Total Differential.** $dz = z_x dx + z_y dy$. The change $\Delta u$ in $u$ caused by a change $\Delta x$ in $x$ and $\Delta y$ in $y$ is

$$\Delta u = u_x \Delta x + u_y \Delta y.$$

**Chain Rule.** If $z = f(x,y)$ and $x = x(r,s)$ and $y = y(r,s)$, then

$$\frac{\partial z}{\partial r} = z_x \frac{\partial x}{\partial r} + z_y \frac{\partial y}{\partial r}$$

$$\frac{\partial z}{\partial s} = z_x \frac{\partial x}{\partial s} + z_y \frac{\partial y}{\partial s}.$$

**Maxima and Minima**

A necessary (but not sufficient) condition that $z = f(x,y)$ has a relative maxima or minima is that $z_x = 0$ and $z_y = 0$. The values $(x_0,y_0)$ at which $z_x = 0$ and $z_y = 0$ are called **critical values.**

*Note*: The minimum of the function $f = x^2 + y^2 + z^2$ occurs at the same point as the minimum of the distance $d = \sqrt{x^2 + y^2 + z^2}$.

The **tangent plane** to the surface $z = f(x,y)$ at $(x_0,y_0,z_0)$ is given by the equation

$$z - z_0 = m_1(x - x_0) + m_2(y - y_0)$$

where $z_0 = f(x_0,y_0)$, $m_1 = z_x(x_0,y_0)$ and $m_2 = z_y(x_0,y_0)$. If the surface is defined by $F(x,y,z) = 0$, see Additional Information.

to $t$ evaluated at a point $P_0$ are the direction numbers of the tangent line to $C$ at $P_0$, we then have the values of the partial derivatives $F_x$, $F_y$, $F_z$ at $P_0$ as the direction numbers of the normal line to the surface at $P_0$. The equations of the **normal line** are then

$$\frac{x - x_0}{F_x(x_0,y_0,z_0)} = \frac{y - y_0}{F_y(x_0,y_0,z_0)} = \frac{z - z_0}{F_z(x_0,y_0,z_0)}.$$

If we now take a different curve lying on the surface and passing through $P_0$, the direction numbers of a line perpendicular to its tangent would be the same partial derivatives. Thus all tangent lines to the surface at $P_0$ are perpendicular to the same normal line and form a plane. The equation of this **tangent plane** is

$$F_x(x_0,y_0,z_0)(x - x_0) + F_y(x_0,y_0,z_0)(y - y_0) + F_z(x_0,y_0,z_0)(z - z_0) = 0.$$

The equation $f(x,y) = 0$ defines a relation between the variables $x$ and $y$. The equation may be differentiated implicitly to obtain an expression for the derivative of $y$ with respect to $x$. (See Chapter 4.) Let us use partial derivatives to write the total differential of $f$ defined by $z = f(x,y) = 0$. It is $df = f_x dx + f_y dy$. Since $f = 0$, the differential $df = 0$, and we have $f_x dx + f_y dy = 0$ which may be solved for the ratios $\frac{dy}{dx} = -\frac{f_x}{f_y}$. This provides us with a second method for implicit differentiation. Consider, for example, the equation $f(x,y) = x^2y^4 + e^{2x} - e^{4y} = 0$. Performing the partial differentiation, we have $f_x = 2xy^4 + 2e^{2x}$ and $f_y = 4x^2y^3 - 4e^{4y}$. Then the derivative of $y$ with respect to $x$ is $\frac{dy}{dx} = -\frac{2xy^4 + 2e^{2x}}{4x^2y^3 - 4e^{4y}} = \frac{xy^4 + e^{2x}}{2x^2y^3 - 2e^{4y}}$.

We can extend this idea to three or more variables. Consider the equation $F(x,y,z) = 0$ as the implicit form of $z = f(x,y)$. The formulas for the **total differentials** of each of these are $dz = z_x dx + z_y dy$ and $dF = F_x dx + F_y dy + F_z dz$. Since $F = 0$ we have $dF = 0$, and we can solve for $dz = -\frac{F_x}{F_z} dx - \frac{F_y}{F_z} dy$. As we compare the two expressions for $dz$, we can conclude that $\frac{\partial z}{\partial x} = -\frac{F_x}{F_z}$ and $\frac{\partial z}{\partial y} = -\frac{F_y}{F_z}$.

**Extrema of a Function of Two Variables.** The necessary condition that a function $f(x,y)$ has a relative maximum or a relative minimum at a critical point $P_0(x_0,y_0,z_0)$ is that $f_x$ and $f_y$ both vanish at $P_0$. This is not a sufficient condition. There are many functions that do not have relative extrema at a critical point; in such cases the point is called a "saddle point." The second derivative test for establishing the extrema at a critical point $P(a,b)$ is to calculate a value for the expression $f_{xx}(a,b)f_{yy}(a,b) - f_{xy}^2(a,b) = N$. Then we have: i) minimum if $N > 0$ and $f_{xx}(a,b) > 0$; ii) maximum if $N > 0$ and $f_{xx}(a,b) < 0$; iii) saddle point if $N < 0$; and iv) the test fails if $N = 0$.

## SOLUTIONS TO THE EXERCISES

I

i. $f(x,y) = 3x^2y^3 - 2x^3y^2$

$f_x = 6xy^3 - 6x^2y^2$

$f_{yx} = 18xy^2 - 12x^2y$

$f_y = 9x^2y^2 - 4x^3y$

$f_{xy} = 18xy^2 - 12x^2y$

ii. $f(x,y) = e^y\cos^2 x - e^x\sin^2 y$

$f_x = e^y(-\sin 2x) - e^x\sin^2 y$

$f_{yx} = e^y(-\sin 2x) - e^x\sin 2y$

$f_y = e^y\cos^2 x - e^x\sin 2y$

$f_{xy} = e^y(-\sin 2x) - e^x\sin 2y$

iii. $f(x,y) = \text{Arctan}\,(3x - 2y)$

$$f_x = \frac{3}{1 + (3x - 2y)^2}$$

$$f_{yx} = -\frac{3(2)(3x - 2y)(-2)}{[1 + (3x - 2y)^2]^2}$$

$$= \frac{12(3x - 2y)}{[1 + (3x - 2y)^2]^2}$$

$$f_y = \frac{-2}{1 + (3x - 2y)^2}$$

$$f_{xy} = \frac{-(-2)(2)(3x - 2y)(3)}{[1 + (3x - 2y)^2]^2}$$

$$= \frac{12(3x - 2y)}{[1 + (3x - 2y)^2]^2}$$

II

i. $u = \ln(x - y),\ x = e^t,\ y = e^{-t}$

$$u_x = \frac{1}{x - y},\quad u_y = -\frac{1}{x - y}$$

$$\frac{dx}{dt} = e^t,\quad \frac{dy}{dt} = -e^{-t}$$

$$\frac{du}{dt} = u_x\frac{dx}{dt} + u_y\frac{dy}{dt}$$

$$= \frac{e^t}{x - y} - \frac{e^{-t}}{x - y} = \frac{e^t - e^{-t}}{x - y}$$

ii. $F = \text{Arctan}\,uv$

$u = \sqrt{x} + \sqrt{y},\ v = \sqrt{x} - \sqrt{y}$

$$F_u = \frac{v}{1 + u^2v^2} = \frac{\sqrt{x} - \sqrt{y}}{1 + (x - y)^2}$$

$$F_v = \frac{u}{1 + u^2v^2} = \frac{\sqrt{x} + \sqrt{y}}{1 + (x - y)^2}$$

$$u_y = \frac{1}{2}y^{-\frac{1}{2}},\quad v_y = -\frac{1}{2}y^{-\frac{1}{2}}$$

$$F_y = \frac{\sqrt{x} - \sqrt{y} - (\sqrt{x} + \sqrt{y})}{2\sqrt{y}[1 + (x - y)^2]}$$

$$= -\frac{1}{1 + (x - y)^2}$$

III

i. $z = 2x^2 - 5y^2,\ P_0(3,-2,-2)$

$z_x = 4x,\ z_x(3,-2,-2) = 12 = m_1$

$z_y = -10y,\ z_y(3,-2,-2) = 20 = m_2$

$z - z_0 = m_1(x - x_0) + m_2(y - y_0)$

$z + 2 = 12(x - 3) + 20(y + 2)$

$z = 12x + 20y + 2$

ii. $z = \text{Arctan}\,\dfrac{y}{x},\ P_0\left(1, 1, \dfrac{\pi}{4}\right)$

$$z_x = \frac{x^2}{x^2 + y^2}\left(-\frac{y}{x^2}\right) = -\frac{y}{x^2 + y^2}$$

$$z_y = \frac{x^2}{x^2 + y^2}\left(\frac{1}{x}\right) = \frac{x}{x^2 + y^2}$$

Then $m_1 = -\frac{1}{2}$ and $m_2 = \frac{1}{2}$.

$$z - \frac{\pi}{4} = -\frac{1}{2}(x - 1) + \frac{1}{2}(y - 1)$$

$$2z + x = y + \frac{\pi}{2}$$

IV

i. $z = 3x^2 + 6xy + 2y^2 - x + y$

$z_x = 6x + 6y - 1 = 0$

$z_y = 6x + 4y + 1 = 0$

$2y = 2,\ y = 1;\ 6x = -5,\ x = -\frac{5}{6}$

# 25 MULTIPLE INTEGRATION AND APPLICATIONS

## SELF-TEST

DIRECTIONS: Write your answers in the numbered regions to the right. To check your answers, turn the page. Study the solutions to the problems you missed, and do the exercises following any problem or group of problems in which you had errors.

In **1-2** evaluate the given integrals.

**1.** $\displaystyle\int_0^{\pi}\int_0^{a(1-\cos\theta)} \rho^2 \sin\theta \, d\rho \, d\theta$

**2.** $\displaystyle\int_1^{2}\int_0^{z}\int_0^{x\sqrt{3}} \frac{x \, dy dx dz}{x^2 + y^2}$

**3.** Use double integration to find the area bounded by $x^2 + y^2 = 18$ and $3y = x^2$.

**4.** Find the volume bounded by the cylindrical surface $x = 1 - y^2$, the plane $z = x$ and the plane $z = 0$.

**5.** Find the surface of the cylinder $x^2 + y^2 = ry$ intercepted by the sphere $x^2 + y^2 + z^2 = r^2$ and in the first octant.

**6.** Find the center of the mass distributed over the area bounded by $4y = x^2$, $y = 0$, $x = 4$ if the density varies as the square of the distance from the origin.

**7.** Find the polar moment of inertia of a mass distributed over the area bounded by $x^2 + y^2 = 2a^2$, $y^2 = ax$, $y = 0$, and the density equal to $kxy$.

1 ________

2 ________

3 ________

4 ________

5 ________

6 ________

7 ________

## SOLUTIONS

1. $$\int_0^{\pi}\int_0^{a(1-\cos\theta)} \rho^2 \sin\theta \, d\rho \, d\theta$$

$$= \int_0^{\pi}\left[\frac{1}{3}\rho^3 \sin\theta\right]_0^{a(1-\cos\theta)} d\theta$$

$$= \frac{a^3}{3}\int_0^{\pi} (1-\cos\theta)^3 \sin\theta \, d\theta$$

$$= \frac{a^3}{3}\left[\frac{(1-\cos\theta)^4}{4}\right]_0^{\pi} = \frac{1}{12}a^3$$

2. $$\int_1^2\int_0^z\int_0^{x\sqrt{3}} \frac{x \, dydxdz}{x^2+y^2}$$

$$= \int_1^2\int_0^z \frac{x}{x}\text{Arctan}\,\frac{y}{x}\Big|_0^{x\sqrt{3}} dxdz$$

$$= \frac{\pi}{3}\int_1^2\int_0^z dxdz = \frac{\pi}{3}\int_1^2 zdz$$

$$= \frac{\pi}{3}\frac{z^2}{2}\Big|_1^2 = \frac{\pi}{3}\left(\frac{3}{2}\right) = \frac{\pi}{2}$$

### EXERCISES I

Evaluate the given integrals.

i. $$\int_0^a\int_0^{\sqrt{a^2-x^2}} (x+y)\,dy\,dx.$$

ii. $$\int_0^{\pi/3}\int_0^{\cos 2z}\int_0^{2yz} \cos\frac{x}{y}\,dxdydz$$

3. $x^2 + y^2 = 18$ and $3y = x^2$
The curves intersect at the points $(-3, 3)$ and $(3, 3)$. See Fig. 25-1.

$$A = \int_{-3}^{3}\int_{x^2/3}^{\sqrt{18-x^2}} dydx$$

$$= \int_{-3}^{3}\left(\sqrt{18-x^2} - \frac{x^2}{3}\right) dx$$

$$= \frac{x}{2}\sqrt{18-x^2}$$

$$+ \frac{18}{2}\text{Arcsin}\,\frac{x}{\sqrt{18}} - \frac{x^3}{9}\Big|_{-3}^{3}$$

$$= \frac{9}{2}\pi + 3$$

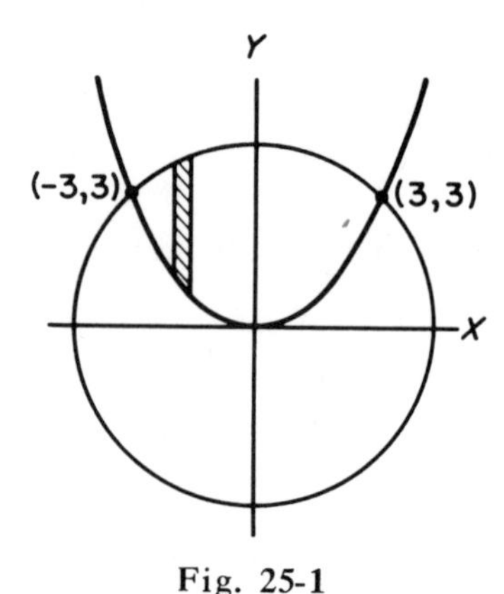


Fig. 25-1

### EXERCISES II

Use double integration to find area:

i. Bounded by $y = \sin x$, $y = \sec^2 x$, $x = 0$ and $x = \frac{\pi}{4}$.

ii. Inside the circle $\rho = 2r\cos\theta$ and outside the circle $\rho = r$.

4. $x = 1 - y^2$, $z = x$, and $z = 0$. Since the parabola $y^2 = 1 - x$ is symmetric about the $X$-axis, we can find the volume in the first octant and double the result. We have $0 \leq x \leq 1 - y^2$ and $0 \leq y \leq 1$ and

$$V = 2\int_0^1\int_0^{1-y^2} x\,dxdy = \int_0^1 (1-y^2)^2 dy$$

$$= y - \frac{2}{3}y^3 + \frac{1}{5}y^5\Big|_0^1 = \frac{8}{15}.$$

5. $x^2 + y^2 = ry$ and $x^2 + y^2 + z^2 = r^2$
Let us consider the projection of the surface on the $Y, Z$-plane and therefore

define the surface in the form $x = f(y,z)$. The needed partial derivatives obtained from the equation of the cylinder are

$$\frac{\partial x}{\partial y} = \frac{r - 2y}{2x} = \frac{r - 2y}{2\sqrt{ry - y^2}}, \quad \frac{\partial x}{\partial z} = 0.$$

The area in the $Y,Z$-plane is bounded by the $Y$ and $Z$ axes and the curve $z^2 + ry = r^2$ which is obtained by eliminating $x$ between the two equations of the solids.

$$A = \int_0^r \int_0^{\sqrt{r^2 - ry}} \left[1 + \frac{(r - 2y)^2}{4(ry - y^2)}\right]^{\frac{1}{2}} dz\,dy$$

$$= \frac{r}{2}\int_0^r \sqrt{r}\, y^{-\frac{1}{2}}\, dy = r^2$$

**EXERCISES III**

i. Find the centroid of the area (i.e., $\delta = 1$) bounded by $y = x$ and $y = 6x - x^2$.

ii. Find the surface of that portion of $3x + 4y + 6z = 12$ whose projection in the $X,Y$-plane is bounded by $x = 0$, $y = 0$, $x + y = 2$.

6. $4y = x^2$, $y = 0$, $x = 4$

$\delta(x,y) = k(x^2 + y^2)$, $y_1 = \dfrac{x^2}{4}$

$$m(A) = k\int_0^4 \int_0^{y_1} (x^2 + y^2)\,dy\,dx$$

$$= k\,4^4\left(\frac{1}{5} + \frac{1}{21}\right) = 4^4\left(\frac{26}{105}\right)k$$

$$M_y = k\int_0^4 \int_0^{y_1} x(x^2 + y^2)\,dy\,dx$$

$$= k\,4^5\left(\frac{1}{6} + \frac{1}{24}\right) = 4^5\left(\frac{5}{24}\right)k$$

$$M_x = k\int_0^4 \int_0^{y_1} y(x^2 + y^2)\,dy\,dx$$

$$= k\,4^4\left(\frac{2}{7} + \frac{1}{9}\right) = 4^4\left(\frac{25}{63}\right)k$$

**ANSWERS**

1. $\frac{1}{12}a^3$
2. $\frac{\pi}{2}$
3. $\frac{9}{2}\pi + 3$
4. $\frac{8}{15}$
5. $r^2$
6. $\bar{x} = \frac{175}{52}$, $\bar{y} = \frac{125}{78}$
7. $I_0 = I_x + I_y = \frac{89}{240}ka^6$

1

2

3

4

5

6

7

$$\bar{x} = \frac{M_y}{m(A)} = \frac{175}{52}, \quad \bar{y} = \frac{M_x}{m(A)} = \frac{125}{78}$$

7. $x^2 + y^2 = 2a^2$, $y^2 = ax$, $y = 0$ and $\delta(x,y) = kxy$

$$I_x = \int_0^a \int_{x_0}^{x_1} kxy^3\,dx\,dy, \text{ with}$$

$$x_0 = \frac{y^2}{a} \text{ and } x_1 = \sqrt{2a^2 - y^2}$$

$$I_x = \frac{5}{48}k\,a^6$$

$$I_y = \int_0^a \int_{x_0}^{x_1} kx^3y\,dx\,dy = \frac{4}{15}ka^6$$

**EXERCISE IV**

i. Find $I_x$, $I_y$, and the associated radii of gyration for the area in the first quadrant bounded by $y = x$ and $y^2 = x^3$.

## BASIC FACTS

The double integral

$$\lim_{\substack{\Delta x \to 0 \\ \Delta y \to 0}} \sum \sum f(x,y) \Delta x \Delta y$$

$$= \int_a^b \int_v^w f(x,y)\, dydx$$

$$= \int_c^d \int_r^s f(x,y)\, dxdy$$

where $v$ and $w$ may be functions of $x$ and $r$ and $s$ may be functions of $y$ is evaluated by performing:

1. the integration on the inner integral, e.g., $\int_v^w f(x,y)\,dy = F(x)$, *holding the outer variable* (e.g., $x$) *constant*;
2. the integration on the resulting integral, e.g., $\int_a^b F(x)\,dx$, and then evaluate between the limits.

**Volume Under a Surface**

$$V = \iint_A f(x,y)\, dA$$

where if $z = f(x,y)$ is the surface, then $A$ is the projection of the surface on the $X, Y$-plane and $dA = dxdy$. The variables and planes may be interchanged.

**Plane Area. Rectangular Coordinates**

$$A = \int_a^b \int_{x_1}^{x_2} dxdy = \int_c^d \int_{y_1}^{y_2} dydx$$

where $x_1 = f(y)$ and $x_2 = g(y)$ or $y_1 = \phi(x)$ and $y_2 = \psi(x)$.

**Mass of an Area, $m(A)$**

$$m(A) = \iint_A \delta(x,y)\, dxdy$$

where $\delta(x,y)$ is the density of the material in $A$.

**Plane Area. Polar Coordinates**

$$A = \iint_R \rho d\rho d\theta = \iint_R \rho d\theta\, d\rho$$

## ADDITIONAL INFORMATION

Consider the surface defined by $z = f(x,y)$ where $f(x,y)$ is a continuous function of the two variables, $x$ and $y$. Let $S$ be a closed area in the $X,Y$-plane and construct a right cylinder with $S$ as the base and the elements parallel to the $Z$-axis. (See Figure 25-2.) This cylinder will intersect the surface $x = f(x,y)$ and enclose the surface area $S'$. Now cover the area $S$ by a set of rectangles of length $\Delta x$ and width $\Delta y$ and erect the set of rectangular parallelepipeds whose bases are the rectangles and whose heights are $z_i$, the smallest value of $z$ on the surface $z = f(x,y)$ for $x$ and $y$ inside the rectangle $\Delta x \Delta y$. The volume of each parallelepiped is $z\Delta x\Delta y$, and if we sum these for all parallelepipeds inside the cylinder we approximate the **volume.** Now as we increase the number of parallelepipeds by taking $\Delta x$ and $\Delta y$ smaller we have the expression

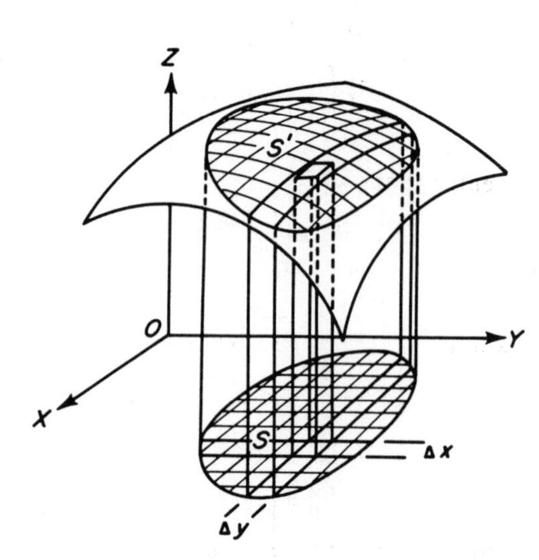


Fig. 25-2

$$V = \lim_{\substack{x \to 0 \\ y \to 0}} \sum \sum f(x,y)\, \Delta x \Delta y = \iint_S f(x,y)\, dxdy$$

as a formula for the volume of the right cylinder with base $S$ and top $S'$. If $f(x,y) > 0$ this is often referred to as the "volume under the surface" for an enclosed area on the surface.

If we had considered a cube of dimensions $\Delta x$ by $\Delta y$ by $\Delta z$, then the volume of this cube is $\Delta V = \Delta x \Delta y \Delta z$. A summation of this volume with respect to $z$ will give the volume of the parallelepipeds discussed above. We thus have the volume as

$$V = \lim_{\substack{x \to 0 \\ y \to 0 \\ z \to 0}} \sum_R \sum \sum \Delta z \Delta y \Delta x = \iiint_R dzdydx.$$

Let us now focus our attention on the region in the $X,Y$-plane. We are then considering a **plane area.** See Figure 25-3. The element of area is a rectangle of dimensions $\Delta x$ and $\Delta y$, that is we have $\Delta A = \Delta x \Delta y$. We can now set up a double summation $\Sigma\Sigma\Delta x\Delta y$ which is defined to be the totality of all such rectangles in the region. The area of the region $S$ can then be defined as $A =$

**Moments of Mass**

Consider a mass distributed over an area $A$ in the $X, Y$-plane and having continuous density $\delta(x,y)$. Then the moments about $x = a$ and $y = b$ are

$$M_a = \iint_A (x - a)\, \delta(x,y)\, dA$$

$$M_b = \iint_A (y - b)\, \delta(x,y)\, dA.$$

**Center of Mass**

$$\bar{x} = \frac{M_y}{m(A)}, \quad \bar{y} = \frac{M_x}{m(A)}$$

where $M_y = \iint_A x\,\delta(x,y)\, dA$

$$M_x = \iint_A y\,\delta(x,y)\, dA$$

and $m(A)$ is the mass in the area $A$.

**Moments of Inertia**

Consider a mass distributed over an area $A$ in the $X, Y$-plane and having a continuous density $\delta(x,y)$. Then the moments of inertia about the $X$-axis, $I_x$, is given by the integral

$$I_x = \iint_A y^2\, \delta(x,y)\, dA$$

and about the $Y$-axis the differential is $x^2 \delta(x,y)\, dA$. If the density is uniform, then $\delta(x,y) = k$. If we consider the moments of inertia about a line $x = a$ or $y = b$, then $x^2$ is replaced $(x - a)^2$ or $y^2$ is replaced by $(y - b)^2$.

The moment of inertia about the origin (*polar moment of inertia*) equals the sum of the moments of inertia about the $X$-axis and $Y$-axis, $I_0 = I_x + I_y$.

**Radius of Gyration,** $r^2 = \dfrac{I}{m}$

$I$ is the moment of inertia about an axis $L$ and $m$ is the total mass of the object.

$\lim_{\substack{x \to 0 \\ y \to 0}} \sum \sum \Delta x \Delta y = \iint_S dxdy$. Considering the previous discussion on volumes under a surface, this is the same as the volume of a right cylinder of unit height erected on $S$; i.e., $f(x,y) = 1$. The summation can be made with respect to $x$ first which covers the horizontal strip shown in Figure 25-3. The second summation would then add all these horizontal strips. However, since $\Sigma\Sigma\, \Delta x \Delta y = \Sigma\Sigma\, \Delta y \Delta x$, the summation could also be made with respect to $y$ first thus covering the vertical strips and then summing these vertical strips. The choice of summation depends upon the definition of the region $S$ and the fact that as we perform the calculation by integration we must have a continuous function and not a relation. Refer again to Figure 25-3 and note that for this configuration we must sum with respect to $y$ first since the curve $y = f_2(x)$ expressed as $x = g_e(y)$ is not a function for $y < c$ and also for $y = f_i(x)$ expressed as $x = g_i(y)$ for $y > d$. The area of the region $S$ shown in Figure 25-3 can be found by a double integration—integrating first with respect to $y$ between the limits $y = f_2(x)$ and $y = f_1(x)$ and then integrating with respect to $x$ between the limits $x = a$ to $x = b$.

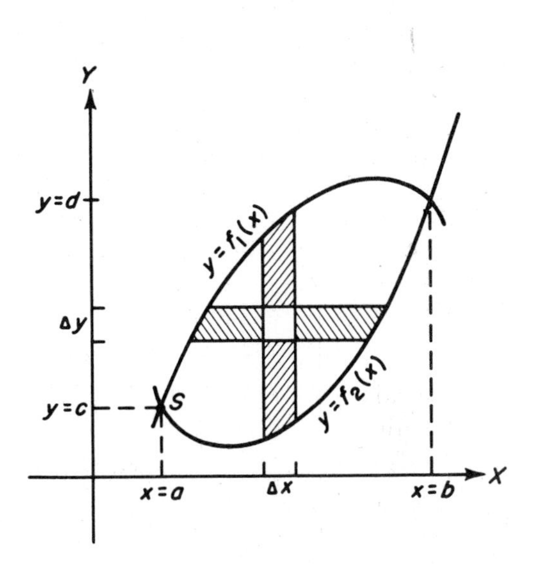


Fig. 25-3

**Surface Area**

$$A = \iint_S \left[1 + \left(\frac{\partial z}{\partial x}\right)^2 + \left(\frac{\partial z}{\partial y}\right)^2\right]^{\frac{1}{2}} dydx = \iint_S \left[1 + \left(\frac{\partial y}{\partial x}\right)^2 + \left(\frac{\partial y}{\partial z}\right)^2\right]^{\frac{1}{2}} dzdx$$

if $S$ is the projection on $X,Y$-plane and $X,Z$-plane, respectively.

**Volume and Mass by Triple Integration**

$$V = \iiint_R dzdydx \quad \text{and} \quad M = \iiint_R \delta(x,y,z)\, dzdydx$$

where $R$ is the region bounded by the surface and $\delta(x,y,z)$ is the density of the solid.

## SOLUTIONS TO THE EXERCISES

### I

i. $$\int_0^a \int_0^{\sqrt{a^2-x^2}} (x+y)\,dydx$$

$$= \int_0^a \left[ \int_0^{\sqrt{a^2-x^2}} (x+y)\,dy \right] dx$$

$$= \int_0^a \left[ xy + \frac{y^2}{2} \right]_0^{\sqrt{a^2-x^2}} dx$$

$$= \int_0^a \left( x\sqrt{a^2-x^2} + \frac{a^2-x^2}{2} \right) dx$$

$$= \frac{2a^3}{3}$$

ii. $$\iiint_V \cos \frac{x}{y}\, dxdydz = K \text{ with}$$

$V$: $x[0, 2yz]$, $y[0, \cos 2z]$, $z[0, \frac{\pi}{3}]$

$$K = \iint_R y \sin \frac{x}{y} \int_0^{2yz} dydz$$

$$= \iint_R y \sin 2z\, dydz$$

$$= \frac{1}{2} \int_0^{\pi/3} y^2 \sin 2z \Big|_0^{\cos 2z} dz$$

$$= \frac{1}{2} \int_0^{\pi/3} \cos^2 2z \sin 2z\, dz$$

$$= -\frac{1}{4}\,\frac{1}{3} \cos^3 2z \Big|_0^{\pi/3} = \frac{3}{32}$$

### II

i. Draw the graphs and determine $0 \le x \le \frac{\pi}{4}$ and $\sin x \le y \le \sec^2 x$.

$$A = \iint_A dydx = \int_0^{\pi/4} (\sec^2 x - \sin x)\, dx$$

$$= \tan x + \cos x \Big|_0^{\pi/4} = \frac{\sqrt{2}}{2}$$

ii. $\rho = 2r \cos \theta$ and $\rho = r$. The points of intersection, $\left(r, \frac{\pi}{3}\right)$ and $\left(r, -\frac{\pi}{3}\right)$ with $r \le \rho \le 2r \cos \theta$.

$$A = \iint_A \rho\, d\rho\, d\theta$$

$$A = \frac{1}{2} \int_{\theta_1}^{\theta_2} (4r^2 \cos^2 \theta - r^2)\, d\theta$$

$$= \frac{1}{2} \left[ 4r^2 \left( \frac{\theta}{2} + \frac{1}{4} \sin 2\theta \right) - r^2 \theta \right]_{-\pi/3}^{\pi/3}$$

$$= r^2 \left( \frac{\pi}{3} + \frac{\sqrt{3}}{2} \right)$$

### III

i. $y = 6x - x^2$ and $y = x$. Draw graphs.

$A = \{P(x,y) \mid 0 \le x \le 5,$
$x \le y \le 6x - x^2\}$

$$A = \iint_A dydx = \int_0^5 (5x - x^2)\, dx = \frac{125}{6}$$

$$M_x = \iint_A y\, dydx = \frac{1}{2} \int_0^5 [(6x-x^2)^2 - x^2]\, dx$$

$$= \frac{1}{2} \int_0^5 (35x^2 - 12x^3 + x^4)\, dx = \frac{5}{6}(125)$$

$$M_y = \iint_A x\, dydx = \int_0^5 (5x^2 - x^3)\, dx$$

$$= \left[ \frac{5}{3} x^3 - \frac{1}{4} x^4 \right]_0^5 = \frac{5}{12}(125)$$

$$\bar{x} = \frac{5\,(125)\,(6)}{12\,(125)} = \frac{5}{2},\ \bar{y} = 5$$

ii. $z = 2 - \frac{1}{2}x - \frac{2}{3}y$, $x + y = 2$

$1 + z_x^2 + z_y^2 = 1 + \frac{1}{4} + \frac{4}{9} = \frac{61}{36}$

$$A = \int_0^2 \int_0^{2-x} \frac{1}{6}\sqrt{61}\, dydx = \frac{\sqrt{61}}{3}$$

### IV

i. $y^2 = x^3$ and $y = x$. Draw figure.

$A = \{P(x,y) \mid 0 \le y \le 1,\ y \le x \le y^{\frac{2}{3}}\}$

$$A = \iint_A dxdy = \left[ \frac{3}{5} y^{\frac{5}{3}} - \frac{1}{2} y^2 \right]_0^1 = \frac{1}{10}$$

$$I_x = \iint_A y^2\, dxdy = \frac{1}{44}$$

$$I_y = \iint_A x^2\, dxdy = \frac{1}{36}$$

$$r_x = \frac{10}{44} = \frac{1}{22}\sqrt{110},\ r_y = \frac{10}{36} = \frac{1}{6}\sqrt{10}$$

**FINAL EXAMINATION** (Answers on page 154)

1. Find $D_x y$ for $y = \ln \tan\left(\frac{x}{2} + \frac{\pi}{4}\right)$.
2. Find $y'$ for $y = \text{Arctan}\,\dfrac{2x}{1 - x^2}$.
3. Find $D_x y$ for $y = (x^2 - x)\sqrt{1 - x}$.
4. Find $\dfrac{dy}{dx}$ for $y = \dfrac{\sqrt{x^2 + a^2}}{a^2 x}$.
5. Find $y' = f(x)$ for $x = \sqrt{1 + \sin y}$.
6. Find $y''$ if $y = xe^{ax}$.
7. Find $y'''$ if $y = x\sqrt{a^2 - x^2}$.
8. Derive the formula for the derivative of $f(x) = u(x)\,v(x)$.
9. Evaluate $\lim\limits_{x \to \infty} \dfrac{2x^3 - 5x^2 + 3x + 7}{3x^3 + 2x^2 - x - 9}$.
10. Evaluate $\lim\limits_{x \to 0} (1 + x)^{\cot x}$.
11. Find the maxima and minima of the function defined by $y = 3x^4 - 4x^3 - 12x^2 + 6$.
12. Find the maxima and minima of the function defined by $y = \dfrac{x + 2}{(x + 1)(x - 2)}$.
13. Find the minimum value of $y = ae^{kx} + be^{-kx}$.
14. Find the points of inflection of $y = \sin x + \cos x$ in the interval $[0, 2\pi]$.
15. If a particle moves according to $s = t^3 + \frac{1}{t}$, find the rate of change of the acceleration at $t = 1$.
16. Find the greatest value of $s$ if
$$s = v_1 t - \tfrac{1}{2} gt^2.$$
17. The legs of an isosceles triangle are each 10 in. long. Find the length of the base if the area is a maximum.
18. Find the dimensions of the rectangle of greatest area that can be inscribed in a circle of radius $r$.
19. Find the velocity and acceleration at $t = 4$ for a particle moving according to the equation
$$s = 4t^{\frac{3}{2}} - 2t^{\frac{1}{2}}.$$
20. A point moves along the curve defined by $x = t - 2t^3$ and $y = t^5 + t^2$. Find the magnitude of the acceleration at $t = \frac{1}{2}$.
21. Find the focus, directrix and length of the latus rectum of $9x + 4y^2 = 0$.
22. Find the equation of the parabola with vertex at $(3, 4)$ and the $Y$-axis as the directrix.
23. Find the equation of a point which moves so that its distance from $(-4, 3)$ plus its distance from $(4, 3)$ equals 12.
24. Find the foci and eccentricity of
    (a) $x^2 + 4y^2 = 8$
    (b) $4x^2 - 9y^2 = 36$.
25. Find the asymptotes for $x^2 - 16y^2 = 16$.
26. Find a standard form equation of the ellipse with $a = 4$ and $(5, 2)$, $(-1, 2)$ as foci.
27. Name the conic defined by
$$x^2 + 4xy + y^2 = 16.$$
28. Find the equation of a line through $(-3, 5)$ and perpendicular to $3x - 5y + 7 = 0$.
29. Transform $x^2 + 4xy + y^2 = 16$ into a standard form.
30. Find the equation of the tangent to the curve defined by $4x^2 + 8x + 13 = 9y^2 - 36y$ at $(2, -1)$.
31. Evaluate $\displaystyle\int \frac{\cos 2x}{3 + \sin 2x}\,dx$.
32. Evaluate $\int (\tan x - \sin x) \cos x\, dx$.
33. Evaluate $\displaystyle\int_{-\infty}^{\infty} \frac{dx}{c^2 + a^2x^2}$.
34. Evaluate $\displaystyle\int_0^{3/2} \frac{dx}{\sqrt{2x - x^2}}$.
35. Evaluate $\displaystyle\int_0^1 x^2 e^x dx$.
36. Find the area bounded by $y^2 = x + 4$ and $y = x - 2$.
37. Find the area bounded by $y = \dfrac{8a^3}{x^2 + 4a^2}$ and the line $x + 2y = 4a$.
38. Find the area bounded by $y = x^3 - x$ and $y = x - x^2$.
39. Find the area in the first quadrant bounded by $y = \sin x$, $y = \cos x$ and $x = 0$.
40. Find the area inside both the circle $x^2 - 2ax + y^2 = 0$ and the parabola $x^2 = ay$; $a > 0$.
41. The cross section of a water gate in a dam is an isosceles trapezoid whose upper base is 12 feet, lower base is 6 feet, and altitude is 6 feet. Find the total pressure on the gate if the surface of the water is 3 feet above the upper base.
42. Find the volume generated when the area bounded by $y(x - 2)^2 = 1$, $y = 0$, $x = 3$, and $x = 4$ is revolved about the $X$-axis.
43. Find the volume generated by revolving the area bounded by $y = \sin x$, $x = 0$, $x = \pi$ about the $Y$-axis.

44. Find the volume generated by revolving the area bounded by $x^2 + 4y^2 = 4$, $y = 0$ about the $Y$-axis.
45. Find the volume generated when the area bounded by $y = 3 + 2x - x^2$ and $y = 0$ is revolved about the $X$-axis.
46. Find the volume of the solid obtained by revolving the region in the first quadrant inside $3x^2 + 4y^2 = 48$ and outside $2y^2 = 9x$ around the $X$-axis.
47. Find the value of $dy$ for the function defined by $y = x^3 - 3x^2$ at $(1,-2)$ if $dx = 0.1$.
48. Find $\Delta y - dy$ at $x = 2$ if $y = x^{-2} - x^{-3}$ and $\Delta x = 1$.
49. Find $\frac{dz}{dt} = f(t)$ if $z = y^2$, $y = \sin x$ and $x = 3t^2$.
50. Find $f(2)$ if $f'(x) = x^2 + 1$ and $f(3) = 10$.
51. Find the solution set $0 \leq x \leq 2\pi$ for the equation $4 \sin x \cos x - 2 \cos x + 2 \sin x = 1$.
52. If $\tan 2\phi = \frac{B}{A - C}$ find $\sin \phi$ if $A = 3$, $B = 4$, $C = 3$.
53. Find the equation of a circle through $(2,-1)$, $(3,4)$, and $(-2,1)$.
54. Find the equation of a line perpendicular to the tangent to the circle $x^2 + y^2 - 3x + 2y = 1$ at $(1,1)$ and passing through the point $(3,-5)$.
55. Find $\arccos \frac{1}{2}$.
56. Find the solution set for $\log x + \log(x-3) = 1$.
57. Find the equation of the line tangent to the curve defined by $x = 2t^2 - 3t$, $y = 3t^2 - 2t$ at $t = 2$.
58. Transform the equation $r^2 \sin 2\theta = 16$ into an equation involving rectangular coordinates.
59. Find the center of curvature at the point $P(2,2)$ for the curve defined by
$$x^3 + xy^2 - 4y^2 = 0.$$
60. Find the radius of curvature at $(0,0)$ for the curve defined by $y = \ln(x + 1)$.
61. Find the length of the arc defined by the equations $x = \frac{a}{2}(2 \cos t - \cos 2t)$, $y = \frac{a}{2}(2 \sin t - \sin 2t)$ from $t = 0$ to $t = \pi$.
62. Expand the function given by $y = \sqrt{1 - x}$ in a Maclaurin series and find its interval of convergence. (Omit the testing of the end points.)
63. Find $u_x$ for $u = (x + y) \sin(x - y)$.
64. Show that $u_{xy} = u_{yx}$ for $u = \frac{x}{x + y}$.
65. Show that $u = \frac{xy}{x + y}$ satisfies $xu_x + yu_y = u$.
66. Find $\mathbf{v}(t)$, $\mathbf{a}(t)$ if $\mathbf{f}(t) = e^t \sin t\, \mathbf{i} - e^t \cos t\, \mathbf{j}$.
67. Find $\mathbf{g}'(t)$ if $\mathbf{g}(t) = \mathbf{u}(t) \times \mathbf{v}(t)$ and the vectors are
$$\mathbf{u}(t) = (t^2 + t)\mathbf{i} - 3t^2\mathbf{j} + e^t\mathbf{k},$$
$$\mathbf{v}(t) = t^2\mathbf{i} + (t^2 - t)\mathbf{j} - e^t\mathbf{k}.$$
68. Find the cosine of the angle between $\mathbf{u}$ and $\mathbf{v}$ if
$$\mathbf{u} = 4\mathbf{i} - 5\mathbf{j} + \mathbf{k} \text{ and } \mathbf{v} = 5\mathbf{i} + 4\mathbf{j} + \mathbf{k}.$$
69. Find the length of the curve defined by $x = t$, $y = \frac{\sqrt{2}}{2} t^2$, $z = \frac{1}{3} t^3$ and $0 \leq t \leq 2$.
70. Find the equation of the plane through $A(2,-1,3)$, $B(1,3,-2)$ and $C(4,-3,1)$.
71. Find the equation of the tangent plane to the surface $2x^2 - y^2 + (z + 2)^2 = 23$ at $P(1,-2,3)$.
72. Find the critical values of $z = (x^2 + x)(y + 1)$.
73. Evaluate $\displaystyle\int_0^{\pi/4} \int_0^{2 \sin \theta} \rho^2 \cos \theta \, d\rho \, d\theta$.
74. Use triple integration to find the volume bounded by the surfaces, $z^2 + y^2 - 4z = 0$, $2z + x = 4$, $y = 0$, $x = 0$. Hint: Use the order of integration $dxdydz$ or $dydxdz$.
75. Find the volume of a cylinder whose base is the area bounded by $x = y^2$ and $y = x^2$ and cut off at the top by the surface $z = 6 + y$.

Determine if the following statements are true or false.

76. If $c > 0$ and $x > y$, then $cy < cx$.
77. If $9x - 3 > 4x + 7$, then $x > 2$.
78. If $a$ and $b$ are positive and $a^2 > b^2$, then $a > b$.
79. If $x = 2$ and $a < b$, then $\frac{b}{x} > \frac{a}{x}$.
80. The equations $r^2 = 9 \tan \theta$ and $x^3 + y^2x = 9y$ have the same graph.
81. The graph of $r^2 + r(\cos \theta + \sin \theta) = 0$ is a circle.
82. The point $\left(-2, \frac{5}{2}\right)$ is the vertex of the parabola $4y^2 - 16x - 20y = 7$.
83. The point $(3,4)$ is the center of the circle $x^2 + y^2 - 6x - 8y = 0$.
84. If $P_1(4,2)$, $P_2(2,6)$, $P_3(-2,2)$, and $P_4(2,0)$, then $\overline{P_1P_4}$ is parallel to $\overline{P_2P_3}$.
85. The graph of $xy = 7$ is a hyperbola.
86. The function defined by $y = \frac{x + 3}{x - 2}$ is discontinuous at $x = 2$.
87. The tangent function is discontinuous at all values of the argument equal to $\left(n + \frac{1}{2}\right)\pi$.

88. The limit of the sum of two functions is the sum of the limits of each function.
89. $\log \sqrt{20} = \frac{1}{3} \log 20$
90. The limit of a constant function as the variable approaches a given value is indeterminate.
91. $\lim_{x \to a} x = a$
92. The domain of the real function defined by $y = 3^{-x}$ is the set of real numbers.
93. If $xy = y^2 - 6$, then at $x = 1$, $y = 3$ and 2.
94. The function $f(x) = \sqrt{9 - x^2}$ is not defined in the real number system if $0 \le x \le 3$.
95. The function $xy - y = 3$ is not defined at $x = 1$.
96. The distance from (4, 2) to the line $y = x + 4$ is $3\sqrt{2}$.
97. If $2^{x+3} = 32$, then $x = 2$.
98. $\cos 2x = \cos^2 x - \sin^2 x$ for all $x$.
99. $\lim_{x \to a} (uv) = \left(\lim_{x \to a} u\right)\left(\lim_{x \to a} v\right)$
100. If $e = 1$, the conic is a parabola.
101. If $N = a^x$, then $x = \log_a N$.
102. The length of the latus rectum of the parabola $2x^2 = 9y$ is 9.
103. The slope of the graph $2y = 4x - 7$ is 4.
104. The graph of $r = a \sin 3\theta$ is a three-leaved rose.
105. $\begin{vmatrix} 3 & 5 \\ 2 & 8 \end{vmatrix} = -14$
106. The lines defined by $2x - 3y = 5$ and the equation $2y - 3x = 5$ are parallel.
107. $\int_0^1 (3x^2 - 6x)\, dx = -2$
108. $\frac{d}{dx} (e^{-x}) = e^{-x}$
109. If $\text{Arcsin}\, x = 0.5$, then $x = 30°$.
110. If $\mathbf{v} = 2\mathbf{i} + 3\mathbf{j}$ and $\mathbf{w} = 3\mathbf{i} + 2\mathbf{j}$, then $\mathbf{v} \perp \mathbf{w}$.
111. $\lim_{x \to 0} \frac{c}{x} = \infty$
112. If $xy = 7$, then $\frac{dy}{dx} = \frac{y}{x}$.
113. If $n$ is a positive integer and $f(x) = x^n$, then $f'(x) = x^{n-1}$.
114. $y = 2x - 2$ is tangent to $y = x^3 - x$ at (1, 0).
115. The minimum of $y = 3 + 2x + 2x^2$ is at $y = \frac{5}{2}$.
116. The instantaneous velocity of a particle that moves according to $s = 2t - 3$ is 2.
117. If $y = x^3$ and $\Delta x = 0.1$, then $dy = 3$ at $x = 1$.
118. If $f'(x) = x^2$ and $f(3) = 0$, then $f(x) = x^3 - 9$.
119. The area bounded by the coordinate axes and $x + y = 1$ is $\frac{1}{2}$.
120. The graph of $x^2 + y^2 - 3x = 5$ is a circle.
121. The graph of $x^2 - y^2 = 7$ is an ellipse.
122. $x = 0$ is an asymptote to $xy = k$.
123. $\sinh^{-1} x = \ln(x + \sqrt{x^2 + 1})$
124. If $N = \sqrt{x}\,(y^{-2})$, then $\ln N = \frac{1}{2} \ln x - 2 \ln y$.
125. If $\ln 4 - \ln 2 = x$, then $e^x = 2$.
126. The tangent to $y = \sin x$ at $x = \frac{\pi}{2}$ is parallel to the $X$-axis.
127. The tangent to $x = t^2$, $y = t^2 + 1$ is $x = t^2$, $y = t^2 + 1$.
128. The area inside the curve $r = 3$ is $9\pi$.
129. If $\mathbf{v} = 4\mathbf{i} - 5\mathbf{j}$ and $\mathbf{w} = 5\mathbf{i} + 4\mathbf{j}$, then the sum $\mathbf{w} + \mathbf{v} = 9\mathbf{i} - 9\mathbf{j}$.
130. If $\mathbf{f}(t) = e^t\mathbf{i} - \cos t\,\mathbf{j}$, then $\mathbf{f}''(t) = e^t\mathbf{i} + \cos t\,\mathbf{j}$.
131. If a curvilinear motion is defined by the equation $\mathbf{f}(t) = (t^2 - t)\mathbf{i} + (t^2 + t)\mathbf{j}$, then $\frac{ds}{dt} = \sqrt{8t^2 - 2}$.
132. If $\mathbf{v} \perp \mathbf{w}$ and $\mathbf{v} = 3a\mathbf{i} - \mathbf{j}$, $\mathbf{w} = a\mathbf{i} + 6a\mathbf{j}$, then $a = 0, 3$.
133. $\int xe^x\, dx = e^x(x + 1) + C$
134. $\int \sin^2 x\, dx = \frac{1}{2} x + \frac{1}{4} \sin 2x + C$
135. $\int \tan^2 x\, dx = \tan x - x + C$
136. $\int \frac{x\,dx}{x^2 - x} = \ln(x - 1) + C$
137. $\int \frac{x + 1}{x^2}\, dx = \ln x + \frac{1}{x} + C$
138. $\int \frac{x\,dx}{x^2 - 1} = \ln(x^2 - 1) + C$
139. $\int \frac{\sqrt{x}\, dx}{1 + x} = 2\sqrt{x} - 2 \arctan \sqrt{x} + C$
140. The mean value of $f(x) = x^2$ in $[1, 4]$ is 7.
141. The moment of the area bounded by $y = \sin x$ and the lines $x = 0$, $x = \frac{\pi}{2}$, and $y = 0$ about the $X$-axis is $\frac{\pi}{8}$.
142. The area bounded by $y = \cos x$, $x = 0$, $x = \frac{\pi}{2}$, and $y = 0$ is 1.
143. The volume generated by revolving the area bounded by $y = \sin 2x$, $x = 0$, $x = \frac{\pi}{4}$, and $y = 0$ about the $X$-axis is $\frac{\pi^2}{8}$.
144. $\lim_{x \to 2} \frac{x^2 - 4}{x^2 + x - 6} = \frac{4}{5}$
145. $\lim_{x \to 0} \frac{\cot 2x}{\cot x} = \frac{1}{2}$
146. $\lim_{x \to 0} \left(\frac{1}{x}\right)^{\sin x} = 0$
147. $\int_1^\infty \frac{dx}{x^3} = \frac{1}{2}$
148. The power series for $f(x) = \sin x$ converges for all $x$.
149. The distance from the origin to a point $P(3, 4, 5)$ is $5\sqrt{2}$.
150. If $u = xy$, $x = e^t$, and $y = e^{-t}$, then $\frac{du}{dt} = 0$.

## ANSWERS

1. $\sec x$
2. $\dfrac{2}{1+x^2}$
3. $\left(\frac{5}{2}x-1\right)\sqrt{1-x}$
4. $-x^{-2}(x^2+a^2)^{-1/2}$
5. $2(2-x^2)^{-1/2}$
6. $ae^{ax}(2+ax)$
7. $-3a^4(a^2-x^2)^{-5/2}$
8. $D_x u(x)v(x) = u(x)D_x v(x) + v(x)D_x u(x)$
9. $\frac{2}{3}$
10. $e$
11. Min at $(-1,1)$ and $(2,-26)$ Max at $(0,6)$
12. Max at $(0,-1)$, Min at $\left(-4,-\frac{1}{9}\right)$
13. $2\sqrt{ab}$
14. $\left(\frac{3\pi}{4},0\right),\left(\frac{7\pi}{4},0\right)$
15. 0
16. $\dfrac{v_1^2}{2g}$
17. $10\sqrt{2}$
18. A square of side $r\sqrt{2}$
19. $v=\frac{23}{2}$, $a=\frac{25}{16}$
20. $\frac{15}{2}$
21. $F\left(-\frac{9}{16},0\right)$, $x=\frac{9}{16}$, $L.R.=\frac{9}{4}$
22. $y^2-8y-12x+52=0$
23. $5x^2+9y^2-54y=99$
24. (a) $F(\pm\sqrt{6},0)$, $e=\frac{1}{2}\sqrt{3}$ (b) $F(\pm\sqrt{13},0)$, $e=\frac{1}{3}\sqrt{13}$
25. $4y=\pm x$
26. $\dfrac{(x-2)^2}{16}+\dfrac{(y-2)^2}{7}=1$
27. Hyperbola
28. $3y+5x=0$
29. $\dfrac{x'^2}{\frac{16}{3}}-\dfrac{y'^2}{16}=1$
30. $9y+4x+1=0$
31. $\frac{1}{2}\ln(3+\sin 2x)+C$
32. $-\cos x-\frac{1}{2}\sin^2 x + C$
33. $\dfrac{\pi}{ac}$
34. $\frac{2\pi}{3}$
35. $e-2$
36. $20\frac{5}{6}$
37. $a^2(\pi-3)$
38. $\frac{37}{12}$
39. $\sqrt{2}-1$
40. $\dfrac{a^2}{12}(3\pi-4)$
41. $306W$
42. $\frac{7\pi}{24}$
43. $2\pi^2$
44. $\frac{8}{3}\pi$
45. $\frac{512}{15}\pi$
46. $13\pi$
47. $-0.3$
48. $-\frac{5}{432}$
49. $6t\sin 6t^2$
50. $\frac{8}{3}$
51. $\frac{\pi}{6},\frac{5\pi}{6},\frac{2\pi}{3},\frac{4\pi}{3}$
52. $\pm\frac{\sqrt{2}}{2}$
53. $11x^2+11y^2-20x-40y=55$
54. $y+4x=7$
55. $\frac{\pi}{3}+2n\pi$, $\frac{5\pi}{3}+2n\pi$
56. 5
57. $y=2x+4$
58. $xy=8$
59. $\left(-\frac{14}{3},\frac{16}{3}\right)$
60. $2\sqrt{2}$
61. $4a$
62. $1-\displaystyle\sum_{n=1}^{\infty}\frac{1\cdot 3\cdot 5\cdots(2n-3)}{n!}\times\left(\frac{x}{2}\right)^n$, $-1<x<1$
63. $\sin(x-y)+(x+y)\cos(x-y)$
64. $u_{xy}=u_{yx}=\dfrac{x-y}{(x+y)^3}$
65. Obtain $u_x$ and $u_y$ and substitute into equation.
66. $\mathbf{v}(t)=e^t(\sin t+\cos t)\mathbf{i}+e^t(\sin t-\cos t)\mathbf{j}$
    $\mathbf{a}(t)=2e^t\cos t\,\mathbf{i}+2e^t\sin t\,\mathbf{j}$
67. $e^t(2t^2+5t+1)\mathbf{i}+e^t(2t^2+5t+1)\mathbf{j}+(16t^3-2t)\mathbf{k}$
68. $\cos\phi=\frac{1}{42}$
69. $\frac{14}{3}$
70. $3x+2y+z=7$
71. $2x+2y+5z=13$
72. $(0,-1)$, $(-1,-1)$
73. $\frac{1}{6}$
74. $\frac{16}{3}$
75. $\frac{43}{20}$
76. True
77. True
78. True
79. True
80. True
81. True
82. True
83. True
84. True
85. True
86. True
87. True
88. True
89. False
90. False
91. True
92. True
93. False
94. False
95. True
96. True
97. True
98. True
99. True
100. True
101. True
102. False
103. False
104. True
105. False
106. False
107. True
108. False
109. True
110. False
111. True
112. False
113. False
114. True
115. True
116. True
117. False
118. False
119. True
120. True
121. False
122. True
123. True
124. True
125. True
126. True
127. True
128. True
129. False
130. True
131. False
132. False
133. False
134. False
135. True
136. True
137. False
138. False
139. True
140. True
141. True
142. True
143. True
144. True
145. True
146. False
147. True
148. True
149. True
150. True

# APPENDIX

## TRIGONOMETRIC FORMULAS

**The Fundamental Identities.**

*The reciprocal relations:*

$$\csc\theta = \frac{1}{\sin\theta};\quad \sec\theta = \frac{1}{\cos\theta};\quad \cot\theta = \frac{1}{\tan\theta}.$$

*The quotient relations:*

$$\tan\theta = \frac{\sin\theta}{\cos\theta};\qquad \cot\theta = \frac{\cos\theta}{\sin\theta}.$$

*The Pythagorean relations:*

$$\sin^2\theta + \cos^2\theta = 1;$$
$$\tan^2\theta + 1 = \sec^2\theta;$$
$$1 + \cot^2\theta = \csc^2\theta.$$

**Sum of Two Angles.**

(1) $\sin(\alpha + \beta) = \sin\alpha\cos\beta + \cos\alpha\sin\beta;$

(2) $\cos(\alpha + \beta) = \cos\alpha\cos\beta - \sin\alpha\sin\beta.$

(3) $\tan(\alpha + \beta) = \dfrac{\tan\alpha + \tan\beta}{1 - \tan\alpha\tan\beta}.$

(4) $\cot(\alpha + \beta) = \dfrac{\cot\alpha\cot\beta - 1}{\cot\alpha + \cot\beta}.$

**Difference of Two Angles.**

(5) $\sin(\alpha - \beta) = \sin\alpha\cos\beta - \cos\alpha\sin\beta.$

(6) $\cos(\alpha - \beta) = \cos\alpha\cos + \sin\alpha\sin\beta.$

(7) $\tan(\alpha - \beta) = \dfrac{\tan\alpha - \tan\beta}{1 + \tan\alpha\tan\beta}.$

(8) $\cot(\alpha - \beta) = \dfrac{\cot\alpha\cot\beta + 1}{\cot\beta - \cot\alpha}.$

**Double Angle Formulas.**

(9) $\sin 2\alpha = 2\sin\alpha\cos\alpha.$

(10) $\cos 2\alpha = \cos^2\alpha - \sin^2\alpha;$
$\qquad = 2\cos^2\alpha - 1;$
$\qquad = 1 - 2\sin^2\alpha.$

(11) $\tan 2\alpha = \dfrac{2\tan\alpha}{1 - \tan^2\alpha}.$

(12) $\cot 2\alpha = \dfrac{\cot^2\alpha - 1}{2\cot\alpha}.$

**Half-Angle Formulas.**

(13) $\sin\dfrac{\theta}{2} = \pm\sqrt{\dfrac{1 - \cos\theta}{2}}.$

(14) $\cos\dfrac{\theta}{2} = \pm\sqrt{\dfrac{1 + \cos\theta}{2}}.$

(15) $\tan\dfrac{\theta}{2} = \pm\sqrt{\dfrac{1 - \cos\theta}{1 + \cos\theta}} = \dfrac{1 - \cos\theta}{\sin\theta} = \dfrac{\sin\theta}{1 + \cos\theta}.$

(16) $\cot\dfrac{\theta}{2} = \pm\sqrt{\dfrac{1 + \cos\theta}{1 - \cos\theta}} = \dfrac{\sin\theta}{1 - \cos\theta} = \dfrac{1 + \cos\theta}{\sin\theta}.$

**Product Formulas.**

(17) $2\sin\alpha\cos\beta = \sin(\alpha + \beta) + \sin(\alpha - \beta).$

(18) $2\cos\alpha\cos\beta = \cos(\alpha + \beta) + \cos(\alpha - \beta).$

(19) $2\sin\alpha\sin\beta = \cos(\alpha - \beta) - \cos(\alpha + \beta).$

(20) $2\sin\beta\cos\alpha = \sin(\alpha + \beta) - \sin(\alpha - \beta).$

**Sums and Differences of Sines and Cosines.**

(21) $\sin x + \sin y = 2\sin\dfrac{x + y}{2}\cos\dfrac{x - y}{2}.$

(22) $\sin x - \sin y = 2\cos\dfrac{x + y}{2}\sin\dfrac{x - y}{2}.$

(23) $\cos x + \cos y = 2\cos\dfrac{x + y}{2}\cos\dfrac{x - y}{2}.$

(24) $\cos x - \cos y = -2\sin\dfrac{x + y}{2}\sin\dfrac{x - y}{2}.$

## FOR OBLIQUE TRIANGLES

*The Law of Cosines:*

$$a^2 = b^2 + c^2 - 2bc\cos\alpha;$$
$$b^2 = a^2 + c^2 - 2ac\cos\beta;$$
$$c^2 = a^2 + b^2 - 2ab\cos\gamma.$$

Solving for the angles, these formulas read

$$\cos\alpha = \frac{b^2 + c^2 - a^2}{2bc};$$

$$\cos\beta = \frac{a^2 + c^2 - b^2}{2ac};$$

$$\cos\gamma = \frac{a^2 + b^2 - c^2}{2ab}.$$

*The Law of Sines:*

$$\frac{a}{\sin\alpha} = \frac{b}{\sin\beta} = \frac{c}{\sin\gamma}.$$

*The Law of Tangents:*

$$\frac{a-b}{a+b} = \frac{\tan\frac{1}{2}(\alpha-\beta)}{\tan\frac{1}{2}(\alpha+\beta)}$$

$$\frac{a-c}{a+c} = \frac{\tan\frac{1}{2}(\alpha-\gamma)}{\tan\frac{1}{2}(\alpha+\gamma)};$$

$$\frac{b-c}{b+c} = \frac{\tan\frac{1}{2}(\beta-\gamma)}{\tan\frac{1}{2}(\beta+\gamma)}.$$

*Half-Angle Formulas:*

$$\tan\frac{\alpha}{2} = \frac{r}{s-a}; \tan\frac{\beta}{2} = \frac{r}{s-b};$$

$$\tan\frac{\gamma}{2} = \frac{r}{s-c};$$

where

$$s = \frac{1}{2}(a+b+c)$$

and

$$r = \sqrt{\frac{(s-a)(s-b)(s-c)}{s}}.$$

*The Area of the Oblique Triangle:*

$$K = \frac{1}{2}\,bc\sin\alpha;\ K = \frac{1}{2}\,ac\sin\beta;\ K = \frac{1}{2}\,ab\sin\gamma.$$

$$K = \sqrt{s(s-a)(s-b)(s-c)}.$$

$$K = \frac{a^2\sin\beta\sin\gamma}{2\sin\alpha};\ K = \frac{b^2\sin\alpha\sin\gamma}{2\sin\beta};$$

$$K = \frac{c^2\sin\alpha\sin\beta}{2\sin\gamma}.$$

## FUNCTIONS OF NEGATIVE ARGUMENTS

*If θ is any positive number, then*

$$\sin(-\theta) = -\sin\theta \qquad \cot(-\theta) = -\cot\theta$$
$$\cos(-\theta) = \cos\theta \qquad \sec(-\theta) = \sec\theta$$
$$\tan(-\theta) = -\tan\theta \qquad \csc(-\theta) = -\csc\theta$$

*for all admissible values of θ.*

### Reference Angle

| $\theta$ | $\alpha$ |
|---|---|
| $0 < \theta < 90°$ | $\alpha = \theta$ |
| $90° < \theta < 180°$ | $\alpha = 180° - \theta$ |
| $180° < \theta < 270°$ | $\alpha = \theta - 180°$ |
| $270° < \theta < 360°$ | $\alpha = 360° - \theta$ |

THEOREM. *Any function of an angle θ, in any quadrant, is numerically equal to the same function of the reference angle for θ; i.e.,*

$$(\text{any function of } \theta) = \pm(\text{same function of } \alpha)$$

The "+" or "−" is determined by the quadrant in which the angle falls.

For $\theta > 2\pi$, we have the relation

$$\text{any function of } \theta = \text{same function of } (\theta - 2n\pi)$$

where $n$ is any integer.

## DEGREES TO RADIANS

| 90° | 120° | 135° | 150° | 180° | 210° |
|---|---|---|---|---|---|
| $\frac{\pi}{2}$ | $\frac{2\pi}{3}$ | $\frac{3\pi}{4}$ | $\frac{5\pi}{6}$ | $\pi$ | $\frac{7\pi}{6}$ |
| 225° | 240° | 270° | 300° | 315° | 330° |
| $\frac{5\pi}{4}$ | $\frac{4\pi}{3}$ | $\frac{3\pi}{2}$ | $\frac{5\pi}{3}$ | $\frac{7\pi}{4}$ | $\frac{11\pi}{6}$ |

## THE HYPERBOLIC FUNCTIONS

$$\sinh x = \frac{e^x - e^{-x}}{2} \qquad \coth x = \frac{e^x + e^{-x}}{e^x - e^{-x}},\quad x \neq 0$$

$$\cosh x = \frac{e^x + e^{-x}}{2} \qquad \operatorname{sech} x = \frac{2}{e^x + e^{-x}}$$

$$\tanh x = \frac{e^x - e^{-x}}{e^x + e^{-x}} \qquad \operatorname{csch} x = \frac{2}{e^x - e^{-x}},\quad x \neq 0$$

$$\sinh^{-1} x = \ln(x + \sqrt{x^2+1}) \quad (\text{all } x).$$
$$\operatorname{Cosh}^{-1} x = \ln(x + \sqrt{x^2-1}) \quad (x \geq 1).$$
$$\tanh^{-1} x = \frac{1}{2}\ln\frac{1+x}{1-x} \quad (x^2 < 1).$$
$$\coth^{-1} x = \frac{1}{2}\ln\left(\frac{x+1}{x-1}\right) \quad (x^2 > 1).$$
$$\operatorname{Sech}^{-1} x = \ln\left(\frac{1}{x} + \sqrt{\frac{1}{x^2} - 1}\right) \quad (0 < x \leq 1).$$
$$\operatorname{csch}^{-1} x = \ln\left(\frac{1}{x} + \sqrt{\frac{1}{x^2} + 1}\right) \quad (\text{all } x \neq 0).$$

### Fundamental Identities

| | |
|---|---|
| $\tanh x = \frac{\sinh x}{\cosh x}$ | $\coth x = \frac{\cosh x}{\sinh x}$ |
| $\operatorname{sech} x = \frac{1}{\cosh x}$ | $\operatorname{csch} x = \frac{1}{\sinh x}$ |
| $\cosh^2 x - \sinh^2 x = 1$ | $\operatorname{sech}^2 x = 1 - \tanh^2 x$ |
| $\operatorname{csch}^2 x = \coth^2 x - 1$ | $\cosh x + \sinh x = e^x$ |
| $\cosh x - \sinh x = e^{-x}$ | |

## THE NUMBER SYSTEM

The **real number system** consists of

- the positive integers (1, 2, 3, . . .)
- zero (0)
- the negative integers (–1, –2, –3, . . .)
- the fractions $\left(\frac{1}{2}, \frac{2}{3}, \frac{7}{5}, \ldots\right)$
- the decimal fractions (.25, 1.03, . . .).

A **rational number** is one which can be expressed in the form $\frac{a}{b}$, $b \neq 0$ with $a$ and $b$ being integers.

An **irrational number** is one which cannot be expressed exactly as the quotient of two integers.

An **even number** is an integer that is exactly divisible by 2, and an odd number is one that is not even.

A **fraction** of the form $\frac{a}{b}$, $b \neq 0$ is an indicated division of $a$ by $b$. The term **ratio** has the same definition.

The letter $\boldsymbol{i}$ is used to denote $\sqrt{-1}$ so that $i^2 = -1$.

The number $\boldsymbol{a + bi}$ where $a$ and $b$ are real numbers is called a **complex number.**

## DERIVATIVE FORMULAS

$D_x c = 0.$

$D_x(u \pm v) = D_x u \pm D_x v.$

$D_x y = D_u y \cdot D_x u.$

$$D_x\left(\frac{u}{v}\right) = \frac{vD_x u - uD_x v}{v^2}.$$

$D_x(cu) = cD_x u.$

$D_x(uv) = uD_x v + vD_x u.$

$D_x u^n = nu^{n-1} D_x u.$

$$D_y x = \frac{1}{D_x y}, \quad (D_x y \neq 0).$$

$D_x \sin u = \cos u\, D_x u.$

$D_x \cos u = -\sin u\, D_x u.$

$D_x \tan u = \sec^2 u\, D_x u.$

$D_x \cot u = -\csc^2 u\, D_x u.$

$D_x \sec u = \sec u \tan u\, D_x u.$

$D_x \csc u = -\csc u \cot u\, D_x u.$

$$D_x \operatorname{Arcsin} u = \frac{D_x u}{\sqrt{1-u^2}}.$$

$$D_x \operatorname{Arccos} u = \frac{-D_x u}{\sqrt{1-u^2}}.$$

$$D_x \operatorname{Arctan} u = \frac{D_x u}{1+u^2}.$$

$$D_x \operatorname{Arccot} u = \frac{-D_x u}{1+u^2}.$$

$$D_x \operatorname{Arcsec} u = \frac{D_x u}{u\sqrt{u^2-1}}.$$

$$D_x \operatorname{Arccsc} u = \frac{-D_x u}{u\sqrt{u^2-1}}.$$

$$D_x \log_a u = \frac{1}{u} \log_a e\, D_x u.$$

$$D_x \ln u = \frac{D_x u}{u}.$$

$D_x a^u = a^u \ln a\, D_x u.$

$D_x e^u = e^u D_x u.$

$D_x(u^v) = vu^{v-1} D_x u + (\ln u)u^v D_x v.$

$D_x \sinh u = \cosh u\, D_x u.$

$D_x \cosh u = \sinh u\, D_x u.$

$D_x \tanh u = \operatorname{sech}^2 u\, D_x u.$

$D_x \coth u = -\operatorname{csch}^2 u\, D_x u.$

$D_x \operatorname{sech} u = -\operatorname{sech} u \tanh u\, D_x u.$

$D_x \operatorname{csch} u = -\operatorname{csch} u \coth u\, D_x u.$

$$D_x \sinh^{-1} u = \frac{D_x u}{\sqrt{1+u^2}} \quad \text{(all values of } u\text{)}.$$

$$D_x \operatorname{Cosh}^{-1} u = \frac{D_x u}{\sqrt{u^2-1}} \quad (u > 1).$$

$$D_x \tanh^{-1} u = \frac{D_x u}{1-u^2} \quad (u^2 < 1).$$

$$D_x \coth^{-1} u = \frac{D_x u}{1-u^2} \quad (u^2 > 1).$$

$$D_x \operatorname{Sech}^{-1} u = \frac{-D_x u}{\sqrt{u^2-u^4}} \quad (0 < u < 1).$$

$$D_x \operatorname{csch}^{-1} u = \frac{-D_x u}{\sqrt{u^2+u^4}} \quad (u \neq 0).$$

## SERIES

$$\sin x = x - \frac{x^3}{3!} + \frac{x^5}{5!} - \cdots + (-1)^{n+1}\frac{x^{2n-1}}{(2n-1)!} + \cdots, \text{ all } x.$$

$$\cos x = 1 - \frac{x^2}{2!} + \frac{x^4}{4!} - \cdots + (-1)^{n+1}\frac{x^{2n-2}}{(2n-2)!} + \cdots, \text{ all } x.$$

$$\tan x = x + \frac{x^3}{3} + \frac{2x^5}{15} + \frac{17x^7}{315} + \cdots, \ x^2 < \frac{\pi^2}{4}.$$

$$e^x = 1 + x + \frac{x^2}{2!} + \cdots + \frac{x^n}{n!} + \cdots, \text{ all } x.$$

$$\ln(1+x) = x - \frac{x^2}{2} + \frac{x^3}{3} - \cdots + (-1)^{n+1}\frac{x^n}{n} + \cdots, \ -1 < x \leq 1.$$

$$\operatorname{Arcsin} x = x + \frac{1}{2}\cdot\frac{x^3}{3} + \frac{1\cdot 3}{2\cdot 4}\cdot\frac{x^5}{5} + \cdots + \frac{1\cdot 3\cdots(2n-3)}{2\cdot 4\cdots(2n-2)}\cdot\frac{x^{2n-1}}{(2n-1)}, \ -1 < x < 1.$$

$$\operatorname{Arctan} x = x - \frac{x^3}{3} + \frac{x^5}{5} - \cdots + (-1)^{n+1}\frac{x^{2n-1}}{2n-1} \pm \cdots, \ -1 \leq x \leq 1.$$

$$\sinh x = x + \frac{x^3}{3!} + \frac{x^5}{5!} + \cdots + \frac{x^{2n-1}}{(2n-1)!} + \cdots, \text{ all } x.$$

$$\cosh x = 1 + \frac{x^2}{2!} + \frac{x^4}{4!} + \cdots + \frac{x^{2n}}{(2n)!} + \cdots, \text{ all } x.$$

$$\sinh^{-1} x = x - \frac{1}{2}\cdot\frac{x^3}{3} + \frac{1\cdot 3}{2\cdot 4}\cdot\frac{x^5}{5} + \cdots, \ -1 < x < 1.$$

## SHORT TABLE OF INTEGRALS

### Elementary Forms

$$\int df(x) = \int f'(x)dx = f(x) + C.$$

$$\int a\,du = a\int du.$$

$$\int (du \pm dv \pm dw + \cdots) = \int du \pm \int dv \pm \int dw \pm \cdots.$$

$$\int u^n\,du = \frac{u^{n+1}}{n+1} + C, n \neq -1$$

$$\int \frac{du}{u} = \log_e u + C.$$

### Forms containing $a + bu$

$$\int (a+bu)^n du = \frac{(a+bu)^{n+1}}{b(n+1)} + C, n \neq -1.$$

$$\int \frac{du}{a+bu} = \frac{1}{b}\log_e(a+bu) + C.$$

$$\int \frac{u\,du}{a+bu} = \frac{1}{b^2}[a + bu - a\log_e(a+bu)] + C.$$

$$\int \frac{u\,du}{(a+bu)^2} = \frac{1}{b^2}\left[\frac{a}{a+bu} + \log_e(a+bu)\right] + C.$$

$$\int \frac{u\,du}{(a+bu)^3} = \frac{1}{b^2}\left[-\frac{1}{a+bu} + \frac{a}{2(a+bu)^2}\right] + C.$$

$$\int \frac{du}{u(a+bu)} = -\frac{1}{a}\log_e\left(\frac{a+bu}{u}\right) + C.$$

$$\int \frac{du}{u^2(a+bu)} = -\frac{1}{au} + \frac{b}{a^2}\log_e\left(\frac{a+bu}{u}\right) + C.$$

$$\int \frac{du}{u(a+bu)^2} = \frac{1}{a(a+bu)} - \frac{1}{a^2}\log_e\left(\frac{a+bu}{u}\right) + C.$$

### Forms containing $a^2 \pm b^2u^2$

$$\int \frac{du}{a^2+b^2u^2} = \frac{1}{ab}\arctan\frac{bu}{a} + C.$$

$$\int \frac{du}{a^2-b^2u^2} = \frac{1}{2ab}\log_e\left(\frac{a+bu}{a-bu}\right) + C.$$

$$\int \frac{du}{b^2u^2-a^2} = \frac{1}{2ab}\log_e\left(\frac{bu-a}{bu+a}\right) + C.$$

$$\int u(a^2 \pm b^2u^2)^n du = \frac{(a^2 \pm b^2u^2)^{n+1}}{\pm 2b^2(n+1)} + C, n \neq -1.$$

$$\int \frac{u\,du}{a^2 \pm b^2u^2} = \frac{1}{\pm 2b^2}\log_e(a^2 \pm b^2u^2) + C.$$

$$\int \frac{du}{u(a^2 \pm b^2u^2)} = \frac{1}{2a^2}\log_e\left(\frac{u^2}{a^2 \pm b^2u^2}\right) + C.$$

### Forms containing $\sqrt{a + bu}$

$$\int u\sqrt{a+bu}\,du = -\frac{2(2a - 3bu)(a+bu)^{\frac{3}{2}}}{15b^2} + C.$$

$$\int u^2\sqrt{a+bu}\,du = \frac{2(8a^2 - 12abu + 15b^2u^2)(a+bu)^{\frac{3}{2}}}{105b^3} + C.$$

$$\int \frac{u\,du}{\sqrt{a+bu}} = -\frac{2(2a-bu)\sqrt{a+bu}}{3b^2} + C.$$

$$\int \frac{u^2du}{\sqrt{a+bu}} = \frac{2(8a^2 - 4abu + 3b^2u^2)\sqrt{a+bu}}{15b^3} + C.$$

$$\int \frac{du}{u\sqrt{a+bu}} = \frac{1}{\sqrt{a}}\log_e\left(\frac{\sqrt{a+bu}-\sqrt{a}}{\sqrt{a+bu}+\sqrt{a}}\right) + C, \text{ for } a > 0.$$

$$\int \frac{du}{u\sqrt{a+bu}} = \frac{2}{\sqrt{-a}}\arctan\sqrt{\frac{a+bu}{-a}} + C, \text{ for } a < 0.$$

### Forms containing $\sqrt{u^2 \pm a^2}$

$$\int (u^2 \pm a^2)^{\frac{1}{2}}du = \frac{u}{2}\sqrt{u^2 \pm a^2} \pm \frac{a^2}{2}\log_e(u + \sqrt{u^2 \pm a^2}) + C.$$

$$\int \frac{du}{(u^2 \pm a^2)^{\frac{1}{2}}} = \log_e(u + \sqrt{u^2 \pm a^2}) + C.$$

$$\int \frac{du}{(u^2 \pm a^2)^{\frac{3}{2}}} = \frac{u}{\pm a^2\sqrt{u^2 \pm a^2}} + C.$$

$$\int \frac{u^2\,du}{(u^2 \pm a^2)^{\frac{1}{2}}} = \frac{u}{2}\sqrt{u^2 \pm a^2} - \frac{\pm a^2}{2}\log_e(u + \sqrt{u^2 \pm a^2}) + C.$$

$$\int \frac{u^2du}{(u^2 \pm a^2)^{\frac{3}{2}}} = -\frac{u}{\sqrt{u^2 \pm a^2}} + \log_e(u + \sqrt{u^2 \pm a^2}) + C.$$

$$\int \frac{du}{u(u^2 + a^2)^{\frac{1}{2}}} = -\frac{1}{a}\log_e\left(\frac{a + \sqrt{u^2 + a^2}}{u}\right) + C.$$

$$\int \frac{du}{u(u^2 - a^2)^{\frac{1}{2}}} = \frac{1}{a}\operatorname{arc\,sec}\frac{u}{a} + C.$$

$$\int \frac{du}{u^2(u^2 \pm a^2)^{\frac{1}{2}}} = -\frac{\sqrt{u^2 \pm a^2}}{\pm a^2u} + C.$$

$$\int \frac{du}{u^3(u^2 + a^2)^{\frac{1}{2}}} = -\frac{\sqrt{u^2 + a^2}}{2a^2u^2} + \frac{1}{2a^3}\log_e\left(\frac{a + \sqrt{u^2 + a^2}}{u}\right) + C.$$

$$\int \frac{du}{u^3(u^2 - a^2)^{\frac{1}{2}}} = \frac{\sqrt{u^2 - a^2}}{2a^2u^2} + \frac{1}{2a^3}\operatorname{arc\,sec}\frac{u}{a} + C.$$

$$\int \frac{(u^2 + a^2)^{\frac{1}{2}}du}{u} = \sqrt{u^2 + a^2} - a\log_e\left(\frac{a + \sqrt{u^2 + a^2}}{u}\right) + C.$$

$$\int \frac{(u^2 - a^2)^{\frac{1}{2}}du}{u} = \sqrt{u^2 - a^2} - a\operatorname{arc\,sec}\frac{u}{a} + C.$$

$$\int \frac{(u^2 \pm a^2)^{\frac{1}{2}}du}{u^2} = -\frac{\sqrt{u^2 \pm a^2}}{u} + \log_e(u + \sqrt{u^2 \pm a^2}) + C.$$

Forms containing $\sqrt{a^2 - u^2}$

$$\int (a^2 - u^2)^{\frac{1}{2}}du = \frac{u}{2}\sqrt{a^2 - u^2} + \frac{a^2}{2}\arcsin\frac{u}{a} + C.$$

$$\int \frac{du}{(a^2 - u^2)^{\frac{1}{2}}} = \arcsin\frac{u}{a} + C.$$

$$\int \frac{du}{(a^2 - u^2)^{\frac{3}{2}}} = \frac{u}{a^2\sqrt{a^2 - u^2}} + C.$$

$$\int \frac{u\,du}{(a^2 - u^2)^{\frac{n}{2}}} = \frac{(a^2 - u^2)^{1-\frac{n}{2}}}{n - 2} + C.$$

$$\int \frac{u^2du}{(a^2 - u^2)^{\frac{1}{2}}} = -\frac{u}{2}\sqrt{a^2 - u^2} + \frac{a^2}{2}\arcsin\frac{u}{a} + C.$$

$$\int \frac{u^2du}{(a^2 - u^2)^{\frac{3}{2}}} = \frac{u}{\sqrt{a^2 - u^2}} - \arcsin\frac{u}{a} + C.$$

$$\int \frac{du}{u(a^2 - u^2)^{\frac{1}{2}}} = -\frac{1}{a}\log_e\left(\frac{a + \sqrt{a^2 - u^2}}{u}\right) + C = -\frac{1}{a}\cosh^{-1}\frac{a}{u} + C.$$

$$\int \frac{du}{u^2(a^2 - u^2)^{\frac{1}{2}}} = -\frac{\sqrt{a^2 - u^2}}{a^2u} + C.$$

$$\int \frac{du}{u^3(a^2 - u^2)^{\frac{1}{2}}} = -\frac{\sqrt{a^2 - u^2}}{2a^2u^2} - \frac{1}{2a^3}\log_e\left(\frac{a + \sqrt{a^2 - u^2}}{u}\right) + C = -\frac{\sqrt{a^2 - u^2}}{2a^2u^2} - \frac{1}{2a^3}\cosh^{-1}\frac{a}{u} + C.$$

$$\int \frac{(a^2 - u^2)^{\frac{1}{2}}du}{u} = \sqrt{a^2 - u^2} - a\log_e\left(\frac{a + \sqrt{a^2 - u^2}}{u}\right) + C = \sqrt{a^2 - u^2} - a\cosh^{-1}\frac{a}{u} + C.$$

$$\int \frac{(a^2 - u^2)^{\frac{1}{2}}du}{u^2} = -\frac{\sqrt{a^2 - u^2}}{u} - \arcsin\frac{u}{a} + C.$$

Forms containing $\sqrt{2au \pm u^2}$

$$\int \sqrt{2au - u^2}du = \frac{u - a}{2}\sqrt{2au - u^2} + \frac{a^2}{2}\arccos\left(1 - \frac{u}{a}\right) + C.$$

$$\int u\sqrt{2au - u^2}du = -\frac{3a^2 + au - 2u^2}{6}\sqrt{2au - u^2} + \frac{a^3}{2}\arccos\left(1 - \frac{u}{a}\right) + C.$$

$$\int \frac{\sqrt{2au - u^2}du}{u} = \sqrt{2au - u^2} + a\arccos\left(1 - \frac{u}{a}\right) + C.$$

$$\int \frac{\sqrt{2au - u^2}du}{u^2} = -\frac{2\sqrt{2au - u^2}}{u} - \arccos\left(1 - \frac{u}{a}\right) + C.$$

$$\int \frac{\sqrt{2au - u^2}du}{u^3} = -\frac{(2au - u^2)^{\frac{3}{2}}}{3au^3} + C.$$

$$\int \frac{du}{\sqrt{2au - u^2}} = \arccos\left(1 - \frac{u}{a}\right) + C.$$

$$\int \frac{du}{\sqrt{2au + u^2}} = \log_e(u + a + \sqrt{2au + u^2}) + C.$$

$$\int \frac{u\,du}{\sqrt{2au - u^2}} = -\sqrt{2au - u^2} + a\arccos\left(1 - \frac{u}{a}\right) + C.$$

$$\int \frac{u^2\,du}{\sqrt{2au - u^2}} = -\frac{(u + 3a)\sqrt{2au - u^2}}{2} + \frac{3a^2}{2}\arccos\left(1 - \frac{u}{a}\right) + C.$$

$$\int \frac{du}{u\sqrt{2au - u^2}} = -\frac{\sqrt{2au - u^2}}{au} + C.$$

$$\int \frac{du}{(2au - u^2)^{\frac{3}{2}}} = \frac{u - a}{a^2\sqrt{2au - u^2}} + C.$$

$$\int \frac{u\,du}{(2au - u^2)^{\frac{3}{2}}} = \frac{u}{a\sqrt{2au - u^2}} + C.$$

Exponential and Logarithmic Forms

$$\int e^{au}du = \frac{e^{au}}{a} + C.$$

$$\int b^{au}du = \frac{b^{au}}{a\log_e b} + C.$$

$$\int ue^{au}du = \frac{e^{au}}{a^2}(au - 1) + C.$$

$$\int u^n e^{au}du = \frac{u^n e^{au}}{a} - \frac{n}{a}\int u^{n-1}e^{au}\,du.$$

$$\int \frac{du}{1 + e^u} = u - \log(1 + e^u) = \log\frac{e^u}{1 + e^u} + C.$$

$$\int \frac{du}{a + be^{pu}} = \frac{u}{a} - \frac{1}{ap}\log(a + be^{pu}) + C.$$

$$\int \frac{du}{ae^{mu} + be^{-mu}} = \frac{1}{m\sqrt{ab}}\tan^{-1}\left(e^{mu}\sqrt{\frac{a}{b}}\right) + C.$$

$$\int \log_e u\,du = u\log_e u - u + C.$$

$$\int u^n\log_e u\,du = u^{n+1}\left[\frac{\log_e u}{n + 1} - \frac{1}{(n + 1)^2}\right] + C, n \neq -1.$$

$$\int \frac{du}{u\log_e u} = \log_e(\log_e u) + C.$$

Trigonometric Forms

$$\int \sin u\,du = -\cos u + C.$$

$$\int \cos u\,du = \sin u + C.$$

$$\int \tan u\,du = -\log_e\cos u + C = \log_e\sec u + C.$$

$$\int \cot u\,du = \log_e\sin u + C.$$

$$\int \sec u\,au = \int \frac{du}{\cos u} = \log_e(\sec u + \tan u) + C$$

$$= \log_e\tan\left(\frac{u}{2} + \frac{\pi}{4}\right) + C.$$

$$\int \csc u\,du = \int \frac{du}{\sin u} = \log_e(\csc u - \cot u) + C$$

$$= \log_e\tan\frac{u}{2} + C.$$

$$\int \sec^2 u\,du = \tan u + C.$$

$$\int \csc^2 u\,du = -\cot u + C.$$

$$\int \sec u\tan u\,du = \sec u + C.$$

$$\int \csc u\cot u\,du = -\csc u + C.$$

$$\int \sin^2 u\,du = \frac{u}{2} - \frac{1}{4}\sin 2u + C.$$

$$\int \cos^2 u\,du = \frac{u}{2} + \frac{1}{4}\sin 2u + C.$$

$$\int \cos^n u\sin u\,du = -\frac{\cos^{n+1}u}{n + 1} + C.$$

$$\int \sin^n u\cos u\,du = \frac{\sin^{n+1}u}{n + 1} + C.$$

$$\int \sin mu\sin nu\,du = -\frac{\sin(m + n)u}{2(m + n)} + \frac{\sin(m - n)u}{2(m - n)} + C.$$

$$\int \cos mu\cos nu\,du = \frac{\sin(m + n)u}{2(m + n)} + \frac{\sin(m - n)u}{2(m - n)} + C.$$

$$\int \sin mu\cos nu\,du = -\frac{\cos(m + n)u}{2(m + n)} - \frac{\cos(m - n)u}{2(m - n)} + C.$$

$$\int \frac{du}{a + b\cos u} = \frac{2}{\sqrt{a^2 - b^2}}\arctan\left(\sqrt{\frac{a - b}{a + b}}\tan\frac{u}{2}\right) + C, \text{ when } a > b.$$

$$\int \frac{du}{a + b\cos u} = \frac{1}{\sqrt{b^2 - a^2}}\log\left[\frac{\sqrt{b - a}\tan\frac{u}{2} + \sqrt{b + a}}{\sqrt{b - a}\tan\frac{u}{2} - \sqrt{b + a}}\right] + C, \text{ when } a < b.$$

$$\int \frac{du}{a + b\sin u} = \frac{2}{\sqrt{a^2 - b^2}}\arctan\left[\frac{a\tan\frac{u}{2} + b}{\sqrt{a^2 - b^2}}\right] + C, \text{ when } a > b.$$

$$\int \frac{du}{a + b\sin u} = \frac{1}{\sqrt{b^2 - a^2}}\log\left[\frac{a\tan\frac{u}{2} + b - \sqrt{b^2 - a^2}}{a\tan\frac{u}{2} + b + \sqrt{b^2 - a^2}}\right] + C, \text{ when } a < b.$$

$$\int \frac{du}{a^2\cos^2 u + b^2\sin^2 u} = \frac{1}{ab}\arctan\left(\frac{b\tan u}{a}\right) + C.$$

$\int e^{au} \sin nu\, du = \dfrac{e^{au}(a \sin nu - n \cos nu)}{a^2 + n^2} + C.$

$\int e^{au} \cos nu\, du = \dfrac{e^{au}(n \sin nu + a \cos nu)}{a^2 + n^2} + C.$

$\int u \sin u\, du = \sin u - u \cos u + C.$

$\int u \cos u\, du = \cos u + u \sin u + C.$

Inverse Trigonometric Functions

$\int \text{arc sin } u\, du = u \text{ arc sin } u + \sqrt{1 - u^2} + C.$

$\int \text{arc cos } u\, du = u \text{ arc cos } u - \sqrt{1 - u^2} + C.$

$\int \text{arc tan } u\, du = u \text{ arc tan } u - \log \sqrt{1 + u^2} + C.$

$\int \text{arc cot } u\, du = u \text{ arc cot } u + \log \sqrt{1 + u^2} + C.$

$\int \text{arc sec } u\, du = u \text{ arc sec } u - \log (u + \sqrt{u^2 - 1}) + C$
$= u \text{ arc sec } u - \cosh^{-1} u + C.$

$\int \text{arc csc } u\, du = u \text{ arc csc } u + \log (u + \sqrt{u^2 - 1}) + C$
$= u \text{ arc csc } u + \cosh^{-1} u + C.$

HYPERBOLIC FUNCTIONS

$\int \sinh u\, du = \cosh u + C,$

$\int \cosh u\, du = \sinh u + C,$

$\int \tanh u\, du = \int \dfrac{\sinh u}{\cosh u} du = \ln \cosh u + C.$

$\int \coth u\, du = \log \sinh u + C.$

$\int \text{sech } u\, du = 2 \tan^{-1} (e^u) + C.$

$\int \text{csch } u\, du = \text{long} \tanh\left(\dfrac{u}{2}\right) + C.$

$\int u \sinh u\, du = u \cosh u - \sinh u + C.$

$\int u \cosh u\, du = u \sinh u - \cosh u + C.$

$\int \text{sech } u \tanh u\, du = -\text{sech } u + C.$

$\int \text{csch } u \coth u\, du = -\text{csch } u + C.$

Integrals leading to inverse hyperbolic functions

$\int \dfrac{du}{\sqrt{a^2 + u^2}} = \sinh^{-1} \dfrac{u}{a} + C \qquad (a > 0),$

$\int \dfrac{du}{\sqrt{u^2 - a^2}} = \text{Cosh}^{-1} \dfrac{u}{a} + C \qquad (u > a > 0),$

$\int \dfrac{du}{a^2 - u^2} = \dfrac{1}{a} \tanh^{-1} \dfrac{u}{a} + C \qquad (u^2 < a^2).$

$\int \dfrac{du}{a^2 - u^2} = \dfrac{1}{a} \coth^{-1} \dfrac{u}{a} + C \qquad (u^2 > a^2).$

Integration by parts

$\int u\, dv = uv - \int v\, du.$

# DICTIONARY-INDEX

**Abscissa:** horizontal coordinate, $x$, of a point, **10**

**Absolute value:** the numerical value of a number indicated by $|a| = a$, if $a \geq 0$; $|a| = -a$, if $a < 0$, **5**

**Absolutely convergent series:** a convergent series each of whose terms is the absolute value of an expression, **125**

**Acceleration:** time rate of change of velocity of a moving body, $a = \dfrac{dv}{dt} = \dfrac{d^2s}{dt^2}$, where $s = f(t)$, **28**

  **normal component:** $a_N = v^2|K|$, **95**

  **tangential component:** $a_T = \dfrac{d^2s}{dt^2}$, **95**

  **vector:** $\mathbf{a}(t) = \mathbf{v}'(t) = \mathbf{f}''(t) = a_x\mathbf{i} + a_y\mathbf{j} + a_z\mathbf{k}$, **95, 137**

**Addition formulas:** a term applied to the formulas for the trigonometric and hyperbolic functions of two or more arguments, **155**

**Alternating series:** a series in which the terms are alternately positive and negative, **125**

**Angle:** if a line segment $OA$ is revolved about $O$ until it takes the position $OB$, the revolution has generated the angle $AOB$, **52**

  **between two lines:** in the plane, the angle $\theta$ is given by $\tan\theta = \dfrac{m_2 - m_1}{1 + m_1m_2}$; in three dimensions, $\cos\theta = \cos\alpha_1 \cos\alpha_2 + \cos\beta_1\cos\beta_2 + \cos\gamma_1\cos\gamma_2$, **59, 130**

  **between vectors:** $\cos\theta = \dfrac{a_1a_2 + b_1b_2 + c_1c_2}{|\mathbf{v}|\,|\mathbf{w}|}$, **94, 136**

  **direction:** the angles, $\alpha$, $\beta$, and $\gamma$, of the line $\overline{OP}$ formed by $\overline{OP}$ and the $X$-, $Y$-, and $Z$-axes, respectively, **130**

  **negative:** an angle formed by revolving a line in a clockwise direction, **52**

  **polar:** the angle between the polar axis and the radius vector, **88**

**Antiderivative:** If $\dfrac{d}{dx}f(x) = F(x)$, then $f(x)$ is the antiderivative of $F(x)$, **40**

**Antilogarithm:** If $\log N = x$, then antilogarithm $x = N$, **70**

**Approximate integration:** approximating the value of a definite integral by special summation formulas; e.g., Simpson's Rule, **113**

**Arc:** the locus of the function defined by $y = f(x)$ in the interval $a < x < b$ if $f$ is continuous, **82**

  **length:** distance along the curve $C$ from a point $P_1(x_1, y_1)$ to a point $P_2(x_2, y_2)$, **82, 136**

  **rectifiable:** any arc of a curve which has length, **82**

**Arccos x:** the number whose cosine is $x$, similarly, for the other trigonometric and hyperbolic functions, **53**

**Area:** the size of a plane region, **47, 148**

  **of a surface:** $\int\int_S (1 + z_x^2 + z_y^2)^{1/2} dydx$, **149**

  **of surface of revolution:** $2\pi \int_C r ds$, **112**

**Associative laws of vectors:** $k(n\mathbf{v}) = (kn)\mathbf{v}$, $\mathbf{u} + (\mathbf{v} + \mathbf{w}) = (\mathbf{u} + \mathbf{v}) + \mathbf{w}$, **94**

**Asymptotes:** a line is said to be an asymptote to a curve if the line and curve get closer together but never touch each other, **11**

**Average velocity:** ratio of the distance to the elapsed time, **28**

**Axes:** the perpendicular lines $OX$ and $OY$ in the rectangular coordinate system, **10, 130**

  **radical:** the line obtained by eliminating the quadratic terms from the equations of two circles, **59**

  **polar:** the fixed line from $O$ and usually extending horizontally to the right, **88**

**Base of a logarithm:** *see* **Logarithm of a number, base of**

**Binomial:** a polynomial of two terms, **5**

**Binomial coefficient:** $\dbinom{n}{r}$, **5**

**Binomial theorem:** **5**

**Binormal vector:** $\mathbf{B}(t) = \mathbf{T}(t) \times \mathbf{N}(t)$, **137**

**Cardioid:** the graph of $r = a(1 - \cos\theta)$, **88**

**Cauchy product of two series:** $\sum_{n=0}^{\infty} c_n x^n$, where $\sum_{k=0}^{n} a_{n-k}b_k = c_n$, **125**

**Center of central conics:** point midway between foci, **64**

**Center of curvature:** coordinates of the center of the circle of curvature, **83**

**Center of mass:** point at which a body will balance on a pin, **112, 113, 149**

**Central conic:** a conic defined by an equation for which $B^2 - 4AC \neq 0$, **65**

**Centroid:** center of mass, **112, 113**

**Chain rule:** If $f(x) = g[u(x)]$, then $\dfrac{df}{dx} = \dfrac{dg}{du}\dfrac{du}{dx}$, **17**

**Circle:** the locus of a point that moves at a constant distance from a fixed point, **59**

  **center of a:** fixed point $C(h, k)$ in the general equation $(x - h)^2 + (y - k)^2 = r^2$, **59**

  **of curvature:** that circle passing through $P$ on a curve, has radius equal to the reciprocal of the curvature, $K$, and whose center lies on the concave side of the curve along the normal through $P$, **83**

  **radius of a:** **59**

**Closed interval:** *see* **Interval, closed**

**Cologarithm:** the logarithm of the reciprocal of the number, $\text{colog}\, N = \log \dfrac{1}{N}$, **70**

**Common chord:** the line joining the two points of intersection of two circles, **59**

**Completing the square:** to change $ax^2 + bx + c$ to the form $\left(x + \dfrac{b}{2a}\right)^2 + c - \dfrac{b^2}{4a}$, **4**

**Concave downward:** a curve which remains below the tangent line to the curve at each point in an interval, **29**

**Concave upward:** a curve which remains above the tangent line to the curve at each point in an interval, **29**

**Conditional inequality:** a statement of inequality which is true for only specific values of the variables, **11**

**Conic section:** a curve whose equation is of the second degree; the path of a point which moves so that the ratio of its distance from a fixed point to its distance from a fixed line is a constant, **64**

**proper:** a conic section whose discriminant is not zero, **64**

**Constant:** a symbol which, throughout a discussion, represents a fixed number; a symbol that varies over a set which consists of only one element, **10**

**of integration:** an arbitrary number added to the function $f(x)$ that is the antiderivative of $f'(x)$, **40**

**Constrained extrema:** the extrema of a function of more than one variable subjected to at least one specified condition, **143**

**Continuity:** a function $f$ is said to be continuous at the number $a$ if $f(a)$ is defined and if $\lim_{x \to a} f(x) = f(a)$, **11**

**for functions of two variables:** the function $f$ is continuous at $(a, b)$ if and only if $f(a, b)$ is defined and $f(x, y) \to f(a, b)$ as $(x, y) \to (a, b)$, **142**

**Convergent:** if a sequence approaches a finite limit, it is said to be convergent, **124**

**series:** if $S = \lim_{n \to \infty} S_n$ exists, the series is said to be convergent, **124**

**Coordinates:** the real numbers which locate a point in a plane or space; **10**

**polar:** the radius vector, $r$, and the polar angle, $\theta$, which locate the point $P(r, \theta)$ in a plane, **88**

**rectangular:** the abscissa, $x$, and the ordinate, $y$, of a point $P(x, y)$, in a plane, **10**

**Cosecant:** the trigonometric function defined by

$$\csc \theta = \frac{\text{radius vector}}{\text{ordinate}} \text{ or } \frac{\text{hypotenuse}}{\text{opposite side}}, \mathbf{52}$$

**Cosine:** the trigonometric function defined by

$$\cos \theta = \frac{\text{abscissa}}{\text{radius vector}} \text{ or } \frac{\text{adjacent side}}{\text{hypotenuse}}, \mathbf{52}$$

**Cotangent:** the trigonometric function defined by

$$\cot \theta = \frac{\text{abscissa}}{\text{ordinate}} \text{ or } \frac{\text{adjacent side}}{\text{opposite side}}, \mathbf{52}$$

**Critical point:** a value of $x$ where $f'(x) = 0$ is a critical point of a function $f$, **29**

**for f(x, y):** a value $(x_0, y_0)$ at which both $f_x$ and $f_y$ vanish is called a critical point of $f$, **143**

**Cross product:** a vector $\mathbf{w} = u \times \mathbf{v}$ that is perpendicular to the plane of $u$ and $v$ and whose magnitude is $|\mathbf{w}| = |\mathbf{u}||\mathbf{v}| \sin \theta$ where $\theta$ is the angle between **u** and **v** and **u,v,w** form a right-hand triple; $\mathbf{u} \times \mathbf{v} = \begin{vmatrix} \mathbf{i} & \mathbf{j} & \mathbf{k} \\ a_1 & b_1 & c_1 \\ a_2 & b_2 & c_2 \end{vmatrix}$, **136**

**Curvature of an arc:** the rate of change of the angle of inclination of the tangent line with respect to the arc length, $K = \frac{d\tau}{ds}$, **82, 137**

**radius of a:** the radius of curvature of an arc at a point is given by $R = \frac{1}{|K|}$, **82**

**Cycloid:** the path of a point on the circumference of a circle that rolls along a straight line, **82**

**Cylinder:** a solid formed by erecting a set of equal parallel lines at each point of a closed curve, **112**

**right:** the parallel lines are perpendicular to the plane of the curve, **112**

**Cylindrical shell:** the solid contained between two concentric right circular cylinders, **112**

**Degree:** a sexagesimal measure of an angle, one degree $= 1° = \frac{1}{360}$ of one revolution, **52**

**Definite integral:** $\int_a^b f'(x)dx = f(b) - f(a)$, **46**

**Density:** mass per unit volume, **149**

**Derivative:** $f'(x) = \lim_{h \to 0} \frac{f(x + h) - f(x)}{h}$, **16**

**formulas for a: 16, 17**

**geometric interpretation of a: 22**

**polar coordinates of a:** if $r = f(\theta)$, then

$$\frac{dr}{d\theta} = \frac{d}{d\theta}[f(\theta)] = r', \mathbf{89}$$

**vector function of a:** $\mathbf{f}'(t) = x'(t)\mathbf{i} + y'(t)\mathbf{j} + z'(t)\mathbf{k}$, **95**

**Differential:** the differential of $y = f(x)$ is defined by $dy = f'(x)h$ where $h$ is an arbitrary increment in $x$ and $h = \Delta x = dx$, **34**

**Differential equation:** the equation $y' = f(x)$ or $dy = f(x)dx$ is called a differential equation of the first order and first degree, **41**

**solution of a:** the relation among the variables that satisfies the differential equation, **41**

**Differential operator:** the symbol $\frac{d}{dx}(\ ) = D_x$ denoting the derivative with respect to the independent variable $x$, **16**

**Differentiation:** the process of finding the derivative, **16**

**implicit:** the process of finding the derivative, $y'$, of $f(x, y) = 0$, **22**

**Direction angles:** the angles between a line and the three coordinate axes, **130**

**Direction cosines:** the cosines of the direction angles, **130**

**Direction numbers:** numbers that are proportional to the direction cosines of a line with the same constant of proportionality, **130**

**Directrix:** the fixed line in the definition of a conic section, **64**

**of an ellipse, 64**

**of a hyperbola, 65**

**of a parabola, 64**

**Disc method for volumes:** a procedure for finding the volume of a solid by integration, **112**

**Discriminant of $Ax^2 + Bxy + Cy^2 + Dx + Ey + F = 0$:** the number defined by the determinant $\begin{vmatrix} 2A & B & D \\ B & 2C & E \\ D & E & 2F \end{vmatrix}$, **64**

**Distance between two points:** if the points are $P_1(x_1, y_1)$ and $P_2(x_2, y_2)$, then $d = \sqrt{(x_2 - x_1)^2 + (y_2 - y_1)^2}$; if the points are $P_1(x_1, y_1, z_1)$ and $P_2(x_2, y_2, z_2)$, then $d = \sqrt{(x_2 - x_1)^2 + (y_2 - y_1)^2 + (z_2 - z_1)^2}$, **130**

**Distance from a point to a line:** $P_1(x_1, y_1)$ to the line $ax + by + c = 0$, $d = \left|\frac{ax_1 + by_1 + c}{\sqrt{a^2 + b^2}}\right|$, **59**

**Distance from a point to a plane: 131**

**Distributive law of vectors:** $\mathbf{u} \cdot (\mathbf{v} + \mathbf{w}) = \mathbf{u} \cdot \mathbf{v} + \mathbf{u} \cdot \mathbf{w}$, **94**

**Divergent:** if a sequence does not approach a limit, it is said to be divergent, **124**

**Division of two series:** this operation is performed in the same manner as long division of two polynomials, **125**
**Domain:** the set of first elements of an ordered pair; the interval of values for the independent variable in a functional relation, **10**
**Dot product:** $\mathbf{v} \cdot \mathbf{w} = |\mathbf{v}|\,|\mathbf{w}| \cos\theta = a_1a_2 + b_1b_2 + c_1c_2$, **94, 136**
**Double-angle formulas:** the formulas for the trigonometric functions of $2\theta$; e.g., $\sin 2\theta = 2\sin\theta\cos\theta$, etc., **155**
**Double integral:** the integral $\int\int f(x, y)dydx$, **148**

**e:** base of the natural logarithms, **70**

$e = \lim_{x \to 0} (1 + x)^{\frac{1}{x}}$, **70**

**Eccentricity:** the constant ratio of the distance to a fixed point and a fixed line in the definition of a conic section, **64**
**Element of a sequence:** the individual members of a sequence, **124**
**Ellipse:** the locus of a point which moves so that the sum of its distances from two fixed points is a constant; the conic section for which the eccentricity is less than one, **64**
  **vertices of an:** the points of intersection of the ellipse and its major axis, **64**
**Ellipsoid:** $\frac{x^2}{a^2} + \frac{y^2}{b^2} + \frac{z^2}{c^2} = 1$, **131**
**Elliptic cone:** $\frac{x^2}{a^2} + \frac{y^2}{b^2} - \frac{z^2}{c^2} = 0$, **131**
**Elliptic paraboloid:** $\frac{x^2}{a^2} + \frac{y^2}{b^2} = \frac{z}{c}$, **131**
**Equation of a line:** a linear equation in two variables, **58**
**Equation of a plane:** a linear equation in three variables, **131**
**Equations:** statements of equality, **4**
  **trigonometric:** equations involving the trigonometric functions, **52**

**Error, absolute:** the numerical difference between the true value and the approximate value, **34**
**Error, percentage:** the relative error times 100, **34**
**Error, relative:** the absolute error divided by the true value, **34**
**Evolute:** locus of the center of curvature at each point on a given curve, **83**
**Exponent:** the number of times the like factor $a$ appears in a multiplication, **5**
  **fractional:** the root of a number, **5**
  **zero:** $x^0 = 1$, **5**
**Exponential functions:** *see* **Functions, exponential, 70**
**Exponents, laws of: 5**
**Extrema:** relative maxima and minima, **28, 143**
**Extreme Value Theorem:** If $f$ is a continuous function defined on the closed interval $[a, b]$, there is at least one value, $x_1$, in $[a, b]$ where $f$ has a largest value and at least one value, $x_2$, where $f$ has a smallest value, **28**

**Factorial:** the product of all the integers from 1 to n inclusive, $n! = 1 \cdot 2 \cdot 3 \cdots n$, **5**
**Fluid pressure:** force per unit area of fluid on a surface; the pressure $p$ at a point $Q$ below the surface of a liquid is given by $p = wh$ where $w$ is the unit weight of the liquid and $h$ is the depth below the surface, **47**
**Focus:** the fixed point in the definition of a conic, **64**

**Function:** a set of ordered pairs such that no two ordered pairs have the same first element; the relation between $x$ and $y$ so that for an assigned value of $x$ a unique value of $y$ is determined, **10**
  **decreasing:** if $f(x_2) < f(x_1)$ for $x_2 > x_1$, **29**
  **derivative of a:** the limit of the ratio of the increment of the function, $\Delta y$, to the increment of the independent variable, $\Delta x$, when $x$ varies and approaches zero as a limit, $\frac{dy}{dx} = \lim_{\Delta x \to 0} \frac{f(x + \Delta x) - f(x)}{\Delta x}$, **16**
  **exponential:** a function of the form $y = a^x$ where $a$ is a constant; "the" exponential function, $y = e^x$, where $e$ is the base of the natural logarithms, **70**
  **increasing:** if $f(x_2) > f(x_1)$ for $x_2 > x_1$, **29**
  **inverse:** the function obtained by interchanging the role of the domain and the range of a given function, **10**
  **range of a:** the set of second elements of an ordered pair; the interval of values for the dependent variable in a functional relation, **10**
  **trigonometric:** *see* **Trigonometric functions**
  **periodic:** a function, not equal to a constant, that repeats its values at a regular interval, **53**
  **transcendental:** all functions that are not algebraic, **70**
**Fundamental Theorem of Integral Calculus:** If the interval $[a, b]$ is divided into $n$ subintervals of lengths $\Delta x_i$ and $f(x)$ is continuous in $[a, b]$, then we have

$$\lim_{n \to \infty} \sum_{i=1}^{n} f(x_i)\Delta x_i = \int_a^b f(x)dx$$

where $x_i$ is a value in the subinterval $\Delta x_i$, **46, 47**

**General equation of the second degree,**
  **in x:** $ax^2 + bx + c = 0$, **10**
  **in x and y:** $ax^2 + bxy + cy^2 + dx + ey + f = 0$, **10**
**Geometric progression:** *see* **Progression, geometric**
**Geometric series, sum of a:** $S = \frac{a}{1 - r}$, **124**
**Graph of an equation:** the totality of all points whose coordinates satisfy the relation defined by the equation, **10**
**Graphical representation:** a pictorial (geometric) representation of an algebraic expression, **10**

**Half-angle formulas:** the formulas expressing the trigonometric functions of the half-angle in terms of the angle, **155**
**Harmonic progression:** *see* **Progression, harmonic**
**Harmonic series:** the sum of the terms in a harmonic progression; the $p$-series with $p = 1$, **124**
**Higher derivative:** the derivative of a derivative, **22**
  **notation for:** $\frac{d^2y}{dx^2} = y''$, $\frac{d^4y}{dx^4} = y^{(iv)} = D_x^4 y$, etc, **22**
**Hyperbola:** the locus of a point which moves so that the difference of its distance from the fixed points is a constant; the conic section for which $e > 1$, **64**
  **vertices of a:** the points of intersection of the hyperbola and its transverse axis, **65**
**Hyperbolic functions: 71**
**Hyperbolic paraboloid:** $\frac{x^2}{a^2} - \frac{y^2}{b^2} = \frac{z}{c}$, **131**

**Hyperboloid**

**of one sheet:** $\frac{x^2}{a^2} + \frac{y^2}{b^2} - \frac{z^2}{c^2} = 1$, **131**

**of two sheets:** $\frac{x^2}{a^2} - \frac{y^2}{b^2} - \frac{z^2}{c^2} = 1$, **131**

**Identity:** a statement of equality which is true for all values of the symbols for which each member of the equality is defined, **53**

**Imaginary number:** *see* **Numbers, imaginary**

**Imaginary unit:** $i = \sqrt{-1}$, **156**

**Implicit differentiation:** the process of finding the derivative of $f(x, y) = 0$, **22**

**Improper integral:** the definite integral in an unbounded interval, $\int_a^\infty f(x)\,dx$; the definite integral of a function which is not continuous at a point (or points) in the interval of integration, **119**

**Inclination of a line:** the angle between a line and the positive $X$-axis, **58**

**Increasing function:** if $f(x_2) > f(x_1)$ for $x_2 > x_1$, then $f$ is an increasing function, **29**

**Increment:** the difference between two chosen values of a variable; $\Delta x = x_2 - x_1$, **16**

**Indefinite integral:** the antiderivatives of a function $f$, $\int f(x)dx$, **40**

**Independent variable:** the variable to which arbitrary values may be assigned, **10**

**Indeterminate forms:** expressions that assume a form of the type $\frac{0}{0}$, $\frac{\infty}{\infty}$, $\infty - \infty$, $\infty \cdot 0$, $0^0$, $\infty^0$, or $1^\infty$ for certain values of the variables, **118**

**Index**

**of a radical:** the number indicating the root to be extracted; in $\sqrt[q]{n}$, the number $q$, **4**

**of summation:** the dummy symbol $i$ in $\sum_{i=1}^{n} a_i$, **46**

**Inequality:** a statement that one quantity is greater than or less than another, $a < b$, **4**

**Infinite limit:** if $f(x) \to \infty$ as $x \to a$, then we have an infinite limit, **11**

**Infinite sequence:** a sequence in which after each term there exists another term, **124**

**Infinite series:** the sum of an infinite sequence, **124**

**absolutely convergent:** if $\sum_{n=1}^{\infty} |u_n|$ converges, then the series $\sum u_n$ is said to be absolutely convergent, **125**

**comparison test:** if $u_n \geq 0$ and $u_n < a_n$ for all $n$ and $\sum a_n$ is convergent, then $\sum u_n$ is convergent; if $a_n > 0$ and $u_n \geq a_n$ for all $n$ and $\sum a_n$ diverges, then $\sum u_n$ diverges, **124**

**convergent:** if $s_n$ is the $n$th partial sum of an infinite series and $\lim_{n\to\infty} s_n$ exists, the series is said to be convergent, **124**

**divergent:** if $\lim_{n\to\infty} s_n$ does not exist, the series is said to be divergent, **124**

**integral test:** 124

**ratio test:** let $\lim_{n\to\infty} \left|\frac{u_{n+1}}{u_n}\right| = A$; if $A < 1$, the series is absolutely convergent; if $A > 1$, the series diverges; if $A = 1$, the test fails, **125**

**Inflection point:** the point on a curve at which the tangent line crosses the curve, the points for which $f''(x) = 0$ and $f'''(x) \neq 0$, **29**

**Initial line:** the starting position of a line segment that is rotated about an end point; one side of an angle, **52**

**Inner product of vectors:** the dot product of two vectors, $v \cdot w = |\mathbf{u}| \cdot |\mathbf{w}| \cos\theta = a_1 a_2 + b_1 b_2$, **94**

**Instantaneous velocity:** limit of average velocity as $\Delta t \to 0$, **28**

**Integral:** the function found from a given differential expression; the antiderivative of a function; $\int f'(x)dx = f(x) + C$, **40**

**definite:** $\int_a^b f'(x)dx = f(b) - f(a)$, **46**

**Integrand:** $f'(x)$ in the expression $\int f'(x)dx = f(x) + C$, **40**

**Integration:** the operation of finding the function which, when differentiated, will yield the integrand, **40**

**constant of:** $C$ in the expression $\int f'(x)dx = f(x) + C$, **40**

**Integration by parts:** $\int u dv = uv - \int v du$, **100**

**Intercept:** the value of the variable for which a curve crosses the coordinate axis, **11**

**x:** the value of $x$ for which a curve crosses the $X$-axis, **11**

**y:** the value of $y$ for which a curve crosses the $Y$-axis, **11**

**Intercept form of a straight line:** $\frac{x}{a} + \frac{y}{b} = 1$, **58**

**Interval:** if $a$ and $b$ are numbers then the interval from $a$ to $b$ is the collection of all the numbers between $a$ and $b$, **4**

**closed:** the interval includes the end points, **5**

**half-open:** the interval includes only one end point, **5**

**open:** the interval does not include either end point, **4**

**Inverse functions:** *see* Function, inverse

**Inverse hyperbolic functions: 71**

**Inverse trigonometric functions:** *see* **Trigonometric functions, inverse**

**Latus rectum:** the line segment between the points of intersection of a line through the focus parallel to the directrix and the conic section, **64**

**Law of cosines:** in any triangle, the square of any side is equal to the sum of the squares of the other two sides minus twice their product times the cosine of their included angle, **155**

**Law of Sines:** in any triangle, any two sides are proportional to the sines of the opposite angles, **155**

**Law of Tangents:** in any triangle, the difference of any two sides divided by their sum equals the tangent of one-half the difference of the opposite angles divided by the tangent of one-half their sum, **155**

**Laws of exponents: 5**

**Lemniscate:** graph defined by $r^2 = a^2 \cos 2\theta$, **88**

**Length of arc:** $s = \int_a^b ds$ where $ds^2 = dx^2 + dy^2$, **82**

**Length of vector:** $|\mathbf{v}| = \sqrt{a^2 + b^2 + c^2}$ where the vector is given by $\mathbf{v} = a\mathbf{i} + b\mathbf{j} + c\mathbf{k}$, **94**

**l'Hopital's Rule:** if $\lim_{x\to a} f(x) = 0$, $\lim_{x\to a} F(x) = 0$ and $\lim_{x\to a} \frac{f'(x)}{F'(x)} = L$, then $\lim_{x\to a} \frac{f(x)}{F(x)} = \lim_{x\to a} \frac{f'(x)}{F'(x)} = L$, **118**

**Limacon:** graph defined by $r = b - a\cos\theta$, $b > a$, **88**

**Limit:** a constant $A$ is said to be a limit of a variable $Z$ if, as the variable changes, $|Z - A| < \epsilon$ for any $\epsilon > 0$ however small, **10**

**of a function:** $\lim_{x \to a} f(x) = L$ if $|f(x) - L| < \epsilon$ whenever $0 < |x - a| < \delta$ for $\epsilon > 0, \delta > 0$, **10**

**Line:** the graph of a linear equation in two variables, **58**

**normal to curve:** the line through $P_0$ which is perpendicular to the tangent line at $P_0$, **23**

**tangent to curve:** if the equation of the curve is $y = f(x)$, then the tangent line at $P_0(x_0, y_0)$ is the line through $P_0$ with slope $f'(x_0)$, **22**

**Line segment:** the portion of a line between two specified points, **58**

**Line in three dimensions: 131, 137**

**Linear equation:** an equation in which the variable occurs only to the first degree, **58**

**ln function:** the logarithmic function to the base $e$, **70**

**Locus:** the totality of all points, and only those points, whose coordinates satisfy the given equation, **10**

**Logarithm of a number:** the exponent of the power to which the base must be raised to equal the number, **70**

**base of:** the positive number shich is raised to a power to equal the argument of the logarithm,

**change of base of a:** $\log_b N = \dfrac{\log_a N}{\log_a b}$, **71**

**common:** the logarithm to the base 10, **70**

**natural:** the logarithms to the base $e$, **70**

**Logarithmic equation:** an equation in which the variable is involved in an expression whose logarithm is indicated, **71**

**Logarithmic function:** the inverse function of the exponential function, $y = \log_a x$, **70**

**Maclaurin series:** the Taylor series with $a = 0$, $f(x) = \sum_{n=0}^{\infty} \frac{1}{n!} f^{(n)}(0)x^n$, **125**

**Mass:** $M = \iiint_R \delta(x,y,z)dxdydx$, **149**

**center $(\bar{x},\bar{y})$ of:** the moment of mass about each axis divided by the total mass, **113**

**moment of:** the moment of a mass about a line is the product of the mass and the perpendicular distance from the center of the mass to the line, **112**

**Maxima:** the relative maximum of a function is the value that is greater than any value immediately preceding or following it, **28, 143**

**necessary condition for:** the first derivative is zero, **29**

**test for:** if $f'(x_0) = 0$ and $f''(x_0) > 0$, then $f$ has a relative maximum at $x_0$, **29, 143**

**Mean, Theorem of the:** if $f$ is continuous function of $x$ in $[a,b]$ and $f'(x)$ exists for each $x$ in $(a,b)$, then there is at least one value $x_1$ in $(a,b)$, such that $f'(x_1) = \dfrac{f(b) - f(a)}{b - a}$, **28**

**Mean value of a function:** $f_m = \dfrac{1}{b - a} \displaystyle\int_a^b f(x)dx$, **112**

**Midpoint of a line segment:** the coordinates of the midpoint of $\overline{P_1P_2}$ are $x = \frac{1}{2}(x_1 + x_2)$, $y = \frac{1}{2}(y_1 + y_2)$, **58**

**Minimum:** the relative minimum of a function is the value that is less than any value immediately preceding or following it, **28, 143**

**necessary condition for:** the first derivative is zero, **29**

**test for:** if $f'(x_0) = 0$ and $f''(x_0) < 0$, then $f$ has a relative minimum at $x_0$, **29, 143**

**Minute:** a measure of an angle equal to $\frac{1}{60}$ of a degree, **52**

**Moment of inertia:** the moment of inertia of a particle of mass $m$ about the axis $L$ is $mr^2$ where $r$ is the perpendicular distance of the object from the axis, **149**

**Moment of a mass:** *see* **Mass, moment of**

**Natural logarithms:** logarithms to the base $e$; also called Napierian logarithms, **70**

**Negative integer:** the integer obtained when a larger positive integer is subtracted from a smaller positive integer, **156**

**Normal:** the line perpendicular to the tangent line at the point on a curve, **23**

**Numbers:**

**even:** a number that has 2 as a factor, **156**

**imaginary:** the number resulting from the extraction of an even root of a negative number, **156**

**irrational:** a number that cannot be expressed exactly as the quotient of two integers ($\sqrt{2}, e, \pi$, *etc.*), **156**

**rational:** a number which can be expressed as the quotient of two integers, **156**

**real:** the set of numbers composed of all the rational and irrational numbers, **156**

**Numerical integration:** technique for finding the integral of a function whose values are given at discrete points, **113**

**Oblique tangent:** one that is not parallel to either coordinate axis, **23**

**Octants:** the division of three-dimensional space caused by three mutually perpendicular planes, **130**

**Open interval:** *see* **Interval, open**

**Ordered pair:** two items in which one has been designated as the first, **10**

**Ordinate:** the vertical coordinate, $y$, of a point $P(x,y)$ in a plane, **10**

**Origin:** the point of intersection of the coordinate axes, **10**

**Orthogonal vectors:** vectors that are perpendicular to each other; $\mathbf{v} \cdot \mathbf{w} = 0$, **94**

**Pappus, Theorems of:** two theorems concerning the volume and area of solids of revolution, **113**

**Parabola:** the locus of points which are equidistant from a fixed point and a fixed line; the conic section for which $e = 1$, **64**

**vertex of a:** the point of intersection of the parabola and its axis, **64**

**Paraboloid:** the surface generated by revolving a parabola about its axis, **131**

**Parallel lines:** two lines which do not intersect; two lines whose slopes are equal, **58**

**Parametric equations:** if $x$ and $y$ are related by $x = f(t)$ and $y = g(t)$, we say that the relation is defined by a set of parametric equations and $t$ is called the parameter, **82**

**Partial derivative:** $f_x(x,y) = \lim_{h \to 0} \dfrac{f(x + h,y) - f(x,y)}{h}$, **142**

**Partial fractions, method of:** the decomposition of a rational function into the sum of simpler expressions, **106**

**Periodic function:** *see* **Function, periodic**

**Perpendicular lines:** two lines which intersect in a right angle; two lines, the product of whose slopes equals $-1$, **58, 130**

**Perpendicular vectors:** two vectors so positioned that they intersect in a right angle; $v \cdot w = 0$, **94**

**Plane:** any three points not on a straight line determine a plane; the surface defined by a linear equation in three variables, **131, 137**

**Point of inflection:** a point on a curve at which $F''(x_0) = 0$ and whose graph is concaved upward on one side and concaved downward on the other side, **29**

**Polar angle:** the angle between the polar axis and the radius vector, **88**

**Polar axis:** the fixed line from a point $O$ and usually extended horizontally to the right, **88**

**Polar coordinates; $P(r,\theta)$:** the radius vector $r = \overline{OP}$ and the polar angle $\theta$, **88**

**arc length in:** $ds = \sqrt{f(\theta)^2 + [f'(\theta)]^2}\, d\theta$, **89**

**area in:** $dA = \frac{1}{2}r^2 d\theta$, **89**

**curvature in:** $K = \dfrac{|r^2 + 2(D_\theta r)^2 - rD_\theta^2 r|}{[r^2 + (D_\theta r)^2]^{3/2}}$, **89**

**Polar moment of inertia:** the moment of inertia with respect to the origin; $I_0 = \int_a^b \int_{y_1}^{y_2} (x^2 + y^2)dydx$, **149**

**Pole:** the origin in polar coordinates, **88**

**Power series:** a series of the form $\Sigma a_i x^i$ where $x$ is a variable and $a_i$ are constants, **125**

**interval of convergence of a:** the totality of values of $x$ for which a power series converges, **125**

**Principal values:** the smallest numerical value of the angles defined by the inverse trigonometric functions, **53**

**Progression:** a sequence of numbers that is formed according to a specific rule, **124**

**arithmetic:** $a, a + d, a + 2d, \ldots$, **124**

**geometric:** $a, ar, ar^2, ar^3, \ldots$, **124**

**harmonic:** a sequence of numbers whose reciprocals form an arithmetic progression, **124**

**Projection of a vector:** $\mathbf{v} = ai + bj$ upon the vector $\mathbf{w} = ci + dj$ is a vector $\mathbf{u} = ci + yj$ where $x = kc$ and $y = kd$ and $k = \pm \dfrac{ac + bd}{\sqrt{c^2 + d^2}}$, **94**

**p-series:** $\sum_{n=1}^{\infty} \dfrac{1}{n^p}$, **124**

**Quadrants:** the four parts into which a plane is divided by the coordinate axes, **10**

**Quadratic formula:** $x = \dfrac{-b \pm \sqrt{b^2 - 4ac}}{2a}$, **5**

**Quadratic in x and y:** $ax^2 + bxy + cy^2 + dx + ey + f = 0$, **64**

**Radian:** the central angle subtended by an arc of a circle equal in length to the radius of the circle; 1 radian $= \dfrac{180}{\pi}$ degrees, **52**

**Radical, $\sqrt{\ }$:** the symbol used to indicate the root of a number, **4**

**Radius of curvature:** $R = \dfrac{[1 + (y')]^{3/2}}{|y''|}$ where the derivatives $y'$ and $y''$ are evaluated at the point $P(x_1, y_1)$ on the curve, **82**

**Radius of gyration:** $r^2 = \dfrac{I}{m}$, **140**

**Radius vector:** the line segment from the origin to a point $P$ in a plane, **88**

**Range:** *see* **Function, range of a**

**Rate of change, average:** the quotient $\Delta z/\Delta x$, **28**

**Ratio test for series:** *see* **Infinite series, ratio test**

**Rational function:** polynomials or fractions formed by dividing one polynomial by another, **106**

**Rationalizing substitutions:** substitutions used to evaluate integrals where the integrand contains fractional powers, **107**

**Rectangular coordinates:** the coordinates of a point in the system formed by two line axes that are perpendicular to each other, **10**

**Rectifiable arc:** any arc of a curve which has length is said to be rectifiable, **82**

**Reference angle: 52**

**Relation between variables:** if two variables are associated in such a way that for a first element there may be one or more second elements in the ordered pairs, **10**

**Relative error:** *see* **Error, relative**

**Relative extrema:** *see* **Maxima** and **Minimum**

**Repeated integral:** the itegrated integration given by $\int\int f(x,y)dydx$, **148**

**Rolle's Theorem:** if $f(x)$ is continuous in the interval $[a,b]$, has a derivative at every point between $a$ and $b$, and is zero at $a$ and $b$, then $f'(x)$ vanishes at some point $x_1$ between $a$ and $b$, **28**

**Rose:** curves defined by $r = a \sin n\theta$, **88**

**Rotation of axes: 65**

**Scalar product of vectors:** *see* **Dot product**

**Secant:** the trigonometric function defined by the ratio $\dfrac{\text{radius vector}}{\text{abscissa}}$ or $\dfrac{\text{hypotenuse}}{\text{adjacent side}}$, **52**

**Second:** a sexagesimal measure of an angle equal to $\frac{1}{60}$ of a minute, **52**

**Second derivative:** the derivative of the first derivative, **22**

**notation for:** $\dfrac{d^2y}{dx^2} = y'' = D_x^2 y = f''(x)$, **22**

**Sequence:** an ordered set of objects so that there is a first, a second, etc., **124**

**Series:** the sum of a sequence, **124**

**Sexagesimal measure:** the units degree, minute, and second used to measure angles, **52**

**Shell, cylindrical:** used in finding the volume of a solid of revolution, **112**

**Simpson's Rule:** numerical approximation of integral

$$\int_a^b f(x)dx = \frac{x}{3}(y_0 + 4y_1 + 2y_2 + 4y_3 + \cdots + 4y_{n-1} + y_n),$$

**113**

**Sine:** the trigonometric function defined by the ratio $\dfrac{\text{ordinate}}{\text{radius vector}}$ or $\dfrac{\text{opposite side}}{\text{hypotenuse}}$, **52**

**Slope:** the tangent of the angle of inclination (the acute angle between a line and the positive $X$-axis), **58**

**Slope of tangent line to a curve at $P(x,y)$:** the derivative of the function defining the curve and evaluated at the point $P(x,y)$, **22**